THE APPLICATION OF THE PRECAUTIONARY PRINCIPLE IN PRACTICE

This overview of the role played by the precautionary principle in international trade law, European law and national law compares how precautionary considerations have been applied in the fields of pesticide regulation and the regulation of base stations for mobile telephones in Sweden, the UK and the USA. A number of problems in the current application of the precautionary principle are identified and discussed. For example, it is shown that a firm reliance on a wide and open-ended precautionary principle may lead to problems with the consistency, foreseeablility, effectiveness and efficiency of measures intended to reduce environmental or health risks. It is suggested that the precautionary principle indeed may be an important tool but in order to be acceptable it must be coupled with strong requirements on the performance of risk assessments, cost–benefit analyses and risk trade-off analyses.

JOAKIM ZANDER is currently responsible for the free movement of goods and services at the European Free Trade Association (EFTA) Surveillance Authority in Brussels. He has previously worked as a legal officer in DG Enterprise of the European Commission and as a political assistant in the European Parliament. He holds a PhD in Law from Maastricht University, the Netherlands, and a Master of Laws Degree (*juris kandidat*) from Uppsala University, Sweden.

THE APPLICATION OF THE PRECAUTIONARY PRINCIPLE IN PRACTICE

Comparative Dimensions

JOAKIM ZANDER

CAMBRIDGE
UNIVERSITY PRESS

CAMBRIDGE
UNIVERSITY PRESS

University Printing House, Cambridge CB2 8BS, United Kingdom

One Liberty Plaza, 20th Floor, New York, NY 10006, USA

477 Williamstown Road, Port Melbourne, VIC 3207, Australia

314-321, 3rd Floor, Plot 3, Splendor Forum, Jasola District Centre, New Delhi - 110025, India

103 Penang Road, #05-06/07, Visioncrest Commercial, Singapore 238467

Cambridge University Press is part of the University of Cambridge.

It furthers the University's mission by disseminating knowledge in the pursuit of education, learning and research at the highest international levels of excellence.

www.cambridge.org
Information on this title: www.cambridge.org/9780521768535

First published 2010

A catalogue record for this publication is available from the British Library

Library of Congress Cataloging in Publication data
Zander, Joakim.
The application of the precautionary principle in practice : comparative dimensions / Joakim Zander.
p. cm.
Includes bibliographical references and index.
ISBN 978-0-521-76853-5
1. Police power. 2. Precautionary principle. 3. Risk. I. Title.
K3289.Z36 2010
344.04′6–dc22
2010021887

ISBN 978-0-521-76853-5 Hardback

For Liisa and Milla

CONTENTS

ACKNOWLEDGEMENTS

This book has been long in the making and it would never have been completed, had I not enjoyed the company and assistance of a large number of people. Consequently, I would like to extend my gratitude to the entire Faculty of Law at Maastricht University, and all its members, for allowing me to carry out the research upon which this book is based and teach at the Faculty between 2003 and 2007. In particular, I would like to thank Professor Ellen Vos for providing supervision of the PhD thesis that formed the basis for this book. Furthermore, I would like to thank my colleagues at the Faculty for their friendship, endless hours of discussion, coffee, hot cocoa and ping-pong. In particular, I would like to mention the study group on risk and regulation, which supplied ample opportunity for the exploration of all angles of the field of risk regulation. Special thanks go to two students at Maastricht University; Anna Fink and Alain Lambert who provided excellent assistance with editing and translation. I would also like to thank Chris Engert in Florence, Italy, for carrying out the language editing of the book.

I would furthermore like to thank Professor Cass Sunstein, then at Chicago Law School, now Administrator of the Office of Information and Regulatory Affairs at the White House, and Professor Ragnar Löfstedt at King's Centre for Risk Management, King's College, London, for receiving me as a visiting researcher and taking the time to discuss the precautionary principle. In this context, I would like to thank The Netherlands Organisation for Scientific Research (NWO) for making this possible by granting me the funds to spend time at Chicago Law School in the autumn of 2005.

I am also grateful for valuable comments provided by members of the PhD assessment committee appointed by Maastricht University and two anonymous reviewers. Obviously, every mistake, omission or blatant misunderstanding in the following pages is mine alone.

On the very eve of the completion of this book, the Treaty on the Functioning of the European Union finally entered into force on 1

December 2009. Although it did not bring about any substantive changes with regard to the subject matter of this book, it did introduce that greatest of joys for any European lawyer: a change in the numbering of the treaty articles. Throughout this book I have thus sought to refer to the current numbering of the articles as they appear in the Treaty on the Functioning of the European Union. Where I have considered it relevant for the understanding of the text, I have kept the old numbering in brackets following the current reference.

Finally, and above and beyond everything else, I would like to thank my wife, Liisa, for putting up with me during the years I have been working on this book. I think that we can both agree that it is not at a moment too soon that this particular train has reached its final destination.

ABBREVIATIONS

AB	Appellate Body
ACP	Advisory Committee on Pesticides
AGNIR	Advisory Group on Non-Ionising Radiation
AIDS	Acquired Immune Deficiency Syndrome
ALARA	as low as reasonably possible
ANSI	American National Standards Institute
AOEL	acceptable operator exposure level
APA	Administrative Procedure Act in the USA
BSE	Bovine spongiform encephalopathy
BST	Bovine Somatotrophin
CAP	Common Agrigultural Policy
CBA	cost–benefit analysis
CEA	cost-effectiveness analysis
CFC	chlorofluorocarbons
CFI	Court of First Instance
CITES	Convention on International Trade in Endangered Species
COM	Committee on Mutagenicity of Chemicals in Food, Consumer Products and the Environment (UK)
COPR	Control of Pesticides Regulations
CRA	comparative risk analysis
CVMP	Committee for Veterinary Medicinal Products
DDT	dichloro-diphenyl-trichloroethane
DecaBDE	decabromodiphenyl ether
DEFRA	Department of Environment, Food and Rural Affairs in the UK
DNA	deoxyribonucleic acid
DSB	Dispute Settlement Body of the WTO
DSU	Dispute Settlement Understanding (WTO)
EAP	Environmental Action Programme
EBCD	ethylenebisdithiocarbamate
EC	European Community
ECJ	European Court of Justice
EEC	European Economic Community
EFTA	European Free Trade Association

EOP	Executive Office of the President of the USA
EPA	United States Environmental Protection Agency
EU	European Union
FACA	Federal Advisory Committee Act in the USA
FAQ	frequently asked questions
FCC	Federal Communications Commission
FDA	Food and Drug Administration in the USA
FDCA	Food, Drug, and Cosmetic Act in the USA
FEPA	Food and Environment Protection Act
FIFRA	Federal Insecticide, Fungicide and Rodenticide Act in the USA
FQPA	Food Quality Protection Act in the USA
GAO	United States Government Accountability Office
GATT	General Agreement on Tariffs and Trade
GM	genetically modified
GMO	genetically modified organisms
HCFCs	hydrochlorofluorocarbons
HEW	United States Department of Health, Education, and Welfare
HPA	Health Protection Agency
HSE	Health and Safety Executive in the UK
ICJ	International Court of Justice
ICNIRP	International Commission for Non-Ionising Radiation Protection
IEEE	Institute of Electrical and Electronics Engineers in the USA
IEGMP	Independent Expert Group on Mobile Phones in the UK
ILGRA	Interdepartmental Liaison Group on Risk Assessment
ITLOS	International Tribunal for the Law of the Sea
KemI	Kemikalieinspektionen (Swedish Chemicals Agency)
LPAs	Local Planning Authorities in the UK
Miljöbalken	The Swedish Environmental Code
MOX	metal oxide
MRL	maximum residue level
NCRP	National Council on Radiation Protection and Measurement
NEPA	National Environmental Policy Act in the USA
NIOSH	National Institute for Occupational Safety and Health in the USA
nimby	not in my back yard
NJA	Nytt Juridiskt Arkiv (Publication containing judgments from the Swedish Supreme Court)
NRPB	National Radiation Protection Board in the UK
OET	Office of Engineering and Technology in the USA
OIRA	Office of Information and Regulatory Affairs in the USA
OMB	Office of Management and Budget in the USA
OSHA	Occupational Safety and Health Administration in the USA
PAN	Pesticide Action Network

PBDE	polybrominated diphenyl ethers
PBL	Plan- och bygglagen (Swedish Planning and Building Act)
PIC	Rotterdam Convention for Prior Informed Consent
PPPR	Plant Protection Products Regulation
PSD	United States Pesticide Safety Directorate
PVC	polyvinyl chloride
QUALYs	Quality Adjusted Life Years
RCHSA	Radiation Control for Health and Safety Act in the USA
REACH	Registration, Evaluation, Authorisation and Restriction of Chemicals
Regeringsformen	The Swedish Constitution
Regeringsrätten	Swedish Supreme Administrative Court
RIA	Regulatory Impact Assessments
RoHS	Restriction of Hazardous Substances Directive
SAP	Scientific Advisory Panel in the USA
SAR	specific absorbtion rate
SCAN	Scientific Committee on Animal Nutrition
SEA	The Single European Act
SPS	WTO Agreement on the Application of Sanitary and Phytosanitary Measures
SSI	Statens Strålskyddsinstitut (formerly named Swedish National Radiation Protection Authority)
SSL	Strålskyddslagen (Swedish Radiation Protection Act)
SSM	Statens Strålsäkerhetsmyndighet (Swedish Radiation Safety Authority)
statliga myndigheter	Swedish system of independent governmental authorities
TBT	Agreement on Technical Barriers to Trade
TETRA	terrestrial trunked radio
TFEU	Treaty on the Functioning of the European Union
UK	United Kingdom
UNCLOS	United Nations Convention on the Law of the Sea
US	United States (of America)
USA	United States of America
WTA	willingness to accept
WTO	World Trade Organisation
WTP	willingness to pay

TABLE OF CASES

International Court of Justice and other International Tribunals

Gabcikovo-Nagymaros Project (Hungry/Slovakia), Judgment, ICJ Reports 1997.

Request for an Examination of the Situation in Accordance with Paragraph 63 of the Court's Judgement of 20 December 1974 in the Nuclear Tests (New Zealand v. *France)*, ICJ Reports 1995 (*French Nuclear Tests II*).

The MOX Plant Case (Ireland v. *UK)*, (Provisional measures), (ITLOS), Order of 3 December 2001.

Southern Bluefin Tuna Cases (New Zealand v. *Japan; Australia* v. *Japan* (Joined Cases)), ITLOS, Order of August 27 1999.

World Trade Organisation (WTO) Appellate Body Reports

Report of the Appellate Body, US – Continued Suspension of Obligations in the EC-Hormones Dispute, WT/DS320/AB/R (*EC-Hormones II*).

Report of the Appellate Body. Japan – Measures Affecting Agricultural Products. WT/DS76/AB/R (*Japan Agricultural Products II*).

Report of the Appellate Body. Japan – Measures Affecting the Importation of Apples. WT/DS245/AB/R (*Japan-Apples*).

Report of the Appellate Body, EC – Sardines. Appellate Body Report, European Communities – Trade Description of Sardines, WT/DS231/AB/R (*EC-Sardines*).

Report of the Appellate Body, Australia – Measures Affecting the Importation of Salmon WT/DS18/AB/R (*Australia-Salmon*).

Report of the Appellate Body, European Communities – Measures Affecting Asbestos and Asbestos-Containing Products. WT/DS135/AB/R (*EC-Asbestos*).

Report of the Appellate Body. United States – Measure Affecting Imports of Woven Wool Shirts and Blouses From India. WT/DS33/AB/R (*US-Wool Shirts and Blouses*)

Report of the Appellate Body, EC Measures Concerning Meat and Meat Products WT/DS26/AB/R (*EC-Hormones I*).

WTO Panel Reports

Panel Report, European Communities - Measures Affecting the Approval and Marketing of Biotech Products, WT/DS291/R, WT/DS292/R WT/DS293/R (*EC-Approval and Marketing of Biotech Products*).

Panel Report, EC Measures Concerning Meat and Meat Products (Hormones) (US) WT/DS26/R (*EC-Hormones I (US)*).

Panel Report, EC Measures Concerning Meat and Meat Products (Hormones) (Canada), WT/DS48/R (*EC-Hormones I (Canada)*).

Panel Report, European Communities - Measures Affecting the Approval and Marketing of Biotech Products, WT/DS291/R, WT/DS292/R WT/DS293/R (*EC-Biotech Products*).

Panel Report, Japan – Measures Affecting the Importation of Apples, WT/DS245/R (*Japan-Apples*).

Panel Report, US-Continued Suspension of Obligations in the EC Hormones Dispute, WT/DS320/R (*Continued Suspension of Obligations-EC Hormones (EC-Hormones II)*).

European Union

European Court of Justice (ECJ)

Case 120/78 *Rewe* v. *Bundesmonopolverwaltung* ('Cassis de Dijon') [1979] ECR 649.

Case 174/82 *Sandoz BV* [1983] ECR 2445.

Case 227/82 *Van Bennekom* [1983] ECR 3883.

Case 94/83 *Heijn* [1984] ECR 3263.

Case 97/83 *Melkunie* [1984] ECR 2367.

Case 240/83 *Waste Oils* [1985] ECR 531.

Case 54/85 *Mirepoix* [1986] ECR 1067.

Case 178/84 *Commission* v. *Germany (Reinhieitsgebot)* [1987] ECR 1227.

Case 331/88 *Fedesa and others* [1990] ECR I-4023.

Case C-405/92 *Etablissement Armand Mondiet SA* v. *Amement Ijslais SARL* [1993] I-6133.

Case C-320/93 *Ortscheit* [1994] ECR I-5243.

Case C-157/94 *Commission* v. *Netherlands* [1997] ECR I-5699.

Cases C-157/96 *The Queen* v. *Ministry of Agriculture, Fisheries and Food, ex parte National Farmers' Union et al.* [1998] ECR I-2211.

Case C-180/96 *United Kingdom* v. *Commission* [1998] ECR I-2265 *(BSE)*.

Case C-341/95 *Gianni Bettati* v. *Safety Hi-Tech* [1998] ECR I-4335.

Case C-120/97 *Upjohn* [1999] ECR I-223.

Case C-6/99 *Association Greenpeace France and Others* v. *Ministere de l'Agriculture et de la Pêche and Others* [2000] ECR I-1651.

Case C-473/98 *Kemikalieinspektionen* v. *Toolex* [2000] ECR I-5681.

Case C-248/99, *France* v. *Monsanto* [2002] ECR I-1.
Case C-3/00 *Denmark* v. *Commission* [2003] ECR I-02643.
Case C-192/01 *Commission* v. *Denmark* [2003] ECR I-9693.
Case C-236/01 *Monsanto Agricoltura Italia and Others* [2003] ECR I-8105.
Case 512/99 *Germany* v. *Commission* [2003] ECR I-00845.
Case C-24/00 *Commission* v. *France* [2004] ECR I-1277.
C-41/02 *Commission* v. *Netherlands* [2004] ECR I-11375.
C-210/03 *Swedish Match AB and Swedish Match AB UK Ltd* v. *Secretary of State for Health* [2004] ECR I-11893.
Cases C-434/02 *Arnold André GmbH & Co. KG* v. *Landrat des Kresies Herford* [2004] ECR I-11825.
Case 132/03 *Codacons and Federconsumatori* [2005] ECR I-4167.
Joined Cases C-15/06 and C-295/06 *European Parliament* v. *Commission*, judgment of 1 April 2008 (not yet reported).
Case C-110/05 *Commission* v. *Italy*, judgment of 10 February 2009 (not yet reported).
Case 142/05 *Mickelsson and Roos*, judgment of 4 June 2009 (not yet reported).
Case C-42/07, *Liga Portuguesa de Futebol Profissional Bwin*, judgment of 8 October 2009 (not yet reported).

Court of the First Instance (CFI)

Case T-120/96 *Lilly Industries* v. *Commission* [1998] ECR II-2571.
Case T-199/96 *Laboratoires pharmaceutiques Bergaderm SA and Goupil* v. *Commission* [1998] ECR II-2805.
Case T-112/97 *Monsanto* v. *Commission* [1999] ECR II-1277.
Case T-74/00 *Artegodan* [2000] II-04945.
Case T-13/99 *Pfizer Animal Health SA* v. *Council of the European Union* [2002] ECR II-03305.
Case T-70/99 *Alpharma Inc.* v. *Council of the European Union* [2002] ECR II-3495.
Case T-177/02 *Malagutti-Vezinhet* v. *Commission* [2004] ECR II-827.
Joined Cases T-366/03 and T-235/04 *Austria* v. *Commission* [2005] ECR II-4005.
Case T-229/04 *Sweden* v. *Commission*, [2007] ECR II-2437.

European Free Trade Association (EFTA) Court

Case E-3/00 *EFTA Surveillance Authority* v. *Norway* [2000–2001] EFTA Court Reports.

United Kingdom

Alfred McAlpine Homes (North) Ltd. v. *Secretary of State for the Environment and Another*, Queen's Bench Division (Crown Office List), 5 April 1993.

Alfred McAlpine Homes (North) Ltd. v. *Secretary of State for the Environment and Another*, Court of Appeal (Civil Division), 3 November 1994.

Associated Provincial Picture Houses Ltd v. *Wednesbury Corporation* [1948] 1 KB 223.

Gateshead MBC v. *Secretary of State for the Environment*, 1 P.L.R., 85, 1994.

Leatch v. *Director-General of National Parks & Wildlife Service and Shoalhaven City Council* [1993] NSWLEC 191 (23 November 1993) (Land and Environment Court of New South Wales) (No. 10376 of 1993).

Newport CBC v. *Secretary of State for Wales*, J.P.L., 377, 1998.

Phillips v. *First Secretary of State and others*, Queen's Bench Division (Administrative Court) [2003] EWHC 2415.

The Queen on the Application of Amvac Chemical UK Limited v. *Secretary of State for Environment, Food and Rural Affairs, Secretary of State for Transport, Local Government and Regions, Food Standards Agency*, High Court of Justice, Queen's Bench Division (Administrative Court) 3 December 2001, 2001 WL 1479891.

Re Cullen, Queen's Bench Division, NIQB 9, 14 February 2005.

R. (on the application of S) v. *Brent LBC* [2002] E.W.C.A Civ 693; [2002] E.L.R., 556.

R. v. *Derbyshire Country Council, ex parte Murray*, Queen's Bench Division (Administrative Court), 6 October 2000.

R. v. *Environment Agency ex parte Turnbull*, Queen's Bench Division (Crown Office List), CO/4848/99, 12 January 2000.

R. v. *Minister for Agriculture ex parte Inquinosa International*, Queen's Bench Division (Crown Office List), 17 January 2000.

R. v. *Ministry of Agriculture, Fisheries and Food ex parte British Agrochemicals Association Ltd.*, English Court of Appeal (Civil Division), 9 November 2001, [2002] 1 C.M.L.R., 8.

R. v. *Secretary of State for Defence, ex parte Smith* (Queen's Bench Division) (1996) 2 WLR 305 B 517.

R. v. *Secretary of State for Trade and Industry ex parte Duddridge and Others*, Queen's Bench Division (Crown Office List), *The Independent*, 4 October 1994.

R. v. *Tandridge D.C. and another, ex parte Al-Fayed*, Queen's Bench Division (Crown Office List) 79 P / C.R., 227, [1999] J.P.L., 825, [1999] 1 P.L.R., 104, *The Times* 28 January 1999, CO/3665/97, 14 January 1999.

T-Mobile (UK) Ltd v. *First Secretary of State*, Court of Appeal (Civil Division), [2005] Env. L.R. 18, C2 and C3.

Trevett v. *Secretary of State for Transport Local Government and the Regions and Others* [2002] E.W.H.C. Admin 2696, 25 November 2002.

United States of America

Associated Fisheries of Maine, Inc. v. *Daley*, (1st Cir. 1997) 127 F.3d 104.

AT&T Wireless PCS, Inc. v. *City Council of Virginia Beach*, (4th Cir. 1998) 155 F.3d 423.

Baltimore Gas and Electric Co. v. *Natural Resources Defence Council*, 462 U.S. 87 (1983).

Banal v. *Freeport-McMoran, Inc.* (5th Cir. 1999) 197 F 3d 161.
Industrial Union Department, AFL-CIO v. *American Petroleum Institute et al.* 448 U.S. 607 (1980).
Cellular Phone Task Force et al., v. *FCC,* (2d Cir. 2000) 205 F.3d 82.
Cellular Telephone Co. v. *The Town Board of Oyster Bay,* (2nd Cir 1999) 166 F.3d 490.
Certified Color Manufacturers Association et al. v. *David Mathews, Secretary of the Department of HEW et al.,* (D.C. Cir. 1976) 465 F.2d 528.
Chevron USA, Inc. v. *Natural Resources Defence Council,* 467 U.S. 87 (1984).
Ciba-Geigy Corp. v. *EPA,* (5th Cir. 1989) 874 F.2d 277.
Corrosion Proof Fittings v. *EPA,* (5th Cir. 1991) 247 F.2d 12–1.
Daubert v. *Merrel Dow Pharmaceuticals,* 509 U.S. 579 (1993).
The Dow Chemical Company v. *William D. Ruckelshaus, Administrator Environmental Protection Agency Defendant-Appellant,* (8th Circuit 1973) 477 F.2d 1317.
EDF v. *EPA,* (D.C. Circuit 1976) 548 F.2d 998.
EDF v. *EPA,* (D.C. Circuit 1972) 465 F.2d 528.
EDF v. *Ruckelshaus,* (D.C. Circuit, 1971) 439 F.2d 584.
EI DuPont de Nemours and Co. v. *Train,* (4th Cir. 1976) 541 F.2d 1018.
Environmental Defence Fund v. *EPA,* (D.C. Cir. 1972) 465 F.2d 528.
Ethyl Corp. v. *EPA,* (D.C. Cir. 1979) 598 F.22d 91.
Fletcher v. *Bealey* 28 Ch.D. 688–700.
FTC v. *Standard Oil Co.,* 449 U.S. 232 (1980).
Gulf South Insulation v. *Consumer Product Safety Commission,* (5th Cir. 1983) 701 F.2d 1137.
Illinois RSA No. 3, Inc. v. *County of Peoria,* 963 F. Supp. 732, (C.D.Ill. 1997).
Industrial Union Department, AFL-CIO v. *American Petroleum Institute,* 448 U.S. 607 (1980).
Lead Indus. Ass'n, Inc. v. *EPA,* (D.C. Cir., 1976) 647 F.2d 1130.
Lee v. *Reilly* (9th Cir. 1992) 968 F. 2d 985.
Love v. *Thomas,* (9th Cir. 1988) 858 F.2d 1347.
Marsh v. *Oregon Natural Resources Council* 490 U.S. 332 (1989).
Massachusetts v. *EPA,* 127 U.S. 1438 (2007).
Midwater Trawlers Cooperative v. *Department of Commerce,* (9th Cir. 2002) 282 F.3d 710.
Motor Vehicle Manufacturers Association v. *State Farm Mutual Auto Insurance, Co,* 463 U.S. 29 (1983).
National Coalition Against the Misuse of Pesticides et al v. *EPA,* (D.C. Cir. 1989) 867 F.2d 636.
National Mining Association v. *Babbitt,* 172 F.3d 906 (D.C. Cir. 1999).
Natural Resources Defense Council, Inc. v. *United States Nuclear Regulatory Comm'n,* (D.C. Cir. 1976) 547 F.2d 633.
Omnipoint Communications Inc. v. *The City of White Plains,* (2nd Cir. 2005) 430 F.3d.
Petersburg Cellular Partnership v. *Boards of Supervisors of Nottoway County,* (4th Cir. 2000) 205 F.3d 688.

Public Citizen v. *Young*, (D.C. Cir. 1987) 831 F.2d 1008.
Reserve Mining Co. v. *EPA*, (8th Cir. 1975) 514 F.2d 492.
Sierra Club v. *Costle*, (D.C. Cir. 1981) 657 F.2d 298.
Sierra Club v. *Marita*, (7th Cir. 1995) 46 F.3d 606.
Sprint Spectrum L.P. v. *Mills*, (2nd Cir. 2002) 283 F.3d 404.
Southwestern Bell Mobile Sys. v. *Todd*, (1st Cir. 2001) 244 F.3d 51.
Tennessee Valley Authority (TVA) v. *Hill*, 437 U.S. 153 (1978).
Texas Oil & Gas Ass'n v. *EPA*, (5th Cir. 1998) 161 F.3d 923.
Vermont Yankee Nuclear Power Corp. v. *Natural Resources Defense Council*, 435 U.S. 519 (1978).

Sweden

Judgment of the Supreme Court in NJA 2004 s. 590.
Judgment of the Supreme Court in NJA 2006 s. 310.
Judgment of the Supreme Administrative Court in case RÅ 1996 Ref. 51.
Judgment of the Supreme Administrative Court in RÅ 1998 Ref. 56.
Judgment of the Supreme Administrative Court in RÅ 1999 Ref. 76.
Judgment of the Supreme Administrative Court in Case no. 3635–1999 delivered on 14 October 1999 (not reported in RA).
Judgment of the Supreme Administrative Court in Case no. 3650–1999 delivered on 14 October 1999, summarised in RÅ 1999 note 217.
Judgment of the Supreme Administrative Court in RÅ 2001 note 6.
Judgment of the Supreme Administrative Court in Case no. 1163–2001 delivered on 12 July 2001 (not reported in RÅ).
Judgment of the Supreme Administrative Court in RÅ 2003 Ref. 43.
Judgment of the Supreme Administrative Court in RÅ 2004 Ref. 14.
Judgment of the Supreme Administrative Court in RÅ 2005 Ref. 2, 6, 7.
Judgment of the Environmental Court of Appeal of 30 March 2005 in case no. M 9408–03 (not reported in MÖD).
Judgment of the Environmental Court of Appeal of 12 May 2005, case no. M 3225–04 (not reported in MÖD).
Judgment of the Environmental Court of Appeal of 12 October 2005 in case no. M 7485–04 (MÖD 2005:55).
Judgment of he Environmental Court at Växjö Tingsrätt of 14 February 2002, case no. M 13/99.
Judgment of Östersunds Tingsrätt of 12 January 2005, case no. M 3123–04.
Judgement of Växjö Tingsrätt of 3 October 2005, case no. M 390–05.

TABLE OF LEGISLATION

Other Sources

International Agreements

African Convention on the Conservation of Nature and Natural Resources, Algiers, 1968.

The 1985 Convention for the Protection of the Ozone Layer: on Substances that Deplete the Ozone Layer, Vienna 1985.

The Agreement for the Implementation of the Provisions of the 1982 UNCLOS Relating to the Conservation and Management of Straddling Fish Stocks and Highly Migratory Fish Stocks, New York 1995

The Convention Concerning the Protection of the World Cultural and Natural Heritage, Paris 1972.

The Convention on International Trade in Endangered Species of Wild Fauna and Flora, Washington DC 1973.

The Convention on the Ban on Import into Africa and the Control of Transboundary Movement and Management of Hazardous Wastes Within Africa, Bamako 1991.

The Convention on the Prior Informed Consent Procedure for Certain Hazardous Chemicals and Pesticides in International Trade. Rotterdam, 1998.

The Convention on Wetlands, Ramsar 1971.

General Agreement on Tariffs and Trade, Geneva 1947.

The Protocol to the 1992 Convention on Biological Diversity on Biosafety, Cartagena de Indias 2000.

The Protocol to the 1992 Framework Convention on Climate Change, Kyoto 1997.

The United Nations Convention on the Law of the Sea, Montego Bay 1982.

The United Nations Framework Convention on Climate Change, New York 1992.

WTO Agreement on Technical Barriers to Trade (TBT Agreement), Marrakesh 1994.

WTO Agreement on the Application of Sanitary and Phytosanitary Measures (SPS Agreement), Marrakesh 1994.

WTO Dispute Settlement Understanding, Marrakesh 1994.

International Resolutions, Declarations and Guidelines

Codex Alimentarius Commission, Procedureal Manual, 15th edition.

Declaration of the United Nations Conference on Environment and Development, Rio de Janeiro 1992.

Declaration of the United Nations Conference on the Human Environment, Stockholm 1972.

The International Commission on Non-Ionizing Radiation Protection, Guidelines for Limiting Exposure to Time-Varying Electric, Magnetic, and Electromagnetic Fields (up to 300 GHz) (ICNIRP Guidelines). Available at: www.icnirp.de/documents/emfgdl.pdf (accessed 1 March 2010).

International Commission on Non-Ionizing Radiation Protection (ICNIRP) Standing Committee on Epidemiology. Epidemiology of Health Effects of Radiofrequency Exposure. *Environmental Health Perspectives*, 112:7 (December 2004).

Ministerial Declaration on Sustainable Development in the ECE Region, Bergen 1990.

The United Nations General Assembly Resolution 37/7 and Annex: World Charter for Nature, 28 October 1982.

EU LEGISLATION AND SOFT LAW

Treaties

Consolidated Version of the Treaty on the Functioning of the European Union, OJ 2008 (115/47).

Consolidated Versions of the Treaty on the European Union and of the Treaty Establishing the European Community, OJ 2002 C 325/1.

Regulations

Council Regulation (EEC) No 2377/90 of 26 June 1990 laying down a Community procedure for the establishment of maximum residue limits for veterinary medicinal products in foodstuffs of animal origin, OJ 1990 L 224/1.

Council Regulation 2821/98/EC amending, as regards the withdrawal of the authorization of certain antibiotics, Directive 70/524/EEC concerning additives in feedingstuffs (1998) OJ L351/4.

Regulation (EC) No. 178/2002 of the European Parliament and of the Council of 28 January 2002 laying down the general principles and requirements of food law, establishing the European Food Safety Authority and laying down procedures in matters of food safety, OJ 2002 L 31/1.

Regulation (EC) No. 1829/2003 of the European Parliament and of the Council of 22 September 2003 on genetically modified food and feed, OJ 2003 L 268/1.

Regulation (EC) No. 1907/2006 of the European Parliament and of the Council of 18 December 2006 concerning the Registration, Evaluation, Authorisation and Restriction of Chemicals (REACH), establishing a European Chemicals Agency, OJ 2006 L 396/1.

Directives

Directive 65/65/EEC of 26 January 1965 on the approximation of provisions laid down by Law, Regulation or Administrative Action relating to proprietary medicinal products, OJ 1965 L 22/369.

Directive 67/548/EEC of 27 June 1967 on the Approximation of the Laws, Regulations, and Administrative Provisions of the Member States Relating to the Classification, Packaging, and Labelling of Dangerous Substances, OJ 1967 L 196/1.

Directive 70/524/EEC of 23 November 1970 concerning additives in feeding-stuffs, OJ 1970 L 270/1.

Directive 75/319/EEC of 20 May 1975 on the approximation of provisions laid down by Law, Regulation or Administrative Action relating to proprietary medicinal products, OJ 1975 L 147/13.

Directive 75/442/EEC of 15 July 1975 on waste, OJ 1975 L 194/39.

Directive 79/409/EEC of 2 April 1979 on the conservation of wild birds, OJ 1979 L 103.

Directive 79/831/EEC of 18 September 1979, amending for the sixth time Directive 67/548/EEC on the approximation of the laws, regulations and administrative provisions relating to the classification, packaging and labelling of dangerous substances, OJ 1979 L 259/10.

Directive 90/219/EEC of 23 April 1990 on the contained use of genetically modified micro-organisms, OJ 1990/L 117/1

Directive 90/220 of 23 April 1990 on the deliberate release into the environment of genetically modified organisms, OJ 1990, L 117/15.

Directive 90/220/EEC of 23 April 1990 on the deliberate release into the environment of genetically modified organisms, OJ L 117/15.

Directive 91/414/EC of 15 July 1991 concerning the placing of plant protection products on the market, OJ L 230/1.

Directive 94/67 of 16 December 1994 on the incineration of hazardous waste, OJ 1994 L 365/34.

Directive 96/51/EC of 23 July 1996, amending Directive 70/524/EEC concerning additives in feeding-stuffs (1996) OJ L235/39.

Directive 96/61 of 24 September 1996 on integrated pollution prevention and control, OJ 1996 L 137/20.

Directive 2001/18/EC of the European Parliament and of the Council of 12 March 2001 on the deliberate release into the environment of genetically modified organisms and repealing Council Directive 90/220/EEC, OJ L 106/1.

Decisions

Council Decision 90/218/EEC of 25 April 1990 concerning the administration of Bovine Somatotrophin (BST) OJ L 116/1.

Council Decision 94/936 of 20 December 1994 amending Decision 90/218/EEC concerning the placing on the market and administration of bovine somatotrophin, OL L 366/19.

Council Decision 96/239/EC of 27 March 1996 on emergency measures to protect against bovine spongiform encephalopathy OJ 1996 L 078/9.

Council Decision 1999/468 of 28 June 1999 laying down the procedures for the exercise of implementing powers conferred on the Commission OJ 1999 L 184/ 23.

Communications of the Commission

Communication of the Commission on Consumer Health and Food Safety, COM/97/83/Final.

Communication from the Commission on the Precautionary Principle, COM 2000/1/ Final.

Communication from the Commission of June 2002 on Impact Assessment, COM/2002/726, adopted 5 June 2002.

Communication from the Commission to the Council and the European Parliament, of 16 March 2005, Better Regulation for Growth and Jobs in the European Union, COM/2005/97 L, Final.

Other

Bulletin EC 6–1990, Conclusions of the Presidency, Point 1.14 and Annex II.

EU Bulletin 12–2000, Chapter 10 Community institutions, point 1.4.75.

Conclusions of the Presidency, European Council at Nice 7–10 December 2000, Annex III, Point 1. Available at: www.europarl.europa.eu/summits/nice1_en.htm (accessed 1 March 2010).

Council Recommendation 1999/519/EC of 12 July 1999 on the limitation of exposure of the general public to electromagnetic fields (0 Hz to 300 GHz), OJ 1999 L 199/59.

Declaration of the Council of the European Communities and of the Representatives of the Governments of the Member States, meeting in the Council on the Programme of action of the European Communities on the Environment, OJ 1973, C 112/1.

Impact Assessment Guidelines, SEC/2005/791 adopted 15 June 2005.

Opinion of the Scientific Committee on Plants on Specific Questions from the Commission Regarding the Evaluation of Paraquat in the Context of Council Directive 91/414/EEC, SCP/PARAQ/002-Final, 16 January 2002.

Short Report of the Meeting of the Standing Committee on the Food Chain and Animal Health (Phytopharmaceuticals Section) held on 3 October 2003 in Brussels. SCFA 4/2003. p. 2. Available at: http://ec.europa.eu/food/committees/regulatory/scfcah/phytopharmaceuticals/summary16_en.pdf (accessed 1 March 2010).

Resolution of the Council and Representatives of the Governments of the Member States, meeting within the Council on the continuation and implementation of an EC policy and action programme on the environment (1987–1992), OJ 1987, C328/1.

Review report for the active substance paraquat, 3 October 2003, SANCO/10382/2002–final. Available at: http://ec.europa.eu/food/plant/protection/evaluation/existactive/list_paraquat.pdf (accessed 1 March 2010).

SWEDISH LEGISLATION, PREPARATORY WORK, AND SO ON

Legislative Acts

Miljöskyddslag (SFS 1969:387).
Förvaltningsprocesslag (SFS 1971:291).
Arbetsmiljölag (SFS 1977:1160).
Förvaltningslag (SFS 1986:223).
Plan- och Bygglag (SFS 1987:10).
Lag (SFS 1988:205) om rättsprövning av vissa förvaltningsbeslut.
Strålskyddslagen (SFS 1988:220).
Lag (SFS 1991:639) om förhandsgranskning av biologiska bekämpningsmedel.
Kommunallagen (SFS 1991:900).
Lag (SFS 1998: 808) om införande av miljöbalken.

Decrees

Förordning (SFS 1985:643) med instruktion för kemikalieinspektionen.
Förordning (SFS 1985:836).om bekämpningsmedel (SFS 1985:836).
Förordning (SFS 1988:295) med instruktion för Statens strålskyddsinsititut.
Förordning (SFS 1988:525) med instruktion för Kemikalieinspektionen
Förordning (SFS 2006:524) med instruktion för Statens strålskyddsinstitut.
Verksförordning (SFS 1995:1322) med instruktion för Statens strålkskyddsinstitut.
Förordning (SFS 1998:947) om bekämpningsmedel.
Förordning (SFS 2006:1010) om växtskyddsmedel.

Preparatory Works

Proposition 1969:28.
Proposition 1973:17.
Proposition 1987/88:69.
Proposition 1994/95:230.
Proposition 1997/98:45.
Proposition 2000/01:130.

SOU 1996:103 Del 1. Miljöbalken: En skärpt och samordnad miljölagstiftning för en hållbar utveckling.
SOU 1996:103 Miljöbalken - Del 1. Miljöbalken: En skärpt och samordnad miljölagstiftning för en hållbar utveckling.

Other

Interview with case handler at the Swedish Ministry of the Environment, 8 September 2004.
Interview with officials at KemI, 9 September 2004.
KemI Memo (P. Bergkvist), PM inför beslut om fortsatt godkännande för EBDC-medel. 9 November 1994 (not published, but available from the author).
KemI Memo (P. Bergkvist), Yttrande over kompletterand material från Rohm and Haas och Elf Atochem Agri angående överklagandena av Kemikalieinspektionens beslut den 9 november 1994 i fråga om bekämpningsmedel innehållande EBDC, 23 August 1995 (not published, but available from the author).
Kommentarer till Statens strålskyddsinstituts allmänna råd (SSI FS 2003:3) om begränsning av allmänhetens exponering för elektromagnetiska fält. Available at: www.ssi.se/forfattning/pdf/2002_3Komment.pdf (accessed 1 March 2010).
Kommentarer till Statens strålskyddsinstituts allmänna råd (SSI FS 2003:3) om begränsning av allmänhetens exponering för elektromagnetiska fält.
Myndigheternas försiktighetsprincip om lågfrekventa electriska och magnetiska fält-en vägledning för beslutsfattare. Available at: www.av.se/dokument/publikationer/adi/adi_477.pdf (accessed 1 March 2010).
Statement by Swedish Foreign Office official Anders Kruse, quoted in Swedish daily newspaper *Svenska Dagbladet*, 24 August 2004. Available at: www.svd.se/dynamiskt/inrikes/did_8027374.asp (accessed 1 March 2010).
Statens strålskyddsinstituts föreskrift (SSI FS 2002:3), Allmänna råd om begränsning av allmänhetens exponering för elektromagnetiska fält (General Advice on Limiting the Public's Exposure to Electomagnetic Fields (hereinafter the General Advice).
Översiktsplan 2003 Nyköpings kommun - Kommunens vilja och ambition, 50 and 88. Available at: www.nykoping.se/op2003/pdf/huvud.pdf (accessed 1 March 2010).

UNITED KINGDOM

Legislative Acts

Health and Safety at Work etc. Act 1974.
Food and Environment Protection Act 1985.
Environment Act 1995.
Health Protection Agency Act 2004.

Regulations and Orders

Control of Pesticides Regulations, SI 1986/1510.
Control of Pesticides (Advisory Committee on Pesticides) Order 1985, SI No. 1985 No. 1516, 1517.
The Town and Country Planning Order 1995, Statutory Instrument 1995 No. 418.
The Management of Health and Safety at Work Regulations 1999.
Plant Protection Products Regulations 2005, SI 1435, s. 4, 5.

Other

Code of Best Practice on Mobile Phone Network Development, Office of the Deputy Prime Minister, 2002.
Guidelines for Environmental Risk Assessment and Management, HMSO 2000. Now updated and available at: www.defra.gov.uk/environment/quality/risk/eramguide/, Sections 3.2–3.5, 3.8, 8.2 (accessed 1 March 2010).
Ibbs Report: 'Improving Management in Government: The Next Steps', A Report to the Prime Minister. HMSO 1988.
Independent Export Group on Mobile Phones, *Mobile Phones and Health*, published on 11 May 2000 (The Stewart Report). Available at: www.iegmp.org.uk/index.htm (accessed 1 March 2010).
Mutagenicity of dichlorvos, COM statement COM/02/S2 – January 2002. Available at: www.advisorybodies.doh.gov.uk/com/dichlorvos.htm (accessed 1 March 2010).
National Radiation Protection Board, *Restrictions on Human Exposure to Static and Time Varying Electromagnetic Fields and Radiation*, Doc. NRPB, 4(5) 1993.
National Radiation Protection Board, *Advice on the 1998 ICNIRP Guidelines for*

Limiting Exposure to Time-varying Electric, Magnetic and Electromagnetic Fields (up to 300 GHz), Doc. NRPB, Vol. 10, No. 2 1999.

National Radiation Protection Board, *Advice on the 1998 ICNIRP guidelines for limiting exposure to time-varying electric, magnetic and electromagnetic fields (up to 300 GHz)*, Doc. NRPB Vol. 15, No. 2 2004. Available at: www.hpa.org.uk/web/HPAwebFile/HPAweb_C/1194947415497 (accessed 1 March 2010).

The Pesticides Safety Directorate's Framework Document. Available at: www.pesticides.gov.uk/corporate.asp?id=232 (accessed 1 March 2010).

The Precautionary Principle: Policy and Application, Report of the Interdepartmental Liaison Group on Risk Assessment (ILGRA) Report. Available at:www.hse.gov.uk/aboutus/meetings/committees/ilgra/minrpt3.pdf (accessed 1 March 2010).

Planning Policy Guidance 8 and Code of Best Practice on Mobile Phone Network Development, Office of the Deputy Prime Minister, 2002.

White Paper, 'This Common Inheritance: Britain's Environmental Strategy', presented to Parliament, September 1990, HMSO, Cm 1200.

UNITED STATES OF AMERICA

Legislative Acts

Administrative Procedure Act 1974, 5 U.S.C. 551 (APA).
The Clean Air Act 1970, 42 U.S.C. 7401.
Data Quality Act, Amendment to the Paperwork Reduction Act, 4 U.S.C. 3501.
Endangered Species Act 1973, 16 U.S.C.
Federal Advisory Committee Act, Public Law 92–463, 5 U.S.C. App. 1.
Federal Food, Drug and Cosmetic Act 1958, 21 U.S.C. 348.
Federal Food, Drug and Cosmetic Act 1996 (FFDCA), 21 U.S.C.
Federal Water Pollution Control Act (Clean Water Act), 33 U.S.C. 1251.
Food Quality Protection Act 1996 (FPQA), Pub L. 104.
The Fungicide, Insecticide, and Pesticide Act, 15 U.S.C., 2601.
National Environmental Protection Act, 42 U.S.C. 4321.
Radiation Control for Health and Safety Act 1968, 42 U.S.C. 263.
The Regulatory Improvement Act 1998, S. 981, 105th Congress, 627.
Telecommunications Act 1996, Pub. L. No. 104–104, 110 Stat. 56.
Toxic Substances Control Act, 7 U.S.C., 136.

Executive Orders

Executive Order No. 12,291, 3 C.F.R. 127(1982).
Executive Order 12,866, 3 C.F.R. 638 (1994).

Other

Evaluating Compliance with FCC Guidelines for Human Exposure to Radiofrequency Electromagnetic Fields, OET Bulletin 65, Edition 97–01, August 1997.

Federal Advisory Committees Follow Requirements, but FCC Should Improve Its Process for Appointing Committee Members, Report to the Chairman, Committee on Government Reform, House of Representatives, General Accounting Office, GAO-05–36, 2004.

Office of Management and Budget, Guidelines for Ensuring and Maximizing the Quality, Objectivity, Utility, and Integrity of Information Disseminated by Federal

Agencies; Notice; Republication, Federal Register / Vol. 67, No. 36 / 22 February 2002.

Office of Management and Budget, Regulatory Analysis, Circular A-4 To the Heads of Executive Agencies and Establishments, 17 September 2003.

The Presidential/Congressional Commission on Risk Assessment and Risk Management, Risk Assessment and Risk Management in Regulatory Decision-making, Final Report Vol. 2, 1997, 2 and 3.

Questions and Answers about Biological Effects and Potential Hazards of Radiofrequency Electromagnetic Fields, OET Bulletin 56, Fourth Edition, August 1999.

Reorganization Plan No. 3 of 1970, Special Message from the President to the Congress About Reorganization Plans to Establish the Environmental Protection Agency and the National Oceanic and Atmospheric Administration, 9 July 1970. Available at: www.epa.gov/history/org/origins/rorg.htm (accessed 1 March 2010).

1

Introduction

The aim of this book is to discuss the role played by the precautionary principle when deciding on regulatory action in relation to risks posed to modern society, and the effects that such action may have for European and international trade. Primarily, this discussion aims to address the issue of how the European Union (EU) currently makes use of the principle, and what lessons the EU institutions and Member States should draw from the use of precautionary considerations in a number of national and international regulatory regimes.

The last 30 years have brought a new awareness of the hidden costs of industrialisation and the welfare state. Suddenly, societies experience the realisation of risks which they did not know that they had created. Such risks include the highly publicised and potentially catastrophic, such as climate change, but also more hidden risks, such as the long-term effects of exposure to chemicals and low-frequency radiation from, for example, mobile phones and electricity networks. Other potential risks are possibly created through the development of new technologies such as gene modification. At the same time, the emergence of highly integrated regional and global markets for trade in goods and services has limited the possibilities for individual states to successfully enact national risk regulatory strategies.

What connects many of the risks that societies are faced with today is the uncertainty which surrounds them. In many cases, the link between an activity and a risk has not been scientifically established. Furthermore, the extent of a risk may be difficult or impossible to ascertain. So far, there has been deep disagreement on how to approach uncertain risks in different practical situations.

Among academic commentators, some claim that uncertain risks are a reasonable price to pay for the advances of society as manifested in the increased quality of life for the individual. The advantages of modern society outweigh its inherent risks. Others resort to pessimism and hold that the state of the world at this moment indeed is so dire that radical measures need to be considered. The German sociologist Ulrich Beck famously wrote in his *World Risk Society*:

> I know of no greater security and no deeper source of creativity than a pessimism which cannot be outbid. Where everything is at stake, everything can and must be re-thought and re-examined.[1]

It is out of this awareness that the notion of precaution in relation to the environment and health has sprung.

One instrument for dealing with uncertain environmental and health risks to society is the precautionary principle, which has been developing as a legal concept in national legal systems since the late 1960s. The precautionary principle affords the decision-maker an opportunity to act before risks have materialised, when only the contours of what might be risks are visible. As the awareness of both perceived and real risks has increased over the years, the importance of 'the precautionary principle' has also increased. It should, however, be pointed out that, in practice, governmental intervention before the emergence of scientific certainty is far from a modern concept. What is unique about the precautionary principle is that it appears to treat precautionary action in a more conceptual fashion, and, in some cases, it aims to elevate precautionary action to the status of legal principle.

Different states make different judgements as to what constitutes risks, and as a consequence the precautionary principle has been applied differently in different states and regions, as well as in different areas of policy within the same state. At the same time, it has remained a loosely defined concept with unclear borders in regional and international legal systems. This makes the precautionary principle controversial in today's world of globalised markets, where states no longer are entirely free to pursue their own risk policies.

At EU level, the application of the precautionary principle takes place primarily at two different levels. First, the European institutions apply it when drafting legislation and deciding on the correct implementation of EU rules by the Member States. Second, the EU Member States often seek to make use of the precautionary principle when issuing national policies by using it as a justification in order to obtain derogations from EU rules. Sometimes, it appears to be used as 'a magic wand', the invocation of which can justify virtually any policy choice. In order to safeguard the consistent and foreseeable development of EU law, it is necessary to

[1] U. Beck, *World Risk Society* (Blackwell Publishers Ltd, 1999), p. 88.

investigate more closely what is required by the entity wishing to rely on this principle.

Furthermore, the precautionary principle plays a role in international trade law. In a number of cases, World Trade Organisation (WTO) Panels and the Appellate Body have elaborated on its role under the WTO Agreements.

When considering international trade issues, it is often argued that the EU is a strong proponent of precautionary action and that the USA is an opponent of it. The EU suggests that it employs one specific precautionary principle, the application of which is guided by certain defined requirements.[2] The USA prefers to use the term precautionary approach, rather than principle, in order to signify that it affords it less legal weight.

When looking at the policies of individual Member States of the EU as well as the national policies of the USA, the picture becomes somewhat more complicated. In certain areas, it appears that there is great disagreement among the Member States of the EU as to what actually constitutes precaution.[3] Some Member States, such as Sweden, take pride in being 'precautionary' in general, with a firm, legally binding precautionary principle in effect in national legislation. It is investigated below whether such a 'strong' reliance on the precautionary principle is, indeed, equally applicable in all policy areas and what the basis for such an approach is. Conversely, Member States such as the United Kingdom that do not have a tradition of precautionary action may apply a stricter precautionary approach in certain defined areas than in others. Here, too, the issue of how and when the precautionary principle actually comes into play arises.

When considering the differences in the application of the precautionary principle, the question arises as to whether it is possible to talk about one single precautionary principle. This book aims to investigate if it is indeed the case that the 'precautionary principle' is defined differently depending upon the area in which it is being applied, and by whom. In this context, how the precautionary principle has been applied in EU and WTO law and in the national legal orders of Sweden, the UK and the USA is discussed and analysed.

[2] Communication from the Commission of 2 February 2002 on the Precautionary Principle, COM (2001) 1.

[3] See, for example, J. Scott and E. Vos, 'The Juridification of Uncertainty: Observations on the Ambivalence of the Precautionary Principle within the EU and the WTO', in Ch. Joerges and R. Dehousse (eds.), *Good Governance in Europe's Integrated Market* (Oxford University Press, 2002).

It is investigated whether precautionary considerations are applied differently in relation to comparable risks in different policy areas within the same state, and between different states. Since the refusal to accept one single, all-encompassing precautionary principle does not, in itself, mean that precautionary action is not taken in certain individual cases, this book uses the term 'precautionary approach' to indicate a vague and open application of precaution, which is not necessarily based on a conceptual precautionary principle. The term 'precautionary principle' is used to describe an institutionalised and systematic use of precaution in one or more policy areas.

This book is based on the presumption that all responsible governments will act out of precaution in certain situations, regardless of whether they are subscribing to 'a precautionary principle' or not. Below, the issue of whether it is possible to identify a common precautionary principle, or common principles for precautionary action, which are applicable in all policy areas and in relation to different states, is analysed.

The aim of this analysis is to examine if a common foundation for treating the precautionary principle as a legal principle exists. Whether such a common conceptual framework is, indeed, necessary, or whether it can be concluded that the application of precautionary measures is better left up to the discretion of the individual states in each case, is also discussed. The primary target for this discussion is the EU, and of particular importance in this regard is the interrelationship between the EU institutions and the Member States.

The first part of this book deals with the reasons for the increased recourse to precaution in environmental and health protection, as well as the legal status of the precautionary principle in European and international trade law. The second part is devoted to the study of two different policy areas in three different countries.

The research builds on the excellent work already carried by several authors with regard to the current legal status of the precautionary principle in international, as well as European, law.[4] This book aims to take the

[4] See, for example, A. Trouwborst, 'Prevention, Precaution, Logic and Law: The Relationship Between the Precautionary Principle and the Preventative Principle in International Law and Associated Questions', *Erasmus Law Review*, 2:2 (2009), 105; M. Fitzmaurice, *Contemporary Issues in International Environmental Law*, (Edward Elgar Publishing, 2009), pp. 1–67; N. de Sadeleer, *Implementing the Precautionary Principle: Approaches from the Nordic Countries, the EU and the USA* (Earthscan, 2007); J. Scott, *The WTO Agreement on Sanitary and Phytosanitary Measures*, (Oxford University Press, 2007); E. Fisher *et al.*, *Implementing the Precautionary Principle*, (Edward Elgar Publishing, 2006); A. Trouwborst, 'The Precautionary Principle in General International Law: Combating

matter a step further and look beyond the inclusion of the precautionary principle as a concept in legal acts and policy documents at international and European levels. Instead, the primary focus is on the practical application of the concept. Of importance in this regard is the study of the relevant case law in the legal systems investigated. Furthermore, the focus is on the concrete application of the principle by a number of states in their national policies. These case studies aim to discuss whether there are differences between how precautionary considerations are applied in comparable situations in certain areas and in certain countries.

The areas investigated are chemicals (pesticides) and mobile phone systems (base stations). These areas have been chosen because they are examples of areas surrounded by varying, but comparable, amounts of uncertainty as to the risks associated with them. At the same time, they are areas in which different states have chosen to apply different approaches to these risks. The areas surveyed include risks to human health posed by food and technology, and risks to the environment. They share the basic trait that they are, to varying extents, involuntary risks: risks that are difficult or impossible to avoid even through vigilant behaviour on the part of the subjects to the risk.

Throughout this book the precautionary principle is treated by breaking it down to a number of constituent parts. This means that in every legal system and in each case study the same topics are evaluated and compared. This includes, in all cases, an assessment of the definition and status of the precautionary principle in the relevant legal system; the type

the Babylonian Confusion', *RECIEL*, 16:2 (2007), 185; M. B. A. van Asselt and E. Vos, 'The Precautionary Principle and the Uncertainty Paradox', *Journal of Risk Research*, 9, (2006), 4; P. van den Bossche and D. Prevost, 'The Agreement on the Application of Sanitary and Phyto-sanitary Measures', in P. McRory, A. Appleton and M. Plummer (eds.), *The World Trade Organisation: Legal, Economic and Political Analysis*, (Springer, 2005), pp. 231–70; K.-H. Ladeur, 'The Introduction of the Precautionary Principle into EU Law: A Pyrrhic Victory for Environmental and Public Health Law? Decision-making Under Conditions of Complexity in Multi-level Political Systems', *CMLR*, 40:6 (2003), 1455; Scott and Vos, 'The Juridification of Uncertainty', 253–86; J. Cameron, 'The Precautionary Principle in International Law', in T. O'Riordan, J. Cameron and A. Jordan (eds.), *Re-interpreting the Precautionary Principle*, (Cameron May, 2001); W. Douma, 'The Precautionary Principle; Its Application in International, European and Dutch Law', Proefshchrift Rijksuniversiteit Groningen, 6 January 2003; D. Grimeaud, Hoofdstuk II, 'The Precautionary Principle in International Environmental and Trade Law', in M. Faure and E. Vos (eds.), *Juridische afpakening van het voorzorgbeginsel: mogelijkheden en grenzen, Achtergrondstudie voor de Nederlandse Gezondheidsraad*, No. A03/03 2003. A. Trouwborst: *Evolution and Status of the Precautionary Principle in International Law*, (Aspen Publishers, 2002); D. Freestone and E. Hey (eds.), *The Precautionary Principle in International Law: The Challenge of Implementation*, (Kluwer Law International, 1996).

and scope of judicial review applied by courts; if a separation of risk assessment and risk management is upheld; an analysis of the burden and level of proof in place for the issuing of precautionary measures; if costs and benefits are considered; if risks are targeted rationally; and if trade-offs are considered. All of the above issues necessarily come into play when issuing precautionary measures, whether they are explicitly considered in a decision or not.

The first question to be answered in this book is if it is possible to find an objectively justifiable precautionary principle which is generally agreed upon by the states and organisations investigated below. If the answer to this question is yes, then this principle will be outlined and analysed. If the answer to this question is no, then the discrepancies will be identified and considered. Furthermore, the issue of whether it is necessary or even desirable, to establish such a system is also considered. Finally, the future role to be played by the precautionary principle is discussed. The main focus of this discussion is the EU and how the Member States and the EU institutions can, and should, make use of the precautionary principle in order to safeguard an efficient and reliable risk reduction strategy.

Pesticides regulation is, to a large extent, regulated at EU level. Below it is discussed that, regardless of this, EU Member States might, in some cases, attempt to use the precautionary principle in order to apply their own risk levels. Two Member States of the EU, Sweden and the United Kingdom, have been especially chosen in order to illustrate the different approaches assumed by Member States to the precautionary principle within the EU. These two states are then compared with the policies adopted by the United States in the same areas.

The reason for choosing to compare Sweden and the UK is that they traditionally represent different European approaches to precaution when it comes to the environment and health protection. Sweden has a long tradition of far-reaching precautionary thinking when it comes to, for example, chemicals, pollution and public health protection.[5] The UK, on the other hand has traditionally applied a less stringent approach. However, recent 'food scares', for example, the BSE (bovine spongiform encephalopathy) crisis, has put a new emphasis on environmental and health concerns.

[5] See, for example, the preparatory works to the original Swedish Chemicals Code of 1973 in Proposition 1973:17, pp. 96 *et seq.*

The United States, on the other hand, has a tradition of precaution when it comes to environmental and health protection in domestic policies,[6] but is today perceived – at least by the EU – as not accepting the precautionary principle within the international arena. Here, the current use of precautionary considerations by the USA is investigated and compared with the approach assumed by Sweden, the UK and the EU.

The discussion on how and when to use precautionary measures is likely to continue for the foreseeable future, especially as the public becomes increasingly aware of the risks to health and the environment. This book aims to contribute to this debate by providing a detailed look at the precautionary principle in a number of areas and states in order to provide some interesting input for the further development of the precautionary principle in the EU and its Member States.

[6] See, for example, D. Vogel, 'Ships Passing in the Night: GMOs and the Politics of Risk Regulation in Europe and the United States', Working Paper No 2001/16, Robert Schuman Centre for Advanced Studies, European University Institute, 2001.

2

Risk and uncertainty: basic concepts and tools for the application of the precautionary principle

2.1 Introduction

When Rachel Carson published her hugely influential book *Silent Spring* in 1962, she laid the foundation upon which the then recently conceived environmental movement would rest.[1] With a language which is as dystopian as it is strangely poetic, Carson framed a rhetoric that would come to be much used by the environmental movement. Her fears and predictions with regard to chemicals, in particular dichloro-diphenyl-trichloroethane (DDT), still linger in the public minds of Europeans and Americans alike. Reports published over time maintaining that some of these fears are exaggerated have had nowhere near the same resonance as her original work.

Silent Spring was followed by other works in the 1970s which warned of the environmental consequences of industrialisation: *Limits to Growth* was published in 1972 and made the spectacular, and much publicised, claims that, by 1990, the world would have run out of gold, silver, mercury and zinc.[2] In *The End of Affluence*, published in 1974, Paul and Anne Ehrlich predicted the dangers following global cooling and the near end of fish in the seven seas.[3] Fortunately, none of these particular claims have materialised. However, in recent years publications have also emerged with the aim of highlighting the risks to human health and the environment that have actually materialised as a consequence of modernisation.[4]

As a response to the above literature, some authors have also emphasised the lack of development and innovation that would have resulted

[1] R. Carson, *Silent Spring* (Houghton Mifflin, 1962).

[2] D. Meadows *et al.*, *The Limits to Growth: A Report for the Club of Rome's Project on the Predicament of Mankind* (Universe Books, 1974).

[3] P. R. Ehrlich and A. Ehrlich, *The End of Affluence: A Blueprint for your Future* (Ballantine Books, 1974).

[4] The European Environmental Protection Agency, 'Late Lessons From Early Warnings: The Precautionary Principle 1896–2000', Environmental Issue Report No. 22, 2002. Available at: http://reports.eea.europa.eu/environmental_issue_report_2001_22/en (accessed 1 March 2010).

had a certain amount of risks not been taken.[5] Certainly, there is no lack of either nightmare scenarios or sunshine stories in the history of risk regulation.

At the beginning of the present century, the precautionary principle has emerged in an increasingly risk-averse Europe as a 'miracle cure' of sorts to the problems of dealing with risks. In allowing for regulation in spite of scientific controversy and the lack of clear scientific evidence, it seems to provide the regulator with the ultimate tool to address any risk deemed necessary. Until recently, the strict interpretation of the precautionary principle has been criticised mainly by American authors.[6] In Europe, this critical discourse has been elaborated on to a more limited extent.[7]

This chapter aims to provide the theoretical background to the fundamental concepts of risk regulation in general and the precautionary principle in particular. Critical to this discussion is the definition of risk and risk perception, scientific uncertainty, how risks are analysed and how the instruments for weighing the costs and benefits of regulation are set up, as well as risk trade-off analysis. Furthermore, some attempts at conceptualising the precautionary principle are briefly summarised.

2.2 Fundamental concepts

2.2.1 *Defining risk*

Throughout the ages, mankind has been subjected to risks, by chance and as a means of surviving, as well as by choice. No doubt a fundamental part of human existence has been preoccupied with considering which risks

[5] See, for example, A. Wildawsky, 'Trial and Error versus Trial Without Error', in J. Morris, *Rethinking Risk and the Precautionary Principle* (Butterworth-Heinemann, 2000), pp. 22–46.

[6] See, for example, J. Morris, *Rethinking Risk and the Precautionary Principle* (Butterworth-Heinemann, 2000); C. Sunstein, *Risk and Reason: Safety, Law and the Environment* (Cambridge University Press, 2004); C. Sunstein, *The Laws of Fear: Beyond the Precautionary Principle* (Cambridge University Press, 2005); G. E. Marchant and K. L. Mossman, *Arbitrary and Capricious: The Precautionary Principle in the European Union Courts* (American Enterprise Institute Press, 2004).

[7] See, for example, F. Furedi, 'Precautionary Culture and the Rise of Possibilistic Risk Assessment', *Erasmus Law Review*, 2:2, (2009), 198; J. C. Hanekamp, 'Neither Accepatble nor Certain – Cold War Antics for 21st Century Precautionary Culture', *Erasmus Law Review*, 2:2 (2009), 221; Ladeur, 'The Introduction of the Precautionary Principle into EU Law', 1455; and L. Bergkamp, 'Understanding the Precautionary Principle (Parts I and II)', *Env. Liability*, 10 (2002), 18–30 and 67–82.

to take and which to avoid, long before this was known as 'risk regulation'. Despite this, the 'mastery of risk', the intention to control or apply one's behaviour in accordance with risks, is considered a uniquely modern concept.[8] The rational science of risk, as we now know it, originally developed as a mathematical puzzle concerned with optimising gambling results.[9] Later it matured within the framework of economics and finance.[10] During the second half of the twentieth century the influence of risk as a concept had spread to almost all areas of society and life. In fact, at present the 'mastering' of risk has become so fundamental for Western societies that influential commentators argue that the age of modernisation has given way to a 'risk society', in which the assessment and management of risks are its overriding concerns.[11]

Generally, the term 'risk' is understood to express something about the likelihood that a hazard will occur.[12] Definitions usually focus on the possibility of humans altering the chain of events leading to the realisation of a hazard. Thus, the relationship between the actions and the consequences is considered from a non-fatalistic viewpoint.[13] Building on previous literature, Jaeger *et al.* have defined the term risk as:

> A situation or event in which something of human value (including humans themselves) has been put at stake and where the outcome is uncertain.[14]

Stirling has formulated it in the following fashion:

> [Risk is] a situation under which it is possible to define all possible outcomes and confidently assign a probability to reflect the likelihoods of each outcome.[15]

[8] P. L. Bernstein, *Against the Gods: The Remarkable Story of Risk* (John Wiley & Sons Inc., 1996), p. 11.

[9] D. A. Moss, *When All Else Fails: Government as the Ultimate Risk Manager* (Harvard University Press, 2002), p. 22 *et seq*.

[10] Bernstein, *Against the Gods*, p. 97 *et seq*.

[11] See, for example, U. Beck, *Risk Society: Towards a New Modernity* (Sage Publishers, 1992), and Beck, *World Risk Society*.

[12] See, for example, The Royal Society, *Risk: Analysis, Perception, Management*, The Royal Society, 1992, p. 2; J. Wiener, 'Precaution in a Multi-Risk World', Duke Law School Public Law and Legal Theory Working Paper No. 23, December 2001, p. 3.

[13] B. Rohrmann and O. Renn, 'Risk Perception Research – An Introduction', in O. Renn and B. Rohrman, *Cross-cultural Risk Perception: A Survey of Empirical Studies* (Kluwer, 2000), p. 14.

[14] C. C. Jaeger, O. Renn, E. A. Rosa and T. Webler, *Risk, Uncertainty and Rational Action* (Earthscan, 2001), p. 17.

[15] A. Stirling, 'On Science and Precaution in the Management of Technological Risk'. An Esto Project, Prepared for the European Commission – JRC Institute Prospective Technological Studies Seville, May 1999 (available from author), p. 40.

A hazard is a situation which can lead to harm, and harm is viewed as a loss to a human being or a population group, or a negative impact on nature.[16] A risk, as described here, can be defined as 'certain' due to the fact that the hazard is known to exist and because mathematical probabilities can be assigned to its occurrence.

Even if the acknowledgement of risks allows a framework within which to reason about the unknown, risks often come 'trailing clouds of vagueness', as Nobel laureate Kenneth Arrow once put it.[17] Sometimes, these 'clouds of vagueness' are so great that they obscure the hazard associated with a risk, or the harm that this hazard may inflict on humans or on nature. In these cases it may not be possible to determine precisely what the hazard entails or the extent (or mere existence) of the harm such a hazard may bring. It is, therefore, no longer possible to determine mathematically the probability of the occurrence of harm and the risk becomes 'uncertain'. The economist John Maynard Keynes once described 'uncertain knowledge' in the following manner:

> I do not mean merely to distinguish what is known for certain from what is only probable. The game of roulette is not subject, in this sense, to uncertainty. [...] The sense in which I am using the term is that in which the prospect of a European war is uncertain, or the price of copper and the rate of interest twenty years hence, or the obsolescence of a new invention. [...] About these matters, there is no scientific basis on which to form any calculable probability whatever. We simply do not know![18]

For the purposes of this book, these risks will be referred to as 'uncertain risks'.

One way to categorise different risks is thus to separate 'certain risks' from 'uncertain risks'. These two groups of risks can be further divided into the subgroups of 'voluntary risks' and 'involuntary risks'. Voluntary risks are those risks to which one actively chooses to subject oneself. Examples of such risks include skydiving and smoking. Involuntary risks are those risks about which one has little or no choice to accept them or not. Examples from this category include contaminated drinking water and a society's use of nuclear power. Some authors hold that people in general consider involuntary risks to be more troublesome

[16] The Royal Society, *Risk: Analysis, Perception, Management*, p. 3.

[17] K. I. Arrow, 'Know a Hawk from a Handsaw', in M. Szenberg (ed.), *Eminent Economists: Their Life and Philosophies* (Cambridge University Press, 1992).

[18] J. M. Keynes, 'The General Theory of Employment', *The Quarterly Journal of Economics*, 51 (1937), 213.

than voluntary risks. The reason for this seems to lie in the fact that there is no choice as to whether to take the risk or not, and therefore levels of accepted risk are lower. Connected to this is the lack of control one has of a situation imposed by involuntary risks. As a consequence of this, people are in general assumed to accept a higher level of risk when they have some way in which they can influence the outcome of a situation.[19]

2.2.2 *The perception of risk*

It has been held that scientific experts no longer play the prominent role in risk regulation that they once did. Increasingly, the media, stakeholders and other scientists comment on, disagree with and debate scientific findings.[20] Some commentators find this to be a positive development and argue in favour of further input by lay people in risk assessments.[21] However, it is important to bear in mind some cognitive problems when thinking about risks.

Non-experts generally do not view risk in an objective and rational manner. Instead, risks tend to be valued differently depending on how they are presented. In general, people tend to gamble more in order to avoid harm than they are willing to gamble for gain, even if the mathematical probabilities are equal. Furthermore, lay people tend to overestimate low probabilities and tend to be generally risk averse.[22] This behaviour leads to what Viscusi has called 'the precautionary paradox', which means that people apply the contradictory behaviour of taking great care to avoid low probability risks, while, at the same time, engaging in higher probability

[19] C. Starr, 'Social Benefit Versus Technological Risk', *Science*, 165 (1969), 1232. For a critique of these findings, see P. Slovic, B. Fischoff and S. Lichtenstein, 'Facts and Fears: Understanding Perceived Risk', in P. Slovic, *The Perception of Risk* (Earthscan, 2000), pp. 146–8. It is held by Slovic *et al.* that the involuntariness of the risks in the Starr study were not necessarily what caused the aversion. Rather it is suggested that the catastrophic nature of many involuntary risks may be the reason behind the attitudes presented. See also, S. Jasanoff, *Science at the Bar*, Harvard University Press, 1995, p. 13, and Sunstein, *Risk and Reason*, note 6, p. 67.

[20] R. Löfstedt, 'Risk Communication and Management in the Twenty-First Century', 7:3 *International Public Management Journal* (2004), 335.

[21] See, for example, M. O'Brien, *Making Better Environmental Decisions: An Alternative to Risk Assessment* (MIT Press, 2000).

[22] D. Kahneman and A. Tversky, 'Prospect Theory: An Analysis of Decision under Risk', *Econometrica*, 47:2 (1979), 263.

risks without much thought.[23] In short, their behaviour is not rational. An expert in a certain field values a risk very differently from someone who is not an expert.[24]

Slovic *et al.* summarised four basic issues that influence lay people's perception of risks. First, cognitive limitations coupled with the anxieties of considering life as a gamble lead to uncertainties being denied, risks being distorted and statements of fact overestimated. Second, the perception of risks is often influenced by the familiarity with certain types of hazards. Third, like experts, lay people tend to put emphasis on the frequencies of death. However, unlike experts, they place great significance on qualitative aspects of risks, such as dread, the likelihood of fatal accidents and catastrophic potential. Finally, Slovic *et al.* concluded that it can be very difficult to change the perception of risk, even when 'evidence' is presented.[25]

Sunstein pointed out three basic tenets that most people use to frame their thinking of risk. First, people tend to think of risks in an all-or-nothing manner. Either the risk is taken, or it is not. However, in most cases, there is a wide spectrum between taking a risk and not taking it. Second, people tend to believe in the 'benevolence of nature'. Thus, in surveys, people state that they do not fear natural pesticides as much as they fear man-made ones. However, in reality, natural pesticides are just as toxic as man-made pesticides.[26] Third, people believe in the possibility of zero risk. As is discussed in more detail below, science, in most cases, does not lead to certain results, and uncertainty is an integral part of all risks. Therefore, an absolute zero-risk approach is utopian at best, unless all risk-taking were halted, which most people would find less than desirable.[27]

Another key aspect of the perception of risk is the public trust in the risk messenger. Since most information about risks is not received through personal experience but through second-hand learning, the confidence of the intermediary in this regard is of crucial importance.[28] Löfstedt holds

[23] K. Viscusi, *Rational Risk Policy* (Oxford University Press, 1998), pp. 11–14 and 25.

[24] See S. Breyer, *Breaking the Vicious Circle: Towards Effective Risk Regulation* (Harvard University Press, 1993), p. 21, Table 4.

[25] P. Slovic, B. Fischoff and S. Lichtenstein, 'Rating the Risks', in: P. Slovic, *The Perception of Risk* (Earthscan, 2000), p. 119. Similar findings are presented in Slovic *et al.*, 'Facts and Fears', pp. 136–53. See also, Slovic, *The Perception of Risk*, pp. 264–75.

[26] Viscusi, *Rational Risk Policy*, pp. 86–8.

[27] Sunstein, *Risk and Reason*, p. 36.

[28] Rohrmann and Renn, 'Risk Perception Research – An Introduction', p. 31.

that the distrust in policy-makers and industry in the 1990s contributed greatly to the public's declining feeling of safety in Europe, although, in reality, society was safer than ever before.[29]

Furthermore, the perception of risk might be influenced by the fundamental values a person subscribes to, or to which cultural group he or she belongs. The perception of risk may also vary according to national backgrounds.[30]

The apparently inherent irrationality in the public perception of risks has led some commentators to state that public opinion that is ill-informed or based upon biases and heuristics should be disregarded by a responsible risk regulator.[31] Some have even stated that current risk regulation that is not solely based upon rational criteria leads to a substantial amount of deaths which could have been prevented had a rational allocation of resources been the determinant factor.[32] Other commentators have held that allowing public opinion to influence the assessment of uncertain risks is essential in order to allow ethics, values and politics to influence scientific conclusions and create 'robust' solutions.[33]

2.2.3 Scientific uncertainty, prevention and precaution

Above, a distinction between 'uncertain' and 'certain' risks was made. However, even when dealing with certain risks, uncertainty comes into play, because even if the hazard is known and the probability of harm is possible to estimate, uncertainty will pertain to the exact time and magnitude of its occurrence, as well as to the exact quality of the measurements upon which the estimation is based.[34]

In risk literature some commentators have defined uncertainty as a situation in which the hazard and harm is known, but it is impossible to assign probabilities to its realisation. Situations in which the outcomes are not known and the probabilities are impossible to assign are then

[29] R. Löfstedt, *Risk Management in Post-Trust Societies* (Palgrave Macmillan, 2005), p. 9.

[30] Rohrmann and Renn, 'Risk Perception Research – An Introduction', pp. 33–8. For a concrete example in this regard, see C. Sunstein, On the Divergent American Reactions to Terrorism and Climate Change, *Colum. L Rev.*, 107 (2007), 503.

[31] Wiener, 'Precaution in a Multi-Risk World', p. 4, and B. Duriodé, 'The True Cost of Precautionary Chemicals Regulation', *Risk Analysis*, 23:2 (2003), 395.

[32] T. O. Tengs, M. E. Adams, J. S. Pliskin, D. G. Safran, J. E. Siegel, M. C. Weinstein and J. D. Graham, 'Five-hundred life-saving interventions and their cost-effectiveness', *Risk Analysis*, 13 (1995), 369–90.

[33] S. O. Funtowicz and J. R. Ravetz, 'Science for the Post-Normal Age', *Futures*, 25:7 (1993), 749.

[34] See, for example, F. Knight, *Risk, Uncertainty and Profit* (Houghton Mifflin Co., 1933).

known as ignorance.[35] Others have focused on a definition of uncertainty as simply a lack of information.[36] However, Walker *et al.* have pointed out that uncertainty can also prevail in situations where information is abundant.[37] More information does not necessarily remove the uncertainty. Walker *et al.* have, instead, proposed a broad definition which holds that uncertainty is 'any departure from the unachievable ideal of complete determinism'.[38] Following such a definition, any situation would be fraught with a varying degree of uncertainty. Complete determinism is utopian. In the most commonly accepted definitions of the precautionary principle, it is clearly established that it applies in cases where 'full scientific proof' is lacking.[39] In order for something to be scientifically proven, it is not required that all uncertainty be resolved. Instead, scientific certainty is generally accepted to be in place when something can be considered to be 95 per cent certain.[40] Thus, the precautionary principle can be considered to apply in the spectrum between 0 and 95 per cent certainty. Where something is scientifically certain, it is likely that the principle of preventive action will be applied.[41] According to this principle, anyone who carries out an activity has the duty to prevent any harm to the environment or to human health which stems from the activity. It is generally understood that this principle applies where a causal link between the activity and the harm is clearly established. The precautionary principle extends the scope of the preventive action to situations in which the causal link is not as clear. Using the above dichotomy, the prevention principle applies when tackling certain risks, and the precautionary principle applies when dealing with uncertain risks.[42]

[35] Stirling, 'On Science and Precaution in the Management of Technological Risk'.

[36] See, for example, S. O. Funtowicz and J. R. Ravetz, *Uncertainty and Quality in Science for Policy* (Kluwer, 1990).

[37] W. E. Walker, P. Harremoes, J. Rotman, J. P van der Sluijs, M. B. A. van Asselt, P. Janssen and M. P. Krayer von Krauss, 'Defining Uncertainty: A Conceptual Basis for Uncertainty Management in Model-Based Decision Support', *Integrated Assessment*, 4:1 (2003), 8.

[38] *Ibid.*

[39] See, below, Chapter 3 in relation to the Rio Declaration.

[40] J. Lemons, K. Schrader-Frechette and C. Cranor, 'The Precautionary Principle: Scientific Uncertainty and Type I and Type II Errors', *Foundations of Science*, 2:2 (1997), 207 and J. Tickner and C. Raffensberger, 'The Precautionary Principle in Action: A Handbook', Science and Environmental Health Network, p. 13. Available at: www.mindfully.org/Precaution/Precaution-In-Action-Handbook.pdf (accessed 1 March 2010).

[41] For a discussion on the status of the precautionary principle in international and European law, see, below, Chapters 3 and 4, respectively.

[42] In relation to this, see also Trouwborst, 'Prevention, Precaution, Logic and Law', 105. He argues that the precautionary principle consumes the prevention principle. The latter

To assist us when thinking in terms of ranges of uncertainties, van Asselt has established a number of degrees of uncertainty, ranging from inexactness to indeterminacy.[43] First, inexactness is the degree closest to certainty in the spectrum. It seems to encompass any uncertainty that can be described adequately in statistical terms. For example, it represents the uncertainty inherent in all data. Measurements of reality can never represent the 'true' value of what is being measured. Inaccuracy or imprecision will always flaw the values established.[44]

The second degree of uncertainty is lack of observations or measurements. This uncertainty is due to a lack of information. With more measurements and research, such uncertainty may possibly be reduced and become scientifically certain.

The third degree of uncertainty can be described as a result of conflicting evidence.[45] Divergent scientific reports exist, which give different answers to the same question. Such divergent opinions may be resolved when additional research is performed and the level of uncertainty may be decreased to inexactness. Similarly, divergent opinions may indicate that no conclusive method exists for evaluating the evidence, and it therefore borders on being immeasurable.

The fourth degree of uncertainty is described as existing due to the fact that the issue is practically immeasurable. It is simply not possible, at least not in the foreseeable future, to obtain the data necessary to reach the desired level of certainty.

The final degree of uncertainty is ignorance, which means that the actual risk is hidden or unexpected and is, therefore, not possible to consider. When considering the other four levels of uncertainty, the risk in

thus lacks relevance. However, this fails to take into account that the impetus for action to counter a risk under the prevention principle is more pronounced and generally less controversial than in the case of the precautionary principle, where the risk is uncertain.

43 M. B. A. van Asselt, *Perspectives on Uncertainty and Risk: The PRIMA Approach to Decision Support* (Kluwer, 2000).

44 Walker *et al.*, 'Defining Uncertainty', 12.

45 M. B. A. van Asselt positions this uncertainty as the second degree (van Asselt, *Perspectives on Uncertainty and Risk*) and as the fourth degree together with Rotmans, in: J. Rotmans and M. B. A. van Asselt, 'Uncertainty Management in Integrated Assessment Modeling: Towards a Pluralistic Approach', *Environmental Monitoring and Assessment* 69:2 (2001), 113. This author finds it most useful to position it between lack of information and immeasurability due to the fact it appears to be a more complex form of uncertainty than lack of information in that it is not necessarily enough to conduct more research to reduce the uncertainty. At the same time, it is not as complex as being described as immeasurable due to the fact that different scientists actually have claims to have measured the issue, albeit having reached diverging conclusions.

question is identified, but its magnitude and probability remain uncertain. With regard to ignorance, the risk itself is 'off screen' and cannot, therefore, be considered.

If this system is applied to the discussion above, the conclusion is that the prevention principle applies in the inexactness phase in van Asselt's scheme, since the causal link here is adequately established, probabilities are possible to assign to the risk, and the remaining uncertainty is merely due to the imperfect nature of all scientific measurements. Ergo, the preventative principle applies in instances of scientific certainty, that is, in cases that are estimated to be 95 per cent certain.

In the second and third phases – lack of information and scientific controversy respectively – the precautionary principle comes into play. Here, the 95 per cent certainty requirement is impossible to fulfil.

2.3 Striking a balance in risk regulation

2.3.1 *The three phases of risk regulation*

2.3.1.1 Risk assessment

Risk assessment is the general term which refers to the scientific element in risk regulation.[46] In some of the literature this phase is referred to as risk estimation, and risk assessment refers to the whole process of making decisions about risks.[47] However, for the purposes of this study risk assessment is understood as the scientific evaluation of risk.

Risk assessments can be quantitative, where the emphasis lies on the expression of numerical results, or they can include qualitative assessments. Broader definitions of risk assessment include the analysis of risk perception, economic considerations and risk trade-off analysis.[48] Here, such considerations will be considered in the risk management phase.

It has been held that the process of risk assessment contains three main parts: (1) the identification of potential outcomes; (2) the estimation of the magnitude of these outcomes; and (3) the probability of the realisation of the outcomes.[49] The aim of the process is to provide the decision-maker in

[46] See, for example, National Research Council, *Risk Assessment in the Federal Government: Managing the Process* (National Academy Press, 1983), p. 17.

[47] See, for example, The Royal Society, *Risk: Analysis, Perception, Management*, p. 3.

[48] National Research Council, *Risk Assessment in the Federal Government*, p. 17.

[49] The Royal Society, *Risk: Analysis, Perception, Management*, p. 3, and J. King, 'Application of Risk Assessment in Policy and legislation in the European Union and in the United

the risk management phase with tangible and accessible data upon which a decision can be based. In the United States four steps are generally emphasised as constituting the risk assessment phase: (1) hazard identification; (2) dose-response assessment; (3) exposure assessment; and (4) risk characterisation.[50] This has also been the view of the European Commission (hereinafter the Commission).[51]

The main reason for maintaining a strict separation between risk assessment and risk management is to avoid policy preferences influencing what should be an 'objective' scientific assessment.[52] However, the question of objective science is a problematical one. The assessment of risks will invariably be influenced by the social interaction between scientists, the reigning scientific and ethical paradigms, and boundary work.[53] At European level, the late recognition of, and subsequently late reaction to, Creutzfeld-Jakob's disease has been attributed largely to the influence of 'British thinking' in the relevant scientific committee. A large number of the members of the committee were British, influential and initially not inclined to favour evidence perceived as negative for the UK.[54] This is but one example of how outside elements influence scientific work. However, recognising this, and striving to minimise non-scientific influence, is necessary to maintain the credibility of scientific advice. Thus, formally separating scientific assessment from political decisions based upon that information is seen as essential by most commentators. As one commentator put it:

> Of course, science has never been value-free, but maybe it should continuously strive to become so and to preclude, rather than to include, external influence.[55]

Kingdom', in P. Calow, *Handbook of Environmental Risk Assessment and Management* (Blackwell Science, 1998), pp. 249 *et seq.*

50 National Research Council, *Risk Assessment in the Federal Government*, pp. 19–20.

51 Communication from the Commission of 2 February 2002 on the Precautionary Principle, COM (2001), 13 and Annex III.

52 *Ibid.*, 33.

53 For a brief summary of this critique, see S. Jasanoff, *The Fifth Branch: Science Advisors as Policymakers* (Harvard University Press, 1990), pp.12–15. See also, Jaeger *et al.*, *Risk, Uncertainty and Rational Action*, pp. 213–21, and K. Shrader-Frechette, 'Scientific Method, Anti-Foundationalism and Public Decision', in E. Löfstedt and L. Frewer (eds.), *The Earthscan Reader in Risk and Modern Society* (Earthscan, 1998), pp. 45–57. For the theoretical foundation for this critique, see T. Kuhn, *The Structure of Scientific Revolutions* (University of Chicago Press, 1962).

54 E. Vos, 'EU Food Safety Regulation in the Aftermath of the BSE Crisis', *Journal of Consumer Policy*, 23 (2000), 231 *et seq.*

55 Duriodé, 'The True Cost of Precautionary Chemicals Regulation', 395.

Other commentators have criticised the use of risk assessments on the basis that they are narrow in scope and risk missing important alternative approaches to risks.[56]

2.3.1.2 Risk management

Risk management has been defined as 'the process of evaluating alternative regulatory actions and selecting among them'.[57] In the risk management phase, the results of the risk assessment are considered when deciding on the appropriate response to a specific risk. Two main reasons have been put forth in support of not allowing the scientific assessment alone to provide the answer to how to respond to a risk. First, as discussed above, science is only one of the components to be considered when deciding on how to respond to a risk. Other factors, such as the costs and benefits of certain regulation, risk trade-offs and public opinion may be as important as the scientific evaluation of the facts underlying a decision. Most activities are associated with varying degrees of certain or uncertain risks, but the resources of all societies are limited, and a selection of which risks to target must be made taking aspects other than the scientific data into account.[58]

Second, the risk management phase needs to ensure a high degree of legitimacy and accountability. Citizens have a right to be able to hold those who make decisions relating to risks accountable for their actions.[59] Scientists are not accountable to the public in the same way that politicians or other public officials are. Therefore, it is argued that it is important that the final decision rests within a structure that can be held accountable within established democratic systems.

Since risk management is disconnected from risk assessment, and there is no universally agreed upon structure for managing risks, different states apply different risk management strategies. It is up to the risk managers of each state to decide upon what level of protection each state is to employ.[60] In doing so, different states choose to attribute varying weight to scientific evidence depending on the area which

[56] See, for example, O'Brien, *Making Better Environmental Decisions.*

[57] National Research Council, *Risk Assessment in the Federal Government*, p. 18.

[58] Viscusi, *Rational Risk Policy*, pp. 84–5.

[59] Weber argued along these lines in relation to bureaucracies needing a structure above them to give them policy guidance. See M. Weber, *The Theory of Social and Economic Organization* (Henderson and Parson, 1947), p. 335.

[60] See, for example, Article 2.1 of the WTO Agreement on the Application of Sanitary and Phytosanitary Measures.

is to be regulated.[61] The precautionary principle is a risk management tool, which can be used by a regulator to enact measures in the face of uncertainty.[62]

2.3.1.3 Risk communication

The final step in the chain of risk analysis has traditionally been assigned to the issue of communicating the assessed and regulated risk. In recent years, the interest in risk communication has increased significantly. There may be several explanations for this, for example, the increase in transparency and right-to-know laws, but the most tangible reason may be the recent 'food scares' experienced in Europe and the subsequent focus on environmental and health issues by the media and the general public.[63] The more concerned the public becomes, the more it seeks to influence the management of risks. The media, sensing the interest in risk-related issues, spurs the interest it sees in citizens by covering risks extensively. However, as mentioned above, the media often simplifies the information it transmits and has problems communicating scientific uncertainty. Thus, the public discourse becomes simplified, and the public may make demands that risk regulators find difficult to fulfil.

The public tends to be 'ambiguity-averse' and demands clear answers to the magnitude and probability of risks.[64] When the regulator cannot provide an unambiguous answer, the degree of trust in the regulator decreases. The regulator, seeking to avoid this, tries to take public opinion into account at an earlier stage in the risk management phase. However, since public opinion may be ill-informed and simplified, the resulting regulation may be equally flawed. Then, the flawed regulation is again picked up by the media and criticised by the public. In this way, what US Supreme Court Justice Breyer has referred to as a 'vicious circle' is created and reinforced.[65]

In such a system as that described above, the issue of communicating risk is absolutely central, since it will determine not only how the public reacts to the risk in question, but also how it reacts to related risks in

[61] This is returned to and investigated in the case studies in the following chapters.

[62] Communication from the Commission of 2 February 2002 on the Precautionary Principle, COM (2001) 1, p. 12.

[63] V. T. Covello, 'Risk Communication', in P. Calow, *Handbook of Environmental Risk Assessment and Management* (Wiley Blackwell, 1998), p. 520.

[64] Bernstein refers to Daniel Ellsberg's work on ambiguity aversion, in Bernstein, *Against the Gods*, p. 280.

[65] Breyer, *Breaking the Vicious Circle*, pp. 50–1.

the future. Slovic has held that, in order to inform about risks effectively, risk communicators must be well educated in the fundamentals of the risk assessments which have been carried out, including its limitations and its unresolved uncertainties. Furthermore, the risk communicator must be aware of the limitations of the public understanding of risks. As discussed briefly above, lay people's perceptions of risks are often flawed and deeply held beliefs are difficult to modify. Moreover, comparing risks statistically to other risks may be a useful way of putting things into perspective.[66]

Fischoff emphasises that risk communication strategies have to be developed on a case-by-case basis, but identifies seven 'developmental stages' for risk communication. These include getting the numbers right, communicating the numbers, explaining the meaning of the numbers, showing that similar risks have been accepted in the past, showing that accepting the risks is actually a good deal, treating the public nicely and, finally, making the public partners in the risk management.[67]

Furthermore, the issue of trust in the risk manager is central to how the risk communication will be received. Löfstedt[68] and Slovic[69] have noted that there is a strong correlation between risks perceived as high and high distrust in risk managers. Furthermore, it is much easier to lose trust than to build it up again. In order to inform and educate the public about risk effectively, the risk communicator must be aware of the level of trust enjoyed by the regulator, as well as be informed about the perception of the risk by the public.[70]

2.3.2 Cost–benefit analysis

Today, most commentators consider some kind of cost–benefit consideration an essential instrument in order to ensure an effective and efficient regulatory approach to risks. This is due to the fact that it is simply not possible to eliminate all risks in societies in which the funds are limited.

66 P. Slovic, 'Informing and Educating the Public about Risk', in P. Slovic, *The Perception of Risk* (Earthscan, 2000), pp. 183–91.

67 B. Fischoff, 'Risk Perception and Communication Unplugged: Twenty Years of Process', *Risk Analysis*, 15:2 (1995), 137.

68 Löfstedt, *Risk Management in Post-Trust Societies*, p. 9.

69 P. Slovic, 'Perceived Risk, Trust and Democracy', in P. Slovic, *The Perception of Risk* (Earthscan, 2000), pp. 317–18.

70 Löfstedt, *Risk Management in Post-Trust Societies*, pp. 4–11.

Thus, risk managers need to be selective.[71] Some commentators have held that regulation that does not take adequate account of the relevant costs leads to wasteful allocation of resources and even unnecessary loss of lives.[72] In some cases, blindness to the costs of regulation has led to exorbitant costs in order to eliminate small or negligible risks. One aspect of this is that, in some cases, a regulator might incur 90 per cent of the costs of regulation in order to eliminate the last 10 per cent of the risk.[73] This is what Breyer has called 'the problem of "the last ten percent"'.[74]

Cost–benefit analysis has been defined as 'a full accounting of the consequences of risk reduction, in both quantitative and qualitative terms'.[75] It seeks to investigate 'whether the expected benefits from a proposed activity outweigh its expected costs'.[76] In the United States the use of cost–benefit analysis as an instrument in the risk management phase has been enshrined in executive orders issued by every president since Reagan, even if it lacks firm support in the law.[77] However, bearing in mind that Republican (Reagan, Bush I and Bush II) as well as Democrat (Clinton) presidents have supported the concept, there seems to be a close political consensus relating to its practice.[78]

In Europe discussions on the use of the cost–benefit instrument are much more recent. A non-binding reference to it is included in the Commission Communication on the Precautionary Principle.[79] Furthermore, EU legislation is subjected to the overriding principle of proportionality,[80] which contains the application of some form of cost–benefit analysis, although it is not defined or formalised.[81]

[71] Viscusi, *Rational Risk Policy*, p. 85.

[72] See, for example, Tengs *et al.*, 'Five-hundred life-saving interventions and their cost-effectiveness'; J. D. Graham, 'Legislative Approaches to Achieving More Protection Against Risk at Less Cost', *U Chi L F* (1997), 13.

[73] Viscusi, *Rational Risk Policy*, p. 99.

[74] Breyer, *Breaking the Vicious Circle*, pp. 11–12.

[75] Sunstein, *Risk and Reason*, p. 106.

[76] Slovic, *The Perception of Risk*, p. 122.

[77] Sunstein, *Risk and Reason*, p. 6. For an early and influential evaluation of the costs associated with regulation in a number of different areas, see R. W. Hahn and J. A. Hird, 'The Costs and Benefits of Regulation: Review and Synthesis', *Yale J. on Reg.*, 8 (1991), 233.

[78] Graham has argued forcefully for Congress to mandate in law the Executive Office of the President to carry out cost–benefit analyses and a clear prioritisation of risks. See Graham, 'Legislative Approaches to Achieving More Protection Against Risk at Less Cost', 7 and 49–57.

[79] COM (2000) 1, p. 18.

[80] Article 5 of the Treaty establishing the European Communities.

[81] See J. Jans, 'Proportionality Revisited', *Legal Issues of Economic Integration*, 27:3 (2000), 239–65.

In order for it to be possible to carry out a cost–benefit analysis, non-economic values, such as environmental protection and human lives, need to be monetised in order for them to be compared with the costs associated with protecting them. For economists this is not a controversial issue since it enables the necessary comparison and gives 'real' value to issues that might otherwise be ignored.[82]

However, the use of cost–benefit analysis has been criticised on the grounds that, when it comes to uncertain risks, there is not sufficient information for a cost–benefit analysis to provide a clear answer.[83] Since the causal link between a risk and an activity may be hidden, or the magnitude of a risk may be impossible to estimate, some variables in the cost–benefit equation might be overlooked, or assigned flawed values. Thus, the result of a cost–benefit analysis may not adequately depict reality, and, according to some, only blur the picture. Instead of 'scrambling' the picture with a cost–benefit analysis, it is argued that it is better to view all the facts individually.[84] Connected to this is the issue that regulators may overestimate technology and processes which are well understood by them and which they may already be promoting.[85]

Others have argued that, far from being value-free, cost–benefit analysis is value-based, especially when it comes to assigning costs and benefits to future events. How much is a human life worth? How do we value the benefits of our children having the opportunity to enjoy untouched forests? In order to carry out a cost–benefit assessment, monetary estimations have to be assigned to these variables. There is obviously no objectively scientific method for doing this, so the regulator falls back on societal values to make hypothetical estimations.[86] Connected to the above are the arguments of critics who hold that cost–benefit analysis is, in general, immoral or unethical, in that it forces human lives and other immeasurable entities to be assigned a monetary value.[87]

[82] K. Viscusi, 'Monetizing the Benefits of Risk and Environmental Regulation', *Fordham Urb. L. J.*, 33 2006, 1003.

[83] Slovic, *The Perception of Risk*, p. 124. For a recent general critique of the use of cost-benefit analysis in regulation, see F. Ackerman and L. Heinzerling, *Priceless: On Knowing the Price of Everything and the Value of Nothing* (The New Press, 2004).

[84] L. Heinzerling, 'Regulatory Costs of Mythic Proportions', *Yale Law Journal*, 107 (1981), 2069.

[85] See, in this regard, J. C. Johnston, 'A Game Theoretic Analysis of Alternative Institutions for Regulatory Cost-Benefit Analysis', *U. Pa. L Rev.*, 150:5 (2002), 1343.

[86] M. Douglas and A. Wildawsky, *Risk and Culture* (University of California Press, 1992), pp. 70–3.

[87] See E. Anderson, *Value in Ethics and Economics* (Harvard University Press, 1993).

It is important to bear the criticisms of cost–benefit analysis in mind when considering the basis for regulation. However, not even the most hardened defenders of the rational regulation of risks consider that cost–benefit analysis should function as the be-all and end-all of risk regulation. Instead, it has been highlighted that it should work together with other analytical tools, such as risk assessment and the understanding of risk perception.[88] It has also been highlighted that, at the end of the day, regulatory decisions regarding risks are not merely technocratic, but of a policy nature. Thus, it is likely that criteria other than the strictly rational criterion will influence the outcome of most decisions.[89]

Cost–benefit analysis for the purposes of this book is not understood as an all-or-nothing formula for providing an unequivocal answer to the question of whether to regulate or not. Instead, it is to be understood as a tool for the decision-maker to use in order to put the benefits as well as the costs of the proposed regulation 'on screen' in order to be able to make an informed choice on regulation.[90]

The transparency offered by a careful cost–benefit analysis can also be successfully used as a tool to explain different regulatory options in a very tangible way to citizens. When rationally comparing costs and benefits with each other, cognitive errors and biases may be corrected. In this way it is hoped that emotional interpretation will not find its way into a formalised weighing of the facts.[91]

2.3.3 Risk trade-off analysis

The procedures to identify hidden risks, weigh risks against each other and rank them according to the order in which they should be dealt with in order to ensure maximum safety for the lowest cost are complementary

[88] Graham, 'Legislative Approaches to Achieving More Protection Against Risk at Less Cost', 20–2, and J. D. Graham and J. K. Hartwell, 'The Risk Management Approach', in J. D. Graham and J. K. Hartwell (eds.), *The Greening of Industry* (Harvard University Press, 1997), pp. 1–2.

[89] Hahn and Hird, 'The Costs and Benefits of Regulation: Review and Synthesis', 259.

[90] Sunstein, *Risk and Reason*, p. 29.

[91] In this regard, it should be noted that some commentators, although agreeing that a weighing of costs and benefits can help correct biases and heuristics, hold that the main problem of risk regulation is that centralised systems lack the political capital to enter into effective bargaining. Therefore, Johnston considers decentralisation of risk regulation to be the most beneficial way to ensure an effective allocation of resources. See J. C. Johnston, 'Paradoxes of the Safe Society: A Rational Actor Approach to the Reconceptualization of Risk and the Reformation of Risk Regulation', *U. Pa. L Rev.*, 151:5 (2003), 747.

to cost–benefit analysis. In employing such approaches, focus is turned away from individual risks and toward a holistic view, encompassing the whole risk spectrum.

The benefits of such approaches are several. First of all, risks do not appear in isolation.[92] The removal, or targeting, of one risk may be accompanied by countervailing risks. A countervailing risk is the risk of an adverse outcome from a measure that was intended to address a target risk. An example of this would be side-effects of medication. The medication targets one risk, but is accompanied by the risk of creating other harm.[93]

Furthermore, the targeting of one risk may result in a risk offset. This means that removing one risk opens up the field to other, potentially more harmful, risks. One example of this is when the EU, after heavy lobbying from Greenpeace, decided to ban polyvinyl chloride (PVC) plastics in children's toys. However, since only the PVC component was targeted, other plastics, about which very little was known, flooded the market. Thus, the positive effect of targeting the risk posed by PVC was offset by the new risks that emerged as a result of it.[94]

When one risk replaces another risk for the same affected people, this is called risk substitution. Finally, when a countervailing risk poses a new potential outcome for a different group of people than those who benefited from the targeting of the first risk, this is known as risk transformation.[95]

If the goal of the regulator is to achieve overall safety for citizens, all of these factors need to be taken into account. It is not sufficient to focus on the targeting of one risk, since this may be accompanied by an increase in other risks for the same or for another group of people. This is especially important, as well as complicated, in cases of uncertainty. Just as not targeting a risk may lead to adverse effects, the actual targeting of a risk may lead to unforeseen or overlooked countervailing effects or risk offsets. As in the case of cost–benefit analysis, thinking in terms of risk trade-offs does not provide a scientifically acceptable, unequivocal miracle cure, but it does allow the regulator to put some of the hidden consequences of regulation, as well as non-regulation, 'on screen'. Once all the

[92] See, for example, Beck, *World Risk Society*, p. 75.

[93] J. D Graham and J. B Wiener, *Risk versus Risk: Tradeoffs in Protecting Health and the Environment* (Harvard University Press, 1995), p. 23.

[94] B. Duriodé, 'Plastic Panics: European Risk Regulation in the Aftermath of BSE', in J. Morris (ed.), *Rethinking Risk and the Precautionary Principle* (Butterworth-Heinemann, 2000), pp. 142–67.

[95] Graham and Wiener, *Risk versus Risk*, p. 25.

effects of regulation are visible, the regulator can choose the 'risk superior' approach to regulating the risk. This means that the regulator aims at the approach that best focuses on the overall risk level, not on the isolated risk. This may include the ranking of risk-reducing policies. This ranking may be based upon cost-effectiveness,[96] but also on how best to ensure overall safety, regardless of costs.[97] This issue is returned to in more detail in the section on the United States, where Regulatory Impact Assessments (RIAs) often function as an objective basis upon which risk assessments are compared side by side.

2.4 Attempts at defining the precautionary principle

No singular 'universally accepted canonical formulation of the precautionary principle' exists today.[98] As is shown throughout the following chapters, a myriad of definitions exist at international level as well as at domestic level. It is telling that, when interviews were performed with civil servants at the Swedish Ministry of the Environment, a strong proponent of early and strict precautionary action, and the question of the definition of the precautionary principle was raised, one civil servant asked which precautionary principle we were talking about, the one included in the Swedish Planning Code or the one included in the Environmental Code.[99] Thus, it would appear that not even in what is, perhaps, the strongest bastion of precautionary action is one singular principle in effect.

Some attempts to define the precautionary principle in the academic literature have been non-specific and have focused on the importance of protecting nature and human health in general.[100] Other views of

[96] Douglas and Wildawsky, *Risk and Culture*, p. 69.

[97] See, in this regard, Viscusi, 'Monetizing the Benefits of Risk and Environmental Regulation', 1034, who holds that a wasteful allocation of funds would not further the health agenda even if the paramount objective in risk regulation would be health and safety, and no consideration would be given to cost.

[98] P. Sandin, 'A Paradox Out of Context: Harris and Holm on the Precautionary Principle', *Cambridge Quarterly of Healthcare Ethics* 15 (2006), 175. In fact, Sandin has, in a work from 1999, identified 19 different versions of the principle. With the increased discussion on the principle since then, it is likely that even more definitions could be identified today. See P. Sandin, 'Dimensions of the Precautionary Principle', *Human and Ecological Risk Assessment*, 5 (1999), 889.

[99] Interview with case handler at the Swedish Ministry of Environment, 8 September 2004.

[100] See, for example, K. Geiser, 'Establishing a General Duty of Precaution in Environmental Protection Policies in the United States: A Proposal', in C. Raffensperger and J. Tickner

the precautionary principle are influenced in one way or another by the wording of Principle 15 of the Rio Declaration, which is returned to in some detail in the following chapter. These definitions tend to be more specific and focus on the serious or irreversible nature of the specific risk incurred, as well as the cost-effectiveness of the chosen measures.[101] In the European Union context, the Commission has defined the precautionary principle as being applicable:

> in those specific circumstances where scientific evidence is insufficient, inconclusive or uncertain and there are indications through preliminary objective scientific evaluation that there are reasonable grounds for concern that the potentially dangerous effects on the environmental, human, animal or plant health may be inconsistent with the chosen level of protection.[102]

Furthermore, according to the Commission, all measures enacted should be proportional, non-discriminatory, consistent, subject to cost–benefit analysis and should also take scientific developments into account. It has been held by Löfstedt and Fairman[103] that an application of the precautionary principle along the lines outlined by the Commission would be acceptable to most policy-makers, but that, in practice, this approach is not always likely to be upheld.[104]

(eds.), *Protecting Public Health and the Environment: Implementing the Precautionary Principle* (Island Press, 1999), pp. xxi and xxiv. Such a general application is also in effect in Swedish legislation, see, below, Chapter 5.1. See also, B. Wahlström, 'The Precautionary Approach to Chemicals Management: A Swedish Perspective', in C. Raffensperger and J. Tickner (eds.), *Protecting Public Health and the Environment: Implementing the Precautionary Principle* (Island Press, 1999), pp. 52–4, and M. Karlsson, 'The Precautionary Principle, Swedish Chemicals Policy and Sustainable Development', *Journal of Risk Research*, 9:4 (2006), 341.

101 See, for example, C. F. Cranor, 'Asymmetric Information, the Precautionary Principle, and Burdens of Proof', in C. Raffensperger and J. Tickner (eds.), *Protecting Public Health and the Environment: Implementing the Precautionary Principle* (Island Press, 1999), pp. 74–99.

102 Commission Communication on the Precautionary Principle, COM (2000) 1, 10.

103 R. Löfstedt and R. Fairman, 'Scientific Peer Review to Inform Regulatory Decision Making: A European Perspective', *Risk Analysis*, 26:1 (2006) 25–31.

104 Written evidence evaluated by House of Lords Select Committee on Economic Affairs. Select Committee on Economic Affairs, 'Government Policy on the Management of Risk', 5th Report of Session 2005–2006, Vol. II: Minutes of Evidence, Authority of the House of Lords, 2006, 217–18. Available at: www.publications.parliament.uk/pa/ld200506/ldselect/ldeconaf/183/183ii.pdf (accessed 1 March 2010). For a critique of the Commission's approach, see J. D. Graham and S. Hsia, 'Europe's Precautionary Principle: Promise and Pitfalls', *Journal of Risk Research*, 5:4 (2002), 371. The issue is returned to in detail below in Section 4.2.5.

Common criticisms of the precautionary principle include claims that it is poorly defined, absolutist and ignores scientific evidence.[105] In the United Kingdom the House of Lords Select Committee on Economic Affairs even went as far as to state that 'ill-defined and ambiguous terms', such as the precautionary principle, 'are generally unhelpful'.[106] Furthermore, it has been held that it does not provide adequate safeguards against politicians using the precautionary principle to take ill-informed decisions in response to public outcries.[107] Some commentators have held that, in order for the precautionary principle not to be misused, it needs to be applied within a formal framework, such as the Commission Communication on the Precautionary Principle.[108]

Sandin holds that there are two basic ways of looking at the precautionary principle: as argumentative or as prescriptive.[109] The argumentative version of the principle deals with which reasons or arguments may be valid when taking decisions under uncertainty. An example of this is Article 15 of the Rio Declaration, which holds that uncertainty cannot be used as a reason for postponing preventive measures.[110] Thus, the uncertainty surrounding a certain risk should not be allowed to be used as an argument in favour of inaction. The prescriptive version, on the other hand, prescribes a certain form of behaviour in cases of uncertainty. This is a stricter version of the principle and functions along the generic lines

[105] See, for example, D. Bodansky, 'Scientific Uncertainty and the Precautionary Principle', *Environment*, 33 (1991), 4, and D. Bodansky, 'Commentary: The Precautionary Principle', *Environment*, 34 (1992), 4–5. For an overview of the conceptual criticism levied against the precautionary principle, as well as a defence against it, see P. Sandin, M. Peterson, S.-O. Hansson, C. Rudén and A. Juthe, 'Five Charges Against the Precautionary Principle', *Journal of Risk Research*, 5:4 (2002), 288.

[106] Select House of Lords Committee on Economic Affairs, 'Government Policy on the Management of Risk', Vol. I, para. 73. Similar sentiments have been expressed by the House of Commons, which have urged the UK government to avoid using the term 'precautionary principle' due to its unclear meaning. See House of Commons Select Committee on Science and Technology, 'Scientific Advice, Risk and Evidence Based Policy Making', 7th Report of Session 2005–2006, Vol. I, Authority of the House of Commons (2006), para. 166. Available at: www.publications.parliament.uk/pa/cm200506/cmselect/cmsctech/900/900-i.pdf (accessed 1 March 2010).

[107] Graham and Hsia, 'Europe's Precautionary Principle'. See also, G. E. Marchant and K. L. Mossman, *Arbitrary and Capricious: The Precautionary Principle in the European Union Courts*, American Enterprise Institute Press, 2004.

[108] Löfstedt, 'Risk Communication and Management in the Twenty-First Century', 335–46, in particular, at 342.

[109] Sandin, 'A Paradox Out of Context', 177.

[110] Article 15 of the Rio Declaration is considered in more detail in Chapter 3.

that 'if there is (1) a threat, which is (2) uncertain, then (3) some kind of action (4) is mandatory'.[111]

Other authors have analysed the different definitions of the versions of the precautionary principle and organised them in schemes ranging from more general non-committing versions to a strict and rigid application of precaution. Wiener and Rogers, for example, have identified three different versions ranging from 'uncertainty does not justify inaction', via 'uncertainty justifies action', to 'uncertainty requires shifting the burden and standard of proof'.[112] The first of these is the mildest version, which merely states that precautionary action may be taken when full scientific certainty is lacking, while the last one is the strictest in that it requires the proponent of an action to show that the action is without risk. All of the above versions are, however, relatively unworkable in that they are silent on the crucial issues of how and what to do when scientific certainty is not in effect.[113]

Karlsson has gone a step further to identify five constituent parts of a strict precautionary principle in chemicals policy. These include worst-case classification, prioritisation based on intrinsic properties, preventive measures (including substitution), maximin decision-making and a reversal of the burden of proof.[114]

For the purposes of this book, the term 'precautionary approach' is used to denote a vague and open application of precaution, which is not necessarily based on a conceptual precautionary principle. The term 'precautionary principle' is used to describe an institutionalised and systematic use of precaution in one or more policy areas. The use of the precautionary principle can be weak, that is, reminiscent of what Sandin refers to as the argumentative version, or strong, which means that the principle is deemed to have more prescriptive characters. No one version is absolute, but consists of various degrees of rigidity. Concrete examples of the above distinctions are dealt with in considerable detail in the following chapters.

The differing definitions and applications of the precautionary principle also raise questions as to what kind of a legal concept the precautionary principle really is. The law is often divided into three categories

111 Sandin, 'A Paradox Out of Context', p. 177.

112 J. Wiener and M. D. Rogers, 'Comparing precaution in the United States and Europe', *Journal of Risk Research*, 5:4 (2002), 317, 320–1.

113 *Ibid.*, 321.

114 Karlsson, 'The Precautionary Principle, Swedish Chemicals Policy and Sustainable Development', 341–4.

of varying characteristics and importance: rules, principles and policy.[115] The name of the concept under discussion here would seem to indicate that it would be a principle of law, but perhaps the issue is not so clear-cut. Dworkin, building on the works of Hart and the legal positivists,[116] has held that a rule is to be defined as a simple function which is to apply similarly in all circumstances. In its simplest form: if (A), then (B).[117] If two rules are in conflict with one another, one will prevail and the other will have to be declared void. Two opposing rules cannot apply to a similar set of circumstances. Principles, on the other hand, are open-ended in character and do not apply in an all-or-nothing fashion. Instead, several principles can apply at the same time, and one can take precedence without the other being considered void or invalid. While rules give direct guidance in concrete situations, principles aim to give instruction in relation to 'justice or fairness or some other dimension of morality'.[118] Finally, policies differ from principles in that they 'set out a standard to be reached, generally an improvement in some economic, political, or social feature of the community'.[119] A policy differs from a principle for different reasons. First, policies are vaguer in character than principles. Where a principle gives a relatively clear direction as to what action should be taken in order to be in compliance with it, a policy states a more general goal to be reached. Second, the content of a principle, according to Dworkin, relates to the concept of justice and how a just legal system is to be maintained. A policy, on the other hand, is not limited to issues of justice and fairness, and can encompass any economic, social or political ambitions.[120]

Where the precautionary principle fits in depends, to a great extent, on how it has been applied in the relevant legal orders. However, even in legal

[115] R. Dworkin, *A Matter of Principle* (Harvard University Press, 1985), pp. 43–5.

[116] See, for example, H. L. A. Hart, *The Concept of Law* (Clarendon Press, 1961).

[117] Dworkin, *A Matter of Principle*, p. 45.

[118] *Ibid.*, pp. 45–6. [119] *Ibid.*, p. 43.

[120] A more nuanced view of the distinction, or lack thereof, between rules, principles and policy has been taken by de Sadeleer, who holds that norms, at least at international level, are open-ended in character. Thus, norms, however they are characterised, function together and are subject to dialogue and negotiations far from the seemingly 'black-letter law' approach by the positivists and, to some extent, the normativists. This post-modern approach would erode the border between the different types of norms. De Sadeleer also introduces a set of norms he names 'directing principles' whose aim it is to spur public policy and allow courts to weigh directly opposing legal interests against each other. The precautionary principle would be an example of such a norm. How this would change the way the precautionary principle is viewed in practical terms is as yet not clear. See N. de Sadeleer, *Environmental Principles: From Political Slogans to Legal Rules* (Oxford University Press, 2002), p. 237.

systems where the principle is codified and endorsed by the courts, it usually remains ambiguous and undefined. Consequently, it is unlikely that the precautionary principle will be considered to apply in the all-or-nothing fashion generally expected from a rule. Instead, it would often have to be considered as either a principle or a policy. With regard to considering it as a principle, there are some concerns. First of all, the precautionary principle does not aim to safeguard the just and fair application of the law, in the manner that traditional legal principles do. Second, it is shown below that there are serious difficulties in establishing what the content and limits of the principle are. Third, even in states where the precautionary principle is heavily relied upon in certain areas, it is entirely left out of the application of the law in other areas.[121] All of this makes it difficult to treat it as a legal principle which technically should apply equally in similar circumstances. Thus, it emerges that the application of the precautionary principle is perhaps best treated as a policy choice. In the following chapters the treatment of the precautionary principle in different states and policy areas is analysed in detail, and this conclusion is returned to within that framework.

2.5 Conclusion

The aim of this chapter has been to give an overview and trace the background to concepts that are central to this book and to risk regulation in general. This is necessary in order to understand the content, purpose and scope of the precautionary principle, which is but one instrument to be used in the broad field of risk regulation.

The concept of uncertainty is essential to the issue of the precautionary principle. In this book, uncertain risks are used to denote risks for which the causal link between the activity and the hazard is unknown. Certain risks are risks for which the causal link is established. It is generally understood that the precautionary principle allows for the regulation of uncertain risks.

Lay people perceive risks in a different manner than experts. For example, biases and heuristics make people overestimate the occurrence of low probability risks. This has led some authors to conclude that opinion based on heuristic errors should be totally disregarded by a responsible government.[122] Others have held that '[f]ar from being egalitarian,

[121] For an in-depth analysis of this, see the case studies below.
[122] Wiener, 'Precaution in a Multi-Risk World', p. 4.

it is real exclusion that begins when prejudice or opinion are taken to be a sound basis for decision-making'.[123] Therefore, in order to safeguard a rational basis for risk regulation, including the application of the precautionary principle, it is generally accepted that the relevant scientific advice should be analysed before regulation is decided upon. Thus, the risk assessment phase should be separated from the risk management phase, in which the decision is taken as to whether, and, if so, how, to regulate a certain activity. In order to ensure that the general public is continuously informed about the analysis and response to societal risks, it is important to communicate regulatory strategies effectively. This helps to increase the trust and the acceptability of the regulatory choices that have been made.

A number of tools are of relevance in the risk management phase, of which the assessment of scientific evidence is but one. Other important sources of information are cost–benefit and cost-effectiveness analyses. Such tools can be designed to be binding on the decision-maker, but they can also be used as sources of information in order to put all the costs and benefits of a particular regulation 'on screen'. Further tools include risk trade-off analyses, which aim to assume a more holistic view of the effects of regulating one specific risk. These tools can be used together in order carry out comparative risk analyses, which allow the decision-maker(s) to choose between, and rank, several risk-reducing strategies. In the following chapter it is investigated whether, and if so, how, these issues are applied in practice.

As stated above, no single definition exists of the precautionary principle. The term has been used to denote everything from a generally 'environmental' approach to specific and binding decision-making rules. The principle has both been praised as a groundbreaking way to protect environment and health and criticised as vague, unscientific and arbitrary. Since no single precautionary principle exists, it is difficult to categorise it as a legal norm, and, due to its inherent ambiguity and its lack of a concrete definition, it is probably best viewed as a policy choice.

[123] Duriodé, 'The True Cost of Precautionary Chemicals Regulation', 395.

3

The precautionary principle in international law

3.1 Introduction

The development and status of the precautionary principle in international environmental law has been chronicled in detail by a number of scholars.[1] The purpose of this chapter is not to reiterate their findings in detail, but merely to highlight a number of the most important and influential international instruments and cases which have shaped the modern application of the precautionary principle. More specifically, the focus of this chapter is on how states can use precautionary considerations to justify trade restrictive measures. In this regard the WTO Agreements are of primary importance.

The role of the precautionary principle under the WTO A greements has been a highly contested issue over the last ten years, and it has been the subject of a number of disputes before the WTO dispute settlement bodies. Generally, the European Union and its Member States have emerged as champions of a precautionary approach which functions as an instrument to justify national or regional regulation in the face of scientific uncertainty and controversy. On the other hand, the United States and other countries have positioned themselves as advocates of a more 'science-based' approach. The stakes involved have often been high and the rhetoric heated.

Like the other chapters of this book, this one aims to go beyond the political statements and the various definitions of the 'precautionary principle', and derivates thereof, to investigate in detail how WTO Panels and

[1] Trouwborst, 'Prevention, Precaution, Logic and Law', 105; Fitzmaurice, *Contemporary Issues in International Environmental Law*, pp. 1–67A; Trouwborst, 'The Precautionary Principle in General International Law', 185–95; Grimeaud, 'The Precautionary Principle in International Environmental and Trade Law'; Douma, 'The Precautionary Principle'; S. Marr, *The Precautionary Principle in the Law of the Sea: Modern Decision Making in International Law* (Martinus Nijhoff Publishers, 2003); Trouwborst, *Evolution and Status of the Precautionary Principle in International Law*; Cameron, 'The Precautionary Principle in International Law'.

the Appellate Body (AB) have come to describe how and when Members of the WTO can act 'precautionary'. The chapter begins with an overview of the relevant provisions in the WTO Agreements which may embody a precautionary principle or approach. Of prime importance here is the WTO Agreement on the Application of Sanitary and Phytosanitary Measures (SPS Agreement), since it is the agreement which so far has been mostly relied on by WTO Members to justify precautionary action. However, the General Agreement on Tariffs and Trade 1994 (GATT 1994) and the Agreement on Technical Barriers to Trade (TBT Agreement) are also discussed.

When the attention is turned to the issue of how Panels and the AB have approached the topic, the SPS Agreement is of primary importance since it is the agreement that so far has been the most commonly relied upon before the dispute settlement bodies in cases concerning the justification of precautionary measures.

This chapter, like the following chapters, identifies a number of constituent parts of the precautionary principle and deals with them separately in order to illustrate all the aspects of relevance when decisions are taken under scientific uncertainty. These aspects include the treatment of issues such as the scope of judicial review, scientific evidence, burden of proof, cost–benefit analysis and public opinion. Finally, this chapter discusses the status and scope of the precautionary principle in the WTO.

3.2 International environmental law

3.2.1 *The precautionary principle in international agreements*

3.2.1.1 Binding international agreements

First appearing in environmental agreements in the 1980s, reference to the precautionary principle has in recent years had what has been described as a 'meteoric rise' in international agreements.[2] However, even though it took until 1985 and the Vienna Convention on Ozone Depleting Substances for the word 'precaution' to find its way into a preamble of an international agreement, precautionary thinking had been influencing earlier international agreements. Among these the 1968 African

[2] D. Cameron and J. Abouchar, 'The Status of the Precautionary Principle in International Law', in D. Freestone and E. Hey (eds.), *The Precautionary Principle in International Law: The Challenge of Implementation* (Kluwer Law International, 1996), p. 29.

Convention on the Conservation of Nature and Natural Resources,[3] the 1971 Ramsar Convention,[4] the 1972 World Heritage Convention,[5] the 1973 CITES[6] and the 1982 UNCLOS[7] can be found, to name a few.[8]

Binding international agreements often stop short of defining precautionary action, and instead make general reference to 'the precautionary principle' or 'a precautionary approach'.[9] However, many agreements make reference to the fact that precautionary measures are designed to be used where full scientific certainty is lacking, and that the lack of such evidence should not be a reason to postpone action.[10] Yet again, some international agreements do not mention the precautionary principle, but are essentially precautionary in function.[11] Furthermore, it should be noted that all binding international agreements where the precautionary principle is mentioned are sectoral in nature.

3.2.1.2 Non-binding international instruments

If the inclusion of precautionary aspects in legally binding international treaties still is relatively scarce, the inclusion of notions of precaution in non-binding international documents is much more common. In fact, the mentioning of precaution in international policy documents is now

[3] African Convention on the Conservation of Nature and Natural Resources, Algiers 1968.

[4] The Convention on Wetlands, Ramsar 1971.

[5] Convention Concerning the Protection of the World Cultural and Natural Heritage, Paris 1972.

[6] The Convention on International Trade in Endangered Species of Wild Fauna and Flora, Washington DC 1973.

[7] The United Nations Convention on the Law of the Sea, Montego Bay 1982.

[8] For a more detailed list of, and in-depth discussion on, early international agreements encompassing precaution, see Trouwborst, *Evolution and Status of the Precautionary Principle in International Law*, pp. 63 *et seq.*

[9] See, for example, the preamble to the Protocol (to the 1985 Convention for the Protection of the Ozone Layer) on Substances that Deplete the Ozone Layer, Vienna 1985 and Article 6 of Agreement for the Implementation of the Provisions of the 1982 UNCLOS Relating to the Conservation and Management of Straddling Fish Stocks and Highly Migratory Fish Stocks, New York 1995.

[10] See, for example, Article 3(3) of the Convention on the Ban on Import into Africa and the Control of Transboundary Movement and Management of Hazardous Wastes Within Africa, Bamako 1991; Article 3(3) of the Framework Convention on Climate Change, New York 1992; Article 6 of the Agreement for the Implementation of the Provisions of the 1982 UNCLOS Relating to the Conservation and Management of Straddling Fish Stocks and Highly Migratory Fish Stocks, New York 1995; and Article 10(6) of the Protocol (to the 1992 Convention on Biological Diversity on Biosafety, Cartagena de Indias 2000.

[11] See, for example, the Protocol to the 1992 Framework Convention on Climate Change, Kyoto 1997.

so widely spread that it is nearly impossible to provide an exhaustive list.[12]

The most influential of the soft law instruments relating to the precautionary principle is Principle 15 of the Rio Declaration, which was agreed upon on 2 April 1992.[13] During the negotiations there had been diverging opinions on the exact wording of Principle 15, but, interestingly, it appears that most states active in the negotiations agreed on the basic idea of including a statement dealing with precaution in one way or another.[14] In its final version, the principle reads:

> In order to protect the environment, the precautionary approach shall be widely applied by States according to their capabilities. Where there are threats of serious or irreversible damage, lack of full scientific certainty shall not be used as a reason for postponing cost-effective measures to prevent environmental degradation.

The wording of Principle 15 appears to be an amalgam of earlier versions of articles in international documents dealing with the precautionary principle. The use of the term 'serious or irreversible damage' as a threshold for the application of a precautionary approach stems from the 20-year older Stockholm Declaration.[15] The use of the wording indicates that some form of minimum level of risk has to be met in order for precautionary measures to be issued. Nothing in the Rio Declaration aims to clarify how this is to be interpreted in practice. Furthermore, the principle adopts the 'triple-negative' definition to state that the lack of scientific certainty shall not be used as a reason for not taking precautionary measures. This model was previously used in the Bergen Declaration[16] and in the 1991 OECD Council Recommendations.[17] Furthermore, the Rio Declaration makes explicit reference to the need for the precautionary measures to be cost-effective. Reference to the weighting of costs and benefits had previously been included in the World Charter for Nature.[18] In the case of the World

12 For a good overview, see Trouwborst, *Evolution and Status of the Precautionary Principle in International Law*, pp. 112–56.

13 Declaration of the United Nations Conference on Environment and Development, Rio de Janeiro 1992.

14 Douma, 'The Precautionary Principle', pp. 77–84.

15 Declaration of the United Nations Conference on the Human Environment, Stockholm, 1972.

16 Ministerial Declaration on Sustainable Development in the ECE Region, Bergen 1990.

17 OECD Council Recommendation C(90)164 on Integrated Pollution Prevention and Control, 31 January 1991.

18 United Nations General Assembly Resolution 37/7 and Annex: World Charter for Nature, 28 October 1982.

Charter for Nature this took the form of a cost–benefit analysis. However, the cost-effectiveness requirement in the Rio Declaration would seem to aim to restrict the regulator of an activity on precautionary grounds to choose the least economically cumbersome way of doing so. The requirement in the World Charter for Nature aims to reverse the burden of proof, away from the regulator and onto the regulated party, to prove that the benefits of the proposed activity outweigh the potential risks. One innovation in the Rio Declaration as compared with previous documents is that it holds that states should apply the precautionary approach according to their own capabilities.

Finally, it should be pointed out that the Rio Declaration makes use of the weaker argumentative version of the precautionary principle, rather than a stronger prescriptive version.[19] Principle 15 of the Rio Declaration is a landmark in the development of the precautionary principle in the international arena. Contrary to many of the previously discussed documents in this chapter, the Rio Declaration is a global and general instrument, which firmly establishes that some form of minimum precautionary approach is widely accepted in the world community. Once again, though, its importance should not be overstated. As has been pointed out, it is weak in character and does not constitute a binding instrument of international law.[20]

3.2.2 *International case law*

International tribunals have been faced with parties basing their arguments on the precautionary principle on a number of occasions. The first time the International Court of Justice was confronted with a case relating to the precautionary principle was in *French Nuclear Tests II* of 1995, where New Zealand asked the Court to investigate the compliance of France with Paragraph 63 of its judgment in *French Nuclear Tests I* of 1974.[21] However, the Court chose to interpret its own jurisdiction rather narrowly and concluded that the object of the second case was not the same as in the original case. Because of these procedural obstacles the Court was not to rule on the substance of the case. However, in the

[19] This issue has been discussed in some detail above in Chapter 2.

[20] For a description of the negotiations leading up to the acceptance of Principle 15 of the Rio Declaration see Douma, 'The Precautionary Principle'.

[21] *Request for an Examination of the Situation in Accordance with Paragraph 63 of the Court's Judgement of 20 December 1974 in the Nuclear Tests* (*New Zealand* v. *France*), ICJ Reports 1995.

three dissents added to the judgment, judges Weeramantry, Palmer and Koroma tentatively discussed the status of the precautionary principle. They all stopped short of declaring that it has any binding effects in the international law. However, Judge Weeramantry held that the principle may be 'gaining increasing support as part of the international law'.[22] Judge Palmer went as far as to say that it 'may' be the case that the precautionary principle forms part of customary international law. In his dissent he discussed at length that cost–benefit considerations would need to be taken into account before issuing precautionary measures.[23]

In the *Gabcikovo-Nagymaros Case* of 1997 the parties agreed to the fact that the precautionary principle posed 'a new requirement' of international environmental law, but there were differences as to the interpretation of the principle.[24] The ICJ decided not to evaluate the content of the precautionary principle, but instead left it to the parties to negotiate the application of the principle.[25]

The precautionary principle ended up before an international tribunal again in the *Southern Bluefin Tuna Cases* of 1999,[26] which was decided upon by the International Tribunal for the Law of the Sea (hereinafter ITLOS) under the United Nations Convention on the Law of the Sea (hereinafter UNCLOS). In the case New Zealand and Australia argued that the precautionary principle would prevent Japan from carrying out its experimental fishing programme for southern bluefin tuna. Japan, on the other hand, held that the programme would enable a more accurate assessment of the relevant fish stocks. In its judgment ITLOS ordered that the experimental programme should be halted and that all parties should adhere to the last agreed-upon limits. The case has been seen by some commentators as proof of the increasing importance of the precautionary principle in international law, and even as an indicator of its status as customary international law.[27] But, as has been pointed out by others,

[22] *Dissenting Opinion of Judge Weeramantry in Request for an Examination of the Situation in Accordance with Paragraph 63 of the Court's Judgement of 20 December 1974 in the Nuclear Tests* (*New Zealand* v. *France*), ICJ Reports 1995, p. 342.

[23] *Dissenting Opinion of Judge Palmer in Request for an Examination of the Situation in Accordance with Paragraph 63 of the Court's Judgement of 20 December 1974 in the Nuclear Tests* (*New Zealand* v. *France*), ICJ Reports 1995, p. 412.

[24] *Gabcikovo-Nagymaros Project (Hungry/Slovakia)*, Judgment, ICJ Reports 1997, p. 7.

[25] *Ibid.*, para. 112.

[26] *Southern Bluefin Tuna Cases* (*New Zealand* v. *Japan*; *Australia* v. *Japan* (Joined Cases)), ITLOS, Order of 27 August 1999.

[27] See, primarily, Cameron, 'The Precautionary Principle in International Law', p. 129, and Trouwborst, *Evolution and Status of the Precautionary Principle in International Law*, pp. 171–4.

ITLOS does not refer to the precautionary principle at all in its judgment in the case.[28] Furthermore, instead of acting out of precaution, it could be argued that ITLOS is merely stating that the management of fish stocks is to be carried out with the agreement of the states concerned. This could be taken to imply that the ITLOS views precaution (or prevention for that matter) not as an objective principle centred on environmental protection, but rather a subjective approach, the content of which may be negotiated between the parties involved.

The above conclusion of ITLOS was confirmed in the *MOX Plant Case* of 2001, which concerned an Irish claim against the UK for failing to investigate properly the impact on the marine environment in the Irish Sea of the establishment of a metal oxide (MOX) plant.[29] Ireland argued, among other things, that the precautionary principle would place the burden of proof on the UK to show that no harm would arise from the establishment of the MOX plant.[30] The UK did not dispute the relevance of the precautionary principle as such, but stated that it considered the risk of harm to be 'infinitesimally small.'[31] As in the *Southern Bluefin Tuna Cases*, ITLOS refrained form making a ruling based upon the precautionary principle, and left it up to the parties to share information on the risks and to communicate with each other in the spirit of 'prudence and caution.'[32] Thus, also in the *Mox Plant Case*, ITLOS left it up to the parties to negotiate the content of and limits to what should constitute precaution.[33]

3.3 WTO law

3.3.1 The precautionary principle and the WTO Agreements

Of primary importance when discussing the legality of precautionary measures in the WTO regime is the WTO Agreement on the Application of Sanitary and Phytosanitary Measures.[34] A sanitary and phytosanitary

[28] See, for example, Douma, 'The Precautionary Principle', p. 172.

[29] *The MOX Plant Case* (*Ireland* v. *U.K.*), (Provisional measures), (ITLOS), Order of 3 December 2001.

[30] *Ibid*., para. 71. [31] *Ibid*., para. 72.

[32] *Ibid*., para. 89.

[33] For a recent summary of the implications of the dispute, see N. Lavranos, 'The Epilogue in the MOX Plant Dispute: An End Without Findings', *EELR*, 18:3 (2009), 180.

[34] For a commentary to the SPS Agreement, see Scott, *The WTO Agreement on Sanitary and Phytosanitary Measures*. See also, van den Bossche and Prevost, 'The Agreement on the Application of Sanitary and Phyto-sanitary Measures', pp. 231–70. For a general overview

measure is any measure that meets one of the four definitions established in Paragraph 1 of Annex A to the SPS Agreement. First, the Agreement applies to measures aimed at protecting animal, plant or human life from risks stemming from the entry of organisms carrying pests or disease. Second, measures aimed at protecting human or animal life against risks arising from additives, contaminants, toxins or disease-causing organisms in foods, beverages or feedstuffs are covered. Third, the Agreement concerns measures aimed at protecting human life or health against risks from diseases carried by animals, plants or products thereof, or from the entry, establishment or spread of pests. Finally, measures aimed at preventing or limiting other damage from the entry, establishment or spread of pests are to be considered SPS measures.[35] Most precautionary measures will have one of these objectives. However, measures not matching this definition will fall under the other agreements, notably the TBT Agreement and/or GATT 1994.[36]

The TBT Agreement applies to technical regulations, standards and conformity assessment procedures.[37] A technical regulation is defined as a document that lays down mandatory product characteristics or their related processes and production methods, including relevant administrative provisions. Technical regulations can, for example, relate to terminology, symbols, packaging, marking or labelling.[38] Standards are defined as documents approved by a recognised body, which provide non-mandatory rules, guidelines or characteristics for common or repeated use for products or related processes and production methods.[39] Finally,

of earlier legal scholarship in the area of the SPS Agreement, see J. Atik and D. A. Wirth, 'Science and International Trade – Third Generation Scholarship', *B.C Int'l & Comp. L. Rev.*, 26 (2003), 171. See also, C. R. Kelly, 'Power, Linkage and Accommodation: The WTO as an International Actor and Its Influence on Other Actors and Regimes', *Berkely J. Int'l L.*, 24 (2006), 79, 110 and note 220.

[35] Article 1, Annex A attached to the SPS Agreement. In Panel Report, European Communities – Measures Affecting the Approval and Marketing of Biotech Products, WT/DS291/R, WT/DS292/R WT/DS293/R, adopted on 21 November 2006 (*EC-Approval and Marketing of Biotech Products*) the Panel elaborated on the definition of SPS measures (see Panel Report, *EC-Approval and Marketing of Biotech Products*, para. 7.149). See also, D. Prevost, 'Opening Pandora's Box: The Panel's Findings in the EC-Approval and Marketing of Biotech Products Dispute', *Legal Issues of Economic Integration*, 34:1 (2007), 73.

[36] For a discussion on how the WTO Agreements relate to each other when the precautionary principle is considered, see N. Bernasconi-Osterwalder *et al.*, *Environment and Trade: A Guide to WTO Jurisprudence* (Earthscan, 2006), pp. 256–64.

[37] TBT Agreement, Articles 1, 2 and 3 of Annex 1.

[38] TBT Agreement, Article 1 of Annex 1.

[39] TBT Agreement, Article 2 of Annex 1.

conformity assessments are procedures which aim to determine compliance with technical regulations or standards.[40]

In situations where measures can be considered to meet the definitions in the SPS Agreement and the TBT Agreement simultaneously, Article 1.5 of the TBT Agreement establishes that the SPS Agreement takes precedence. That this is the case has also been upheld in a number of Panel Reports.[41]

Whereas the SPS Agreement and the TBT Agreement apply to the certain categories of measures outlined above, GATT 1994 applies to all measures affecting trade in goods, regardless of their content and purpose. As was also discussed above, the TBT Agreement and the SPS Agreement are mutually exclusive. However, the SPS Agreement and GATT 1994 may apply simultaneously. As is stated in Article 2.4 of the SPS Agreement, measures which fulfil the obligations of the SPS Agreement shall be presumed to be in compliance with GATT obligations, especially the provisions of Article XX(b). Indeed, one of the motives behind the SPS Agreement was to clarify and flesh out the Article XX(b) exception in GATT 1994. However, the SPS Agreement goes further than GATT 1994 in that it imposes a new and comprehensive set of rules.[42] In the light of the above, it is unlikely that provisions of the SPS Agreement and GATT 1994 would clash. However, should this nonetheless be the case, the SPS Agreement prevails. This follows from the General Interpretative Note to Annex 1A of the WTO Agreement, which holds that in the event of a conflict between GATT 1994 and another multilateral trade agreement on trade in goods, the latter shall prevail.[43]

The SPS Agreement applies to SPS measures 'which may, directly and indirectly, affect international trade' according to its Article 1.1. This makes the scope of the SPS Agreement broader than GATT 1994, which

[40] TBT Agreement, Article 3 of Annex 1.

[41] Panel Report, *EC Measures Concerning Meat and Meat Products (Hormones) (US)* WT/DS26/R (Panel Report, *EC-Hormones (US)*), para. 8.29 and Panel Report, *EC Measures Concerning Meat and Meat Products (Hormones) (Canada)*, WT/DS48/R (Panel Report, *EC-Hormones (Canada)*), para. 8.32 and Panel Report, *EC-Approval and Marketing of Biotech Products*, para. 7.167.

[42] P. van den Bossche, *The Law and Policy of the World Trade Organization*, 2nd edn (Cambridge University Press, 2008), Section 8.4.3.2.

[43] In this regard, van den Bossche holds that it would be 'logical, in a dispute involving an SPS measure, to begin by examining the measure under the *SPS Agreement*.' Van den Bossche, *The Law and Policy of the World Trade Organization*, 1st edn, p. 463. See also, C. Button, *The Power to Protect: Trade, Health and Uncertainty in the WTO* (Hart, 2004), pp. 9–43. This is also supported by the Panel in Panel Report, *EC-Hormones (US)*, para. 8.42, Panel Report, *EC-Hormones (Canada)*, para. 8.45.

concerns specific obligations. GATT 1994 obligations include respect for the non-discrimination principles, that is, the most-favoured-nation principle in Article I, and, perhaps of particular importance when the precautionary principle is concerned, the national treatment obligation in Article III. They also include the rules on market access, which comprise the elimination of quantitative restrictions in favour of tariffs.[44]

Furthermore, the presumption concerning the legality of measures in the SPS Agreement is opposite to the presumption in GATT 1994. Article 2.1 of the SPS Agreement explicitly states that Members have a right to take SPS measures. In GATT 1994 derogating measures are exceptions to the rules on non-discrimination and equal treatment. At the outset it is thus up to the party wishing to challenge an SPS measure to show that the measure is not in conformity with the regime laid down in the agreement, rather than for the party issuing the measures to show that they are in conformity with the agreement.[45] In contrast, under GATT 1994, a challenging Member would have to show prima facie that a measure taken by another Member is not in conformity with the obligations relating to non-discrimination or market access. It would then be for the challenged Member to show that the measures fit under the exceptions in Article XX of GATT 1994.

The preferred level of protection that is to function as a basis for SPS measures is internationally agreed-upon standards.[46] Members may either base or conform their measures to such standards. Measures conforming to international standards are to be deemed as proportionate and in conformity with other relevant WTO provisions.[47] However, Members are allowed to base their measures on the wish to maintain or introduce a higher level of protection than that provided for in international standards if they fulfil the criteria laid down in Article 5.[48]

Article 5.1 makes it clear that SPS measures are to be based on a risk assessment, the definition of which is provided in paragraph 4 of Annex A of the SPS Agreement:

44 GATT, Article XI and XIII.

45 See a detailed discussion below in this chapter under Sections 3.3.2.4 and 3.3.2.5.

46 In this regard, the AB has held that 'the object and purpose of Article 3 is to promote the harmonisation of the SPS measures of Members on as wide a basis as possible, while recognising and safeguarding at the same time the right and duty of Members to protect the life and health of their people. AB Report in *EC-Hormones I*, para. 177.

47 SPS Agreement. Article 3.2. For a discussion on the role of international standards under the SPS Agreement, see Scott, *The WTO Agreement on Sanitary and Phytosanitary Measures*, pp. 243–80.

48 SPS Agreement, Article 3.3.

> The evaluation of the likelihood of entry, establishment or spread of a pest or disease within the territory of an importing Member according to the sanitary or phytosanitary measures which might be applied, and of the associated potential biological and economic consequences; or the evaluation of the potential for adverse effects on human or animal health arising from the presence of additives, contaminants, toxins or disease-causing organisms in food, beverages or feedstuffs.

As can be deduced from the wording of the above, two slightly different risk assessments are prescribed depending on the origin of the risk in question. With regard to pests or diseases, the 'likelihood' of the risk needs to be assessed together with the potential biological and economic consequences. However, with regard to risks present in food, beverages or feed, the 'potential for adverse effects' is to be assessed. No reference is made to the economic consequences of the proposed measures.[49]

Article 5.2 further elaborates on what should be included in a risk assessment. This includes the available scientific evidence, processes and production methods, inspection and sampling methods, relevant existing specific diseases or pests, areas which are free of diseases or pests, relevant ecological and environmental conditions, as well as quarantines or other treatments.

An exception from the obligation to base measures in a risk assessment is included in Article 5.7:

> In cases where relevant scientific evidence is insufficient, a Member may provisionally adopt sanitary or phytosanitary measures on the basis of available pertinent information, including that from the relevant international organizations as well as from sanitary or phytosanitary measures applied by other Members. In such circumstances, Members shall seek to obtain the additional information necessary for a more objective assessment of risk and review the sanitary or phytosanitary measure accordingly within a reasonable period of time.

This means that a measure may provisionally be maintained or introduced even though it has not been adequately backed up by scientific evidence provided that 'scientific evidence is insufficient'.[50] Article 5.7 is the clearest reflection of the precautionary principle in the SPS Agreement. However, the principle is not limited to this article. This issue is returned to below in the following sections.

[49] For a discussion on these requirements see van den Bossche, *The Law and Policy of the World Trade Organization*, 2nd edition, Section 8.4.5.1. See also, below in this chapter in Sections 3.3.2.3 and 3.3.2.4.

[50] This discussion is returned to in detail below in Section 3.3.2.4 on scientific evidence.

In the event that a measure does not meet the definition of an SPS measure, the TBT Agreement may be relevant. The TBT Agreement restates some of the basic principles found in GATT 1994, for instance the most-favoured-nation principle and the principle of national treatment.[51] Where the purpose of the SPS Agreement is to allow for the introduction of trade-restrictive SPS measures in certain cases, the TBT Agreement establishes clearly that the presumption is that like products should be treated equally. However, like GATT 1994, the TBT Agreement allows Members to introduce or maintain technical regulations as long as they do not create 'unnecessary obstacles to international trade'. Furthermore, they must be 'necessary to fulfil a legitimate objective, taking account of the risks non-fulfilment would create'.[52] Objectives that can be considered necessary include human health or safety, animal or plant life or health, and the environment.[53] The list provided in Article 2.2 is not exhaustive.[54] A trade-restrictive measure that is introduced, and considered necessary, must be assessed continuously to ensure that it remains necessary.[55] In this sense the requirements in the TBT Agreement correspond to those in the SPS Agreement. Necessity is a requirement in both, as well as the requirement to dismantle the measures when the situation changes.[56] Further similarities include the recognition of international standards as a basis for the introduction of measures.[57] If a TBT measure is based on such a standard and meets one of the objectives in Article 2.2, it is presumed not to create an unnecessary obstacle to international trade. The TBT Agreement does not go as far as identifying which international standards are appropriate, but it does state that Members 'shall use them, or the relevant parts of them' unless the international standards are to be considered an ineffective or inappropriate way of fulfilling the policy objectives prescribed in Article 2.2. If this is the case, Members may as an exemption, depart from the international standards.[58]

Where the SPS Agreement only allows for measures to be taken after a thorough scientific risk assessment,[59] the TBT Agreement appears to

[51] TBT Agreement, Article 2.1. [52] *Ibid.*, Article 2.2.

[53] *Ibid.*

[54] Van den Bossche, *The Law and Policy of the World Trade Organization*, 1st edn, p. 460.

[55] TBT Agreement, Article 2.3.

[56] For an in-depth discussion on the necessity requirement in the WTO regime see Bernasconi-Osterwalder *et al.*, *Environment and Trade*, pp. 149–201.

[57] TBT Agreement, Article 2.4. [58] TBT Agreement, Article 2.4.

[59] SPS Agreement, Articles 5.1 and 5.2.

be more lenient in its appreciation of which measures may be allowed. Article 2.2 only states that:

> In assessing such risks, relevant elements of consideration are, inter alia: available scientific and technical information, related processing technology or intended end-uses of products.[60]

This signifies that available scientific information is an important aspect to be considered when determining risks under the TBT Agreement. However, there is no strict obligation to rely on scientific information under the TBT Agreement. As was shown above, the SPS Agreement places great weight on an assessment based in 'scientific principles' and obliges a Member maintaining or introducing SPS measures to justify this with 'sufficient scientific evidence'.[61] By stating that 'relevant elements' of the assessment of risks include 'inter alia' scientific and technological information, and so on, Article 2.2 of the TBT Agreement seems to indicate that issues other than strictly scientific ones possibly could be considered when assessing risks.

Furthermore, no corresponding article to Article 5.7 of the SPS Agreement is to be found in the TBT Agreement. Therefore, no clear distinction appears to be made between measures taken under scientific certainty and scientific uncertainty. It thus appears that precautionary measures could be taken under Article 2.2 of the TBT Agreement and that these measures would not necessarily have to be of interim character as is the case in the SPS Agreement.[62] Clearly, such measures would have to respect the requirement in Article 2.2 of not being more trade-restrictive than necessary, taking into account the risks non-fulfilment would create. Room is left for the precautionary principle in the broad sense in the preamble, where it is stated that no country should be prevented from taking measures necessary to protect human, animal or plant life or health or the environment, at levels it considers appropriate.[63]

As was established above, when deciding on cases relating to the precautionary principle, Panels would generally begin by considering the SPS Agreement, and only turn to the TBT Agreement or GATT 1994 if the

[60] TBT Agreement, Article 2.2. [61] SPS Agreement, Article 2.2.

[62] However, as mentioned above, all measures taken under the TBT Agreement need to be updated if the circumstances change in accordance with Article 2.3. This requirement appears to be weaker than the corresponding requirement relating to precautionary measures in the SPS agreement. Primarily, Article 5.7 of the SPS agreement obliges the Member applying SPS measures to carry out research in order to end the state of uncertainty. This will be monitored by the DSB. Article 2.3 appears to be far more general in character.

[63] Bernasconi-Osterwalder *et al.*, *Environment and Trade*, pp. 257–8.

provisions in that agreement do not appear applicable.[64] Both are covered by the Interpretative Note to Annex 1A and should be considered before turning to GATT 1994.[65] Thus a Member cannot choose to rely on the agreement most suitable for its needs. Instead the general principle of *lex specialis* applies.

Under GATT 1994 measures breaching GATT obligations need to be justified under the exceptions listed in Article XX in order to be allowed. As far as the precautionary principle would be argued under GATT 1994, it would have to be justified by one of the exceptions listed in the Article. The list there has generally been interpreted to be exhaustive.[66] The most relevant parts read:

> Subject to the requirement that such measures are not applied in a manner which would constitute a means of arbitrary or unjustifiable discrimination between countries where the same conditions prevail, or a disguised restriction on international trade, nothing in this Agreement shall be construed to prevent the adoption or enforcement by any contracting party of measures:
> [...]
> (b) necessary to protect human, animal or plant life or health;
> [...]
> (d) necessary to secure compliance with laws or regulations which are not inconsistent with the provisions of this Agreement, including those relating to customs enforcement, the enforcement of monopolies operated under paragraph 4 of Article II and Article XVII, the protection of patents, trade marks and copyrights, and the prevention of deceptive practices;
> [...]
> (g) relating to the conservation of exhaustible natural resources if such measures are made effective in conjunction with restrictions on domestic production or consumption.

3.3.2 *The role of the precautionary principle in WTO dispute settlement*

3.3.2.1 The definition of the precautionary principle

The precautionary principle evidently lacks definition or even explicit support in the WTO Agreements.[67] Some authors have held that there are

64 Panel Report, *EC-Hormones* (Canada), para 8:45 and Panel Report, *EC-Hormones* (US), para. 8.42.

65 See, above, note 42.

66 See, for example, J. Scott, 'The European Regulation of GMOs and the WTO', *Colum. J. Eur. L.*, 9 (2003), 213, 230.

67 See Scott, *The WTO Agreement on Sanitary and Phytosanitary Measures*, p. 127 and Button, *The Power to Protect*, pp. 120–2.

equally no rules in the WTO Agreements which would prohibit the application of a precautionary principle or approach.[68] However, as is discussed in detail below, the apparently strict reliance on scientific evidence in the SPS Agreement makes it difficult to see prima facie how the precautionary principle would find reflection in the WTO Agreements, outside of Article 5.7.[69] This, in combination with the fact that the principle was not considered to be part of international customary law, has been stressed by the Appellate Body (AB) as a reason not to base rulings on it alone.[70] However, the AB has also pointed out that the principle 'finds reflection' in Article 5.7 of the SPS Agreement and that 'there is no need to assume that Article 5.7 exhausts the relevance of a precautionary principle', especially since it is also reflected in the sixth paragraph of the preamble and Article 3.3 of the SPS Agreement.[71] In any case, the precautionary principle cannot be considered to override the strict requirements on scientific evidence and risk assessment in Articles 2.2, 5.1 and 5.2 of the SPS Agreement.[72]

Even if the requirement of a risk assessment is strict, this does not mean that the AB does not recognise the fact that the assessment of a risk may not provide conclusive evidence of the existence of risk. Therefore the AB holds that the term 'sufficient scientific evidence', which is laid down in Article 2.2 of the SPS Agreement as the basic requirement for the enactment of all measures under the agreement, does not only refer to mainstream scientific opinion.[73] In cases of unresolved scientific controversy, which is one instance of scientific uncertainty,[74] Members may base their measures on the opinions of the scientific minority.[75] This is especially justified in cases where the risk involved is perceived as life-threatening and imminent.[76] As was shown previously, scientific uncertainty is not a static or monolithic concept, but can exist in various degrees and can be due to several different causes.[77] That this is the case was also recognised by the AB in *EC-Hormones II*.[78] This creates a situation where Members,

[68] Bernasconi-Osterwalder *et al.*, *Environment and Trade*, p. 256.

[69] However, a more thorough analysis of the relevant case law suggests that a possibility of taking precautionary measures may exist outside of Article 5.7.

[70] AB Report, *EC-Hormones I*, paras. 123 and 124. This was also repeated more recently in Panel Report, *EC-Approval and Marketing of Biotech Products*, para. 7.89.

[71] AB Report, *EC-Hormones I*, para. 124. [72] *Ibid.*, para. 125.

[73] *Ibid.*, para. 194. [74] See, above, Section 2.3.1.

[75] AB Report, *EC-Hormones I*, para. 194. [76] *Ibid.*

[77] See, above, Sections 2.3.1 and 2.3.2.

[78] Report of the Appellate Body, US – Continued Suspension of Oblgiations in the EC-Hormones Dispute, WT/DS320/AB/R (*EC-Hormones II*), para. 679.

in cases of scientific uncertainty due to scientific controversy, may issue precautionary measures without seeking recourse in Article 5.7, as long as other basic SPS requirements are abided by. In *Japan-Apples* the Panel acknowledged this, and held that 'scientific uncertainty' and 'insufficient scientific evidence' are not interchangeable concepts.[79] Furthermore, in *Japan-Apples, EC-Approval and Marketing of Biotech Products* and *EC-Hormones II*, the AB and the Panels showed that they view insufficient scientific evidence as a product of lack of quantitative as well as qualitative evidence.[80]

It might be argued that this shows that the AB has made use of a more sophisticated approach to scientific uncertainty than it sometimes is given credit for. Furthermore, it indicates that it does not view the precautionary principle in the WTO Agreements as limited to Article 5.7 of the SPS Agreement. This was also supported in *EC-Hormones I* where the AB's conclusions can be interpreted to mean that 'the precautionary principle is also reflected in other provisions (than Article 5.7) of the SPS Agreement'. As mentioned above, the AB specifically alluded to Article 3.3 and the preamble.[81]

What is also clear is that the requirement of a risk assessment is absolute outside of Article 5.7, and that the risk assessment must yield a minimum of scientific evidence for the application of precautionary measures to be allowed. Thus, outside of Article 5.7 the lack of scientific evidence is not what triggers the use of the precautionary principle. On the contrary, some scientific evidence must be the basis for the measure. The level of proof which is required by the Member wishing to maintain precautionary measures will be returned to shortly. The conclusion as regards the content of Article 5.7 must be that it does not exhaust the application of the precautionary principle. The only thing it does is to establish one carefully circumscribed situation – lack of data – where, temporarily, measures do not need to be backed up by a risk assessment.

Members who do not wish to base their precautionary measures on the results of a risk assessment must thus rely on Article 5.7. In order to

[79] Panel Report, *Japan – Measures Affecting the Importation of Apples*, WT/DS245/R (Panel Report, *Japan-Apples*), para. 184.

[80] AB Report, Japan – Measures Affecting the Importation of Apples. WT/DS245/AB/R (*Japan-Apples*), para. 179; Panel Report, *EC-Approval and Marketing of Biotech Products*, Annex K; and AB Report, *EC-Hormones II*, para. 530. The matter is returned to in more detail below in relation to the section on scientific evidence and the level of proof.

[81] Bernasconi-Osterwalder *et al.*, *Environment and Trade*, p. 264. Scott also finds that the AB in *EC-Hormones I* saw a role for the precautionary principle outside of Article 5.7. See Scott, *The WTO Agreement on Sanitary and Phytosanitary Measures*, p. 128.

do this, they must fulfil four requirements found in Article 5.7 and confirmed by the AB in *Japan Agricultural Products II* and more recently by in *EC-Hormones II*.[82] First, the measures must be imposed in respect of a situation where 'relevant scientific information is insufficient'. Second, they must be adopted 'on the basis of available pertinent information'. Third, additional information must be sought in order to conduct 'a more objective assessment of risk'. Finally, the measures must be of temporary nature, and should be reviewed within a reasonable period of time. The requirements are cumulative in nature and it is enough for one of them not to be present for a measure to fall outside the scope of Article 5.7.[83]

It should be pointed out that the Panel in *EC-Approval and Marketing of Biotech Products* stated that relying on Article 5.7 is an autonomous right of Members, and does thus exempt them from other obligations in the SPS Agreement, such as Articles 2.2 and 5.1.[84] This issue is returned to in more detail below.

3.3.2.2 Judicial review

In WTO dispute settlement, Panels have been empowered by the agreements to take an active part in fact-finding, consultation with and appointment of scientific or technical experts and substantive review of scientific evidence put before them.[85] All of this aims to provide for an 'objective assessment of the facts'.[86] How deep a Panel needs to go in order to assess the evidence is not clear. One author has claimed that the Panel in *EC-Hormones I* positioned itself somewhere in between a deferential review and a *de novo* assessment of the facts: '[A] Panel should use all of

[82] Report of the Appellate Body. Japan – Measures Affecting Agricultural Products. WT/DS76/AB/R (*Japan Agricultural Products II*). paras. 89 and 90 and AB Report, *EC-Hormones II*, para. 676.

[83] AB Report, *Japan Agricultural Products II*, para. 89 and Panel Report, *EC-Hormones II*, para 7.594 and 7.826.

[84] Panel Report, *EC-Approval and Marketing of Biotech Products* para. 7.3007. See also, Prevost, 'Opening Pandora's Box', 83–4.

[85] GATT 1994 Annex II: Understanding on Rules and Procedures Governing the Settlement of Disputes (DSU), Article 13 and SPS Agreement, Article 11. For a detailed discussion on the judicial review in health matters under the WTO Agreements, see H. Horn and P. C. Mavroidis, 'Burden of Proof in Environmental Disputes in the WTO: Legal Aspects', *EELR*, 18:2 (2009), 112. See also Button, *The Power to Protect*, Chapters 5–7 and D. Palmeter, 'The WTO Standard of Review in Health and Safety', in G. A. Bermann and P. C. Mavroidis, *Trade and Human Health and Safety*, Cambridge University Press 2006, pp. 224–35.

[86] GATT 1994 Annex II: Understanding on Rules and Procedures Governing the Settlement of Disputes (DSU), Article 11.

the information at its disposal, including the evidence presented by the two parties, third party submissions, expert witnesses the Panel chooses to call, or any other relevant sources.'[87] This approach gives a considerable freedom to the Panels in determining on what evidence it will base its decisions. Bohanes has held that '[i]t is difficult to overestimate the implications of this: standards of review affect the entire balance of power in a supranational organization, where sovereign members yield powers of interpretation of an agreement to a tribunal'.[88] Croley and Jackson hold that the issue of judicial review was one of the most contended during the Uruguay Round negotiations, which led to the creation of the WTO.[89]

In order for the Panels to adequately assess the often complex and technical information put before them, the agreements provide for the appointment of experts to assist them.[90] In all cases where Panels have made use of expert advice, individual experts rather than expert review groups have been appointed. The reason for this appears to be that Panels have wished to avoid appointing a 'panel within the panel' and the fact that co-ordinating the work in expert groups may be excessively time-consuming.[91]

The AB, on the other hand, is prohibited from revising the factual conclusions reached by the Panels.[92] In *EC-Sardines*, *EC-Hormones I* and *Japan-Apples* the AB stated that:

[87] C. Bisgaard, 'Assessing the Standard of Review for Trade-Restrictive Measures in the Sanitary and Phytosanitary Agreement', in E. Brown Weiss and J. H. Jackson, *Reconciling Environment and Trade* (Transnational Publishers, 2001), p. 367.

[88] J. Bohanes, 'Risk Regulation in WTO Law: A Procedure-based Approach to the Precautionary Principle', *Colum. J. Transnat. L.*, 40 (2001–2002), 323, 346.

[89] S. P. Croley and J. H. Jackson, 'WTO Dispute Procedures, Standard of Review and Deference to National Governments', *AM. J. Int'l L.*, 90 (1996), 193, 211.

[90] See, in particular, Articles 11.2 of the SPS Agreement and 13.2 of the Dispute Settlement Understanding in this regard.

[91] J. Pauwelyn, 'The Use of Experts in WTO Dispute Settlement', *ICLQ*, 51 (2002), 325. Pauwelyn generally sees the use of experts as very beneficial for the swift and rational decision-making in Panel proceedings and he even argues for an extended use of scientific experts, including the appointment of Panel members and secretariat with scientific training (*ibid.*, 363–4). See also, J. Pauwelyn, 'Expert Advice in WTO Dispute Settlement', in G. A. Bermann and P. C. Mavroidis, *Trade and Human Health and Safety* (Cambridge University Press, 2006), pp. 235–56. However, other authors are more sceptical about the appropriateness of placing such great weight on scientific evidence. See for example, R. Howse, 'Democracy, Science, and Free Trade: Risk Regulation on Trial at the World Trade Organization', *Mich. L. Rev.*, 98 (2000), 2329. For a general critique of the reliance on scientific advice in public decision-making, see Jasanoff, *The Fifth Branch*.

[92] Article 17.6 of the DSU holds that 'An appeal shall be limited to issues of law covered in the panel report and legal interpretations developed by the panel.'

> Determination of the credibility and weight properly to be ascribed to (that is, the appreciation of) a given piece of evidence is part and parcel of the fact finding process and is, in principle, left to the discretion of a panel as the trier of facts.[93]

Only if an interpretation of the Panel puts its 'objective assessment of the facts' in question is it going to be reviewed by the AB. However, a simple misinterpretation of the facts by a Panel is not going to make the AB revise its judgment. Instead the AB has held that the Panel must have 'deliberately disregarded', 'refused to consider', 'wilfully distorted' or 'misrepresented' the evidence before it for this to be the case.[94] Thus, the discretion of the Panels when interpreting scientific evidence is very broad. The discretion of the AB, on the other hand, is limited to considering formal aspects of the judgment and to making an assessment *in rem* only regarding the question of whether the Panels have intentionally made a biased assessment of the facts.[95]

The consequence of this deep review by the Panels in relation to the precautionary principle is that science appears to become the arbiter in many cases.[96] This issue is analysed in detail below.

In *EC-Hormones II* the AB elaborated considerably on the limits of the judicial review under the SPS Agreement. In its ruling, the AB considered the Panel to have gone beyond the powers attributed to it when it reviewed the compliance of the EC with the ruling of the AB in *EC-Hormones I*. It stated that the purpose of consulting experts under the Dispute Settlement Understanding (DSU) is not to enable the Panel to carry out its own risk assessment but rather to enable it to consider if the risk assessment performed by the relevant Member 'sufficiently warrants' the SPS measure at issue.[97] The reasoning provided by the AB in this regard deserves, despite its relative length, to be quoted in full:

> The Panel should have first looked at the European Communities' risk assessment. It should then have determined whether the scientific basis

[93] AB Report, European Communities – Trade Description of Sardines, WT/DS231/AB/R (*EC-Sardines*), para. 300; AB Report *EC-Hormones I* para. 132; AB Report *Japan-Apples*, para. 138.

[94] AB Report, *EC-Hormones I*, para. 133. For a critique of the objective standard of review performed by the AB see Button, *The Power to Protect*, pp. 163–91.

[95] For a discussion on the review permitted under Article 17.6 of the DSU see van den Bossche, *The Law and Policy of the World Trade Organization*, 1st edn, pp. 250–2.

[96] Bohanes, for example, has held that the judicial review is one of the main problems when applying the precautionary principle; see Bohanes, 'Risk Regulation in WTO Law', 345–6.

[97] AB Report, *EC-Hormones II*, para. 592.

> relied upon in that risk assessment came from a respected and qualified source. The Panel should have sought assistance from the scientific experts in confirming that it had properly identified the scientific basis underlying the European Communities' risk assessment or to determine whether that scientific basis originated in a respected and qualified source. The Panel should also have sought the experts' assistance in determining whether the reasoning articulated by the European Communities on the basis of the scientific evidence is objective and coherent, so that the conclusions reached in the risk assessment sufficiently warrant the SPS measure. Instead, the Panel seems to have conducted a survey of the advice presented by the scientific experts and based its decisions on whether the majority of the experts, or the opinion that was most thoroughly reasoned or specific to the question at issue, agreed with the conclusion drawn in the European Communities' risk assessment. This approach is not consistent with the applicable standard of review under the *SPS Agreement*.[98]

Thus it appears that the standard of review advocated by the AB in *EC-Hormones II* aims to establish that the risk assessment comes 'from a respected and qualified source', 'has properly identified the scientific basis' and is 'objective and coherent, so that the conclusions reached in the risk assessment sufficiently warrant the SPS measure'. This appears to be a standard which is rather similar to the 'hard look' doctrine applied by federal courts in the United States.[99]

3.3.2.3 Risk assessment and risk management

The AB has held that the requirement of a risk assessment stated in Article 5.1 of the SPS Agreement is strict and that it aims to ensure that the general goal in Article 2.2 is met.[100] The relevant part of Article 2.2 states that an SPS measure must be 'necessary', 'based on scientific principles' and 'not maintained without sufficient scientific evidence'.[101]

In relation to pests and diseases, the AB in *Australia-Salmon* established that a risk assessment under Article 5.1 must address three requirements. First, it must identify the risk and the biological and economic consequences the realisation of it would bring. Second, the likelihood of the realisation must be established. Finally, the likelihood of the realisation of

[98] *Ibid.*, para. 598. [99] See below, Section 7.1.4.

[100] AB Report, *EC-Hormones I*, para. 180.

[101] In Australia – Measures Affecting the Importation of Salmon WT/DS18/AB/R (*Australia-Salmon*) (paras. 137–8), the AB elaborated on the relationship between Article 5.1 and the more general Article 2.2. It concluded that if Article 5.1 is breached, it can be presumed that Article 2.2 is also breached. There has not yet been a case where this presumption has not been found to be true. For a discussion on this see Bernasconi-Osterwalder *et al.*, *Environment and Trade*, pp. 84–91.

the risk must be established according to the SPS measures which might be applied.[102] The third requirement means that it is not enough to evaluate the situation in relation to SPS measures already in place, but other possible measures must be considered. The AB has later emphasised that this is an important factor in the risk assessment.[103] The three requirements stressed in *Australia-Salmon* are based on the definition of risk assessment included in Annex A to the SPS Agreement.[104]

Concerning food-borne risks, Panels and the Appellate Body have stressed two requirements which should be addressed when carrying out a risk assessment under paragraph 4 of Annex A of the SPS Agreement. First, the adverse effects to human or animal health stemming from the food, beverage or feeding-stuff need to be identified. Second, the 'potential' occurrence of these risks needs to be assessed.[105]

Furthermore, the AB has found it of importance that the assessed data is specific enough. In *EC-Hormones I* the data presented by the EC in support of its measures was not based on the hormones at issue per se, but rather on a whole category of hormones. This the Panel as well as the AB considered to be 'relevant but do not appear to be sufficiently specific to the case at hand'.[106] It thus seems that the AB did find the data to be of use, but too general in scope to justify the EC measures.[107] The same line of reasoning was adopted by the AB in *Japan-Apples* and in *EC-Hormones II*.[108]

When assessing the data, the Member is not bound to a specific assessment of the risk but remains free to choose the appropriate methodology. The methodology chosen must be in conformity with Paragraph 4 of Annex A to the SPS Agreement.[109] Article 5.2 of the SPS Agreement lists a number of topics that should be covered by a risk assessment. This list should, however, not be considered as exhaustive. Instead it is indicative of issues that may be of relevance.[110] Nothing should prevent a Member

[102] AB Report, *Australia-Salmon*, para. 120.

[103] AB Report, *Japan-Apples*, para. 209.

[104] Paragraph 4. Annex A attached to the SPS Agreement. See also, D. Winickoff *et al.*, 'Adjudicating the GM Food Wars: Science, Risk, and Democracy in World Trade Law', *Yale J. Int'l Law*, 30 (2005), 81 at footnote 173. See also AB Report, *EC-Hormones II*, para. 525.

[105] Panel Report, *EC-Hormones I (US)*, para. 8.98; Panel Report, *EC-Hormones I (Canada)*, para. 8.101; and AB Report, *EC-Hormones I*, para. 184–6.

[106] AB Report, *EC-Hormones I*, para. 200.

[107] See also, Scott, *The WTO Agreement on Sanitary and Phytosanitary Measures*, p. 94.

[108] AB Report, *Japan-Apples*, footnote 372 to para. 202 and AB Report, *EC-Hormones I*, paras 562–6.

[109] AB Report, *Japan-Apples*, para. 205.

[110] AB Report, *EC-Hormones II*, para. 535.

from including other aspects that more accurately depict the relevant risk 'in the real world where people live and work and die'.[111]

As discussed above, the AB in *EC-Hormones I* emphasised that a risk assessment should not necessarily be limited to the mainstream scientific opinion. Instead it stressed that divergent views on a risk may indicate that the issue is surrounded by scientific uncertainty. However, it also stressed the importance of a clear link between the divergent opinions and the risk at issue.[112] In *EC-Hormones II* the AB emphasised that '[a]lthough the scientific basis need not represent the majority view within the scientific community, it must nevertheless have the necessary scientific and methodological rigour to be considered reputable science.'[113]

The Panel in the *EC-Hormones I* case made a distinction between risk assessment and risk management. It held that risk assessment is a strictly scientific undertaking and not a 'policy' exercise.[114] Risk management, on the other hand, it described as 'non-scientific'.[115] On appeal the AB did not agree with this reasoning. It found no textual basis in the SPS Agreement for separating the two stages; Article 5 only pertains to the issue of risk assessment. The AB therefore denounced the 'restrictive notion of risk assessment' that it considered the Panel to be advocating.[116] This seems to imply that the AB is open to considering 'risk assessment' to be less restrictive than the traditional science-based interpretation that the Panel had advocated.[117] One commentator has held that this conclusion of the AB leaves the governments 'great freedom in risk management even where exercise of this freedom may inhibit trade'.[118] In *Continued*

[111] AB Report, *EC-Hormones I*, para. 187. The AB repeated this reasoning in AB Report, *EC-Hormones II*, paras. 526–8.

[112] AB Report, *EC-Hormones I*, para. 194. [113] AB Report, *EC-Hormones II*, para. 591.

[114] Panel Report, *EC Hormones (US)*, para. 8.94 and Panel Report, *EC Hormones (Canada)*, para. 8.97.

[115] Panel Report, *EC Hormones (US)*, para. 8.95 and Panel Report, *EC Hormones (Canada)*, para. 8.97.

[116] AB Report, *EC-Hormones I*, para. 181.

[117] For a discussion on the differing approaches assumed by the Panel and the AB in *EC-Hormones I* in relation to the content of risk assessments, see J. Scott, 'Of Kith and Kine (and Crustaceans): Trade and Environment in the EU and WTO'. Available at: www.jeanmonnetprogram.org/papers/99/990307.html (accessed 1 March 2010). The Panel had first established a clear separation between scientific risk assessment and the decision-making taking place in the risk management phase. The AB found that such a clear separation was not to be found in the SPS Agreement and argued that risk assessment can include broader considerations than strictly scientific analysis (*EC-Hormones I*, para. 187). However, as is seen in the following, the requirement on scientific evidence is still paramount.

[118] Bohanes, 'Risk Regulation in WTO Law', 341.

Suspension of Obligations-EC Hormones, the Panel once again sought to draw a line of demarcation between risk assessment and issues it considered belonging to the risk management process. While recognising that the AB in *EC-Hormones I* had considered there to be no textual basis in the SPS Agreement for a generally applicable separation of risk assessment and risk management, the Panel sought instead to limit risk assessment to the definition set forth in Annex A(4) of the SPS Agreement.[119] On appeal, the AB considered this restrictive notion of risk assessment to be as erroneous as the one considered by the AB in *EC-Hormones I*.[120]

It could be argued that the view assumed by the Panels in *EC-Hormones I* and *EC-Hormones II* ensures a more systematic approach with clearer implications than the conclusions the AB came to. It could therefore be argued that it would have made sense for the AB to build on the findings of the Panels and primarily confine risk management choices to Article 3.3 of the SPS Agreement. It would then have been clear that the setting of the appropriate level of protection is a risk management decision, while the establishment of the scientific basis for that decision is an eminently scientific enterprise.[121] Although there is no explicit support for this in the WTO Agreements, it is, as is elaborated on elsewhere in this book, a generally accepted distinction in the literature as well as in the administrative practices of most states. Furthermore, it is the preferred method advocated in the Codex Alimentarius.[122]

Connected to the above, in *EC-Hormones II*, the AB elaborated on the issue of if a Member's chosen level of protection can be allowed to influence the risk assessment process. It came to the conclusion that this could indeed be the case.[123] The AB stated:

> There may be circumstances in which the appropriate level of protections chosen by a Member affects the scope or method of the risk assessment. This may be the case where a WTO Member decides to adopt an SPS measure based on an international standard because it seeks to achieve a higher level of protection. In such a situation, the fact that the WTO Member has chosen to set a higher level of protection may require it to perform certain research as part if its risk assessment that is different

[119] Panel Report, US-Continued Suspension of Obligations in the EC Hormones Dispute, WT/DS320/R, paras. 7.519 and 7.520 (*Continued Suspension of Obligations-EC Hormones*).

[120] AB Report, *EC-Hormones II*, para. 545.

[121] For a further discussion on the separation of risk assessment and risk management see above, Section 2.3.1.

[122] Codex Alimentarius Commission, Procedureal Manual, 15th edition, p. 45.

[123] AB Report, *EC-Hormones II*, para. 534 and 685–6.

> from the parameters considered and the research carried out in the risk assessment underlying the international standard.[124]

However, the AB went on to state that this fact 'may not affect the rigour or objective nature of the risk assessment' and it 'does not pre-determine the results of the risk assessment'.[125] Thus, the AB appears to be of the opinion that the chosen level of protection is central to determining the types of scientific testing a certain product should be subjected to, but it should not be allowed to prejudice the objectivity of such tests. This would indeed appear to be a necessary conclusion in order to ensure that Members have a real possibility to decide on the appropriate level of protection. However, the AB could have reached the same end while maintaining a separation between risk assessment and risk management.

When carrying out a risk assessment, Article 5.1 holds that Members shall take into account risk assessment techniques developed by relevant international organisations. In *EC-Hormones II* the USA held that the failure of the EC to strictly adhere to the risk assessment techniques suggested in the Codex Alimentarius constituted a breach of Article 5.1.[126] However, the Panel concluded that it was not of primary importance that the party invoking protective measures follow all the steps outlined in the Codex Alimentarius. Instead it considered it enough that the relevant risk assessment techniques 'had been taken into account' during the assessment.[127]

As mentioned above, the only time the strict risk assessment requirement may be wavered is in cases where Article 5.7 applies. According to the article, precautionary measures may be applied 'on the basis of pertinent information' in cases where 'relevant scientific evidence is insufficient.' This must be compared to the wording of Article 2.2 where it is held that measures cannot be maintained or introduced 'without sufficient evidence'. The AB has considered the latter as meaning that measures may not be maintained where there is no 'rational or objective relationship' between the measure and the relevant scientific evidence.'[128] Thus, it describes a situation where enough scientific evidence exists in order to carry out a risk assessment, but the results of the risk assessment do not support the conclusions drawn by the party wishing to maintain or introduce the measures.[129]

[124] *Ibid.*, para. 534. See also paras. 685–6. [125] *Ibid.*

[126] Panel Report, *EC-Hormones II*, paras. 7.450–7.454.

[127] *Ibid.*, paras. 7.467–7.469. [128] AB Report, *Japan-Apples*, para. 147.

[129] For a further discussion on what it means for a measure to be based on a risk assessment see Scott, *The WTO Agreement on Sanitary and Phytosanitary Measures*, pp. 104–10.

With regard to Article 5.7 the AB has held that the aim is to cater for a situation where 'the body of available scientific evidence does not allow, in quantitative or qualitative terms, the performance of an adequate assessment of risks as required under Article 5.1 and as defined in Annex A to the SPS Agreement.'[130] What the AB here refers to is a situation where a risk assessment is impossible to carry out due to the lack of scientific evidence. The same conclusion was echoed by the AB in *EC-Hormones II*, where it held that:

> Article 5.7 is concerned with situations where deficiencies in the body of scientific evidence do not allow a WTO Member to arrive at a sufficiently objective conclusion in relation to risk.[131]

In such cases, and only in such cases, measures can be maintained without support in a risk assessment. However, it can be discussed whether a risk assessment must not at least be attempted in support of measures. One could be forgiven for asking how a Member would know that relevant data was lacking before it has made an attempt at evaluating it.

3.3.2.4 Scientific evidence and the introduction of precautionary measures

This analysis has revealed that there are two ways of maintaining or introducing precautionary measures under the SPS Agreement. First, measures can be maintained in a situation where a risk assessment is carried out and a rational and objective relationship between the scientific evidence and the precautionary measures is identified. In this situation scientific uncertainty may exist due to diverging opinions or insufficient data to make the risk scientifically certain.

Second, measures can be maintained in a situation where a Member can claim that the scientific data available is so insufficient that it renders the performance of a risk assessment impossible. In this section it is investigated what the party wishing to maintain or introduce measures needs to show in the two respective situations.

As was discussed above, the Panel is the trier of the facts under the DSU. The AB performs a limited review, restricted to interpreting the legality of measures. It will only revise the factual findings of the Panel if it finds that the Panel has made an intentionally biased assessment of the facts. The Panel on the other hand goes to great lengths to interpret the evidence presented to it and to determine that measures are in conformity

[130] AB Report, *Japan-Apples*, para. 179.

[131] AB Report, *EC-Hormones II*, para. 677.

with Articles 2.2 and 5.1 of the SPS Agreement. Four requirements found in the Articles need to be fulfilled by the Member wishing to maintain measures outside of Article 5.7 of the SPS Agreement.[132] First, a risk assessment must be carried out.[133] Second, the measures must be shown to be necessary with regard the goal they aim to achieve. Third, measures must be based on scientific principles. Finally, the measures may not be maintained without 'sufficient scientific evidence'. The last two of these requirements are dealt with here.

In *EC-Hormones I*[134] the Panel found that the main problem with the scientific evidence brought before it by the EC was that it was not specific enough.[135] The *EC-Hormones* case related to a number of specific hormones used as growth promoters in cattle. However, most of the evidence brought by the EC did not relate to these specific hormones, but rather to entire categories of hormones.[136] Such studies were considered by the Panel to be relevant, but not 'sufficiently specific to the case at hand', and therefore did not fulfil the requirement in paragraph 4 of Annex A of the SPS Agreement.[137]

The more specific evidence brought by the EC was limited.[138] Particularly damning for the EC was that its own scientific expert had come to the conclusion that no scientific evidence indicating the harmfulness of the hormones existed.[139] A dissenting scientific opinion by one expert was considered. He held that there was a measurable risk associated with the hormones. However, the Panel found that the risk proposed by the expert, 0–1 in a million, was not enough to signify that an 'identifiable risk' existed and that the use of the hormones for growth promotion purposes was safe.[140] The Panel disagreed with the opinion for two main reasons. First, it considered the fact that the risk proposed referred to the total amount of oestrogen in cattle, not to the amount connected to growth promoters. Second, it held that a risk of such low probability is not even a scientifically identified risk.[141]

[132] AB Report, *EC-Hormones I*, paras. 179 and 180.

[133] See the discussion above in Section 3.3.2.2.

[134] The case was brought before a Panel by Canada and the USA, which held that the EC legislative regime prohibiting beef treated with a number of hormones from entering its market was not in conformity with the SPS Agreement, notably Articles 2.2 and 5.1. For a detailed discussion on the case, see Bernasconi *et al.*, *Environment and Trade*, pp. 268–78 and Scott, 'Of Kith and Kine (and Crustaceans)'.

[135] Panel Report, *EC Hormones (US)*, paras. 8.129 and 8.134.

[136] *Ibid.*, para. 8.127.

[137] AB Report, *EC Hormones I*, para. 199.

[138] Panel Report, *EC Hormones (US)*, para. 8.119–8.122

[139] *Ibid.*, para. 8.123.

[140] *Ibid.*, para. 8.124.

[141] *Ibid.*, note 331.

As was discussed in the previous section, on appeal the AB was critical of the narrow definition the Panel had given to the term risk assessment. It was also critical of the Panel's use of the term 'probable' since it considered it to imply 'a higher degree of threshold of potentiality or possibility'.[142] This the AB found to lead to a situation where a quantitative risk assessment would have to yield as a result a risk of a certain magnitude.[143] This would not be commensurable with the SPS Agreement. Instead, the AB considered that the establishment of a possible cause and effect chain would be sufficient to justify SPS measures.[144] It should be noted that the risks in *EC-Hormones I* were food-borne and therefore the second tier of the definition of risk assessment in paragraph 4 of Annex A of the SPS Agreement was considered by the Panels and the AB. Therefore, the conclusions of the AB with regard the level of proof required relates to the establishment of 'potential' risk, as opposed to the 'likelihood' requirement in the first tier of the definition relating to pest or disease. It appears that the difference in the wording should be taken to indicate a less stringent level of proof in cases relating to food-borne risks than in cases of pests and disease.[145]

Even if the AB in *EC-Hormones I* went to some lengths to point out that it construed the requirement of risk assessment and scientific evidence to be somewhat different (and perhaps more lenient) than the Panel had, in reality the ruling by the AB is more ambiguous. In fact, it appears that the difference between the approaches adopted has more to do with linguistics than with any functional discrepancies. Like the Panel, the AB holds that 'theoretical uncertainty' as to the risk is not what is to be assessed under Article 5.1.[146] Scott has interpreted this as meaning that '[a] hypothetical possibility of risk, arising from the failure of scientists to demonstrate with absolute certainty that there is no risk either now or in the future, is not capable of rationally grounding an SPS measure'.[147] Thus, the reliance on scientific evidence is strongly emphasised also by the AB. And since the performance of comprehensive risk assessments is the most common, if not only, way to scientifically prove the existence of a risk, the performance of risk assessments are also endorsed by the AB.

In *Japan-Apples* the USA challenged measures introduced by Japan to protect itself from the introduction of fire-blight through the importation

142 AB Report, *EC-Hormones I*, para. 184. 143 *Ibid.*, para. 186.

144 For a discussion on this, see Scott, 'Of Kith and Kine (and Crustaceans)'.

145 Van den Bossche, *The Law and Policy of the World Trade Organization*, 2nd edn, Section 8.4.5.1.

146 AB Report, *EC-Hormones I*, para. 186.

147 Scott, 'Of Kith and Kine (and Crustaceans)'.

of apples from the USA.[148] Fire-blight is a bacterium, which causes apples to whither and darken. The risk thus meets the first tier definition in paragraph 4 of Annex A of the SPS Agreement, and what needed to be assessed is the 'likelihood' of the risk rather than its 'potential'.

Japan held that its risk assessment had yielded enough evidence to maintain its protective regime against the USA. As a secondary motion it relied on Article 5.7, in case the Panel would find that the scientific evidence was not enough to justify the measures under Article 5.1.

Two issues needed to be examined by the Panel to determine the risk to Japan of fire-blight spreading through the importation of apples. First, it had to be established if mature and symptomless fruit (the only type exported by the USA to Japan) could harbour fire-blight bacteria. Second, if this be the case, could such apples carry the bacteria and spread it to other fruit after it had entered Japan?

To begin with, the Panel and the AB turned their attention to the issue of whether the measures could be said to be supported by 'sufficient scientific evidence' in accordance with Article 5.1 and 2.2 of the SPS Agreement. In order to understand the scientific evidence before it, the Panel relied on the advice of appointed experts to clarify the detailed scientific issues.[149] The evidence consisted of relevant scientific reports and monographs in combination with the empirical knowledge obtained over many years of trading with apples.

The Panel found that the bulk of the scientific evidence relating to the infection and infestation of mature and symptomless apples on which Japan relied was 'relatively confused, difficult to interpret or even unconvincing'[150] and that the results they predicted were 'extremely rare' at best.[151]

The second stage of this survey dealt with the issue of the cause and effect relationship between the potential prevalence of fire-blight in mature symptomless apples and the spread of fire-blight to other apples. Japan's own risk assessment had failed to find support for such a link[152] but Japan held that it would be a matter of 'common sense' that transmission could be completed through rain, insects, and so on.[153] The Panel was not convinced by this argument. It cited experts in the field who held that large-scale experiments had failed to show a link. Furthermore the

[148] For another analysis of the *Japan-Apples case*, see Bernasconi-Osterwalder *et al.*, *Environment and Trade*, pp. 288–91.

[149] See, for example Panel Report, *Japan-Apples*, paras. 8.125 and 8.169.

[150] *Ibid.*, para. 8.125. [151] *Ibid.*, para. 8.131.

[152] *Ibid.*, para. 8.147. [153] *Ibid.*, para. 8.146.

historic *in vivo* evidence suggested the risk to be 'negligible' and defined this as a likelihood of between zero and one in a million.[154] Experts also held that the risk of travellers to Japan bringing with them infected or infested apples was higher than that associated with wholesale import.[155] The only precautionary measure endorsed by the experts was that apples from heavily fire-blight-infested orchards should not be exported, even if mature and symptom-free.[156]

Finally, the Panel held that fruit other than 'mature and symptomless' apples may harbour the bacteria, and that such bacteria may survive the transport to Japan.[157] However, the USA emphasised that such apples were not exported to Japan and that rigorous tests were in place to ensure that this would continue to be the case. In the light of this, the Panel held that the risk of such apples entering Japan would purely be due to errors in handling or illegal actions. These risks were acknowledged by the experts, but were deemed to be 'small' or 'debatable'.[158]

On the basis of this, the Panel concluded that 'sufficient scientific evidence' had not been produced in favour of the SPS measures, and these were thus in violation of Article 2.2 of the SPS Agreement.[159] The AB further clarified this on appeal and reiterated its position in *Japan-Agricultural Products II*, where it stated that the absence of 'an actual causal link' between the protective measure and the scientific evidence indicates that 'sufficient scientific evidence' is lacking in support of the measures.[160] A 'causal link' must in turn be interpreted as establishing a 'rational or objective relationship'.[161]

Having concluded that sufficient scientific evidence did not exist, the Panel turned to investigate if the measures could be maintained in the absence of scientific evidence under Article 5.7. Here, it began by separating the meaning of the wording of Article 5.7 from that of Article 2.2. The requirement that 'relevant scientific evidence is insufficient' in Article 5.7 is not to be equated with Article 2.2's statement that measures may not be maintained 'without sufficient scientific evidence'.[162] The Panel concluded:

> The current 'situation', where scientific studies as well as practical experience have accumulated for the past 200 years, is clearly not the type of

[154] *Ibid.*, paras. 8.146 and 8.149. [155] *Ibid.*, para. 8.150.
[156] *Ibid.*, para. 8.151. [157] *Ibid.*, paras. 8.154–8.157.
[158] *Ibid.*, para. 8.161. [159] *Ibid.*, para. 8.199.
[160] AB Report, *Japan Agricultural Products II*, para. 83.
[161] AB Report, *Japan-Apples*, para. 162. [162] Panel Report, *Japan-Apples*, para. 8.215.

> situation Article 5.7 was intended to address. Article 5.7 was obviously designed to be invoked in situations where little, or no, reliable evidence was available on the subject matter at issue. With regard to fire blight, not only a large quantity but a high quality of scientific evidence has been produced over the years that describes the risk of transmission of fire blight through apple fruit as negligible. Moreover, this is evidence in which the experts have expressed strong and increasing confidence. We therefore are of the view that the first condition of the first sentence of Article 5.7 is not met.[163]

Thus, it appears that the Panel viewed Article 5.7 as something to which recourse can be had in cases of scientific uncertainty due to the lack of observations and scientific data. If such data is not insufficient in the context of Article 5.7, the requirements of Article 2.2 must be met. The conclusion of the above is that 'sufficient scientific evidence' in Article 2.2 is intended to mean 'sufficient scientific evidence in favour of the SPS measure', whereas 'insufficient scientific evidence' in Article 5.7 refers to scientific evidence relating to the subject matter in general.[164] In relation to this, some authors have held that the scope of Article 5.7 should be allowed to be broader since scientific evidence is perceived as rarely leading to entirely reliable or conclusive results.[165] However, it should be noted that it does not appear to be referring to reliability or conclusiveness in absolute terms, but rather it uses them as relative terms to distinguish situations falling under Article 5.7 from those falling under the strict risk assessment requirement in Articles 2.2 and 5.1. This distinction was also emphasised by the Panel in *EC-Approval and Marketing of Biotech Products*.[166]

Furthermore, and importantly, the AB has clearly stated that, in order to invoke Article 5.7, a Member would not have to show a complete lack

[163] *Ibid.*, para. 8.219.

[164] Bohanes also addresses this issue and he stresses the point that since no Member as of yet has successfully invoked Article 5.7, it is unclear what will be required to be shown in order to do so. It is here argued that only the lack of scientific findings can motivate temporary measures under Article 5.7. See Bohanes, 'Risk Regulation in WTO Law', pp. 338–9. See also, the discussion on sufficient scientific evidence in Scott, *The WTO Agreement on Sanitary and Phytosanitary Measures*, pp. 86–91.

[165] Winickoff *et al.*, 'Adjudicating the GM Food Wars', 113–14. The authors argue that the greater the disparity of the scientific advice, and the greater the lack of consensus on its validity, the greater the possibility would be to rely on Article 5.7. However, this is in contrast to the conclusion of the argument made by this author, who sees Article 5.7 not as a 'safe haven' but simply as an exemption from the requirement of performing risk assessments in urgent cases where this is difficult or impossible. Thus, the AB has rightly employed a narrow interpretation of the article since the overarching requirement of performing a risk assessment should not be eroded.

[166] Panel Report, *EC-Approval and Marketing of Biotech Products*, para. 7.2995.

of scientific findings, but the article applies where the body of available scientific evidence does not allow, in quantitative or qualitative terms, the performance of an adequate assessment of risks as required under Article 5.1.[167] Every situation has to be decided on a case-by-case basis.[168] In relation to this, in *EC-Hormones II* the Panel stated that it did not consider the purpose of Article 5.7 to be to 'lower' the level of proof a Member has to meet in order to introduce or maintain protective measures. Rather it considered Article 5.1 and Article 5.7 to apply to the separate situations outlined above.[169] The Panel in *EC-Hormones II* also stated that the intended level of protection of the Member introducing or maintaining SPS measures under Article 5.7 is irrelevant when considering whether the measures can be allowed. What triggers the use of the article is not the intended level of protection, but the lack of available scientific evidence.[170]On appeal, the AB took issue with this interpretation of the Panel. It stated that the chosen level of protection can be highly influential in determining if the available scientific evidence is sufficient or not. For example, the AB stated that the existence of an international standard creates a presumption that there is sufficient scientific evidence in support of the level of protection indicated by the standard. However, if a Member wishes to maintain a higher level of protection than that offered by the standard it is free to do so. In such a situation it is not evident that the scientific evidence on which the standard is based is sufficient, or even entirely relevant, for the assessment of the risk. However, the AB stressed that the chosen level of protection cannot be allowed to predetermine the outcome of the risk assessment, and the risk assessment must remain a rigorous and objective process.[171]

In *EC-Hormones II* the Panel also addressed the issue of whether available scientific evidence can 'become insufficient' if it has previously been held that the available evidence was sufficient. The Panel held that this could indeed be the case, but only following the emergence of 'a critical mass of new evidence and/or information that calls into question the fundamental precepts of previous knowledge so as to make relevant, previously sufficient, evidence now insufficient.'[172] A 'critical mass' of information and scientific evidence would have to be 'such that they are at the origin of a change in the understanding of a scientific issue'.[173] Having

[167] AB Report, *Japan-Apples*, para. 179. This statement was also reiterated more recently in Panel Report, *EC-Hormones II*, para. 7.608.

[168] AB Report, *Japan-Apples*, para. 185.

[169] Panel Report, *EC-Hormones II*, para. 7.605.

[170] *Ibid.*, para. 7.612.

[171] AB Report, *EC-Hormones II*, paras. 685–7.

[172] *Ibid.*, para. 7.648.

[173] *Ibid.*, note 782 at para. 7.648.

stated this, the Panel addressed all the hormones at issue in the case and stated that sufficient scientific evidence existed in order to perform a risk assessment with regard to all of them.[174] Also on this point the AB disagreed with the Panel. The AB found that a critical mass doctrine along the lines outlined by the Panel indicated that a ' paradigm shift' would have to be at hand for scientific evidence to become insufficient. This the AB found to be a too inflexible approach, which would create too high a threshold.[175] Instead the AB stated that it would be enough that ' new scientific evidence casts doubts as to whether the previously existing body of scientific evidence still permits of a sufficiently objective assessment of risk'.[176]

As was argued above, the precautionary principle can also function under Articles 2.2 and 5.1, and these articles do not require conclusive evidence to be invoked.[177] The only thing that Article 5.7 adds to the equation is that in urgent cases where little or no scientific evidence is available, Members should not be prevented from issuing temporary protective measures. In all other cases a risk assessment should be performed. However, as has been repeatedly accounted for above, the risk assessment does not have to yield a scientifically certain result for measures to be upheld. Thus the precautionary principle can be invoked outside of Article 5.7 as long as a link is established between the evidence and the measures invoked.[178] This assertion is also supported by Button, who holds that most risk assessments are uncertain to one degree or another, and that this uncertainty does not mean that measures cannot be maintained under Articles 2.2 and 5.1 of the SPS Agreement.[179] Scott also sees the possibilities of maintaining precautionary measures outside of Article 5.7, as long as a risk assessment has been carried out. She even holds that the official view of the EU with regard to risk assessments preceding the application of the precautionary principle implies that the view of the European institutions may not be a far cry from that of the WTO Agreements.[180] Van

[174] The Panel summarised these findings at *ibid.*, paras. 7.831–7.837.

[175] AB Report, *EC Hormones II*, paras. 703–6. [176] *Ibid.*, para. 703.

[177] If the requirement had been that a fact would have had to be shown conclusively in order to function as a basis for measures under Article 5.1, minority opinions, for example could not be relied upon. An instance where a reliable scientific minority opinion exists is by definition an area where the evidence must be seen as uncertain and thus inconclusive. See AB Report, *EC-Hormones I*, para. 194, and the discussion above in Section 3.3.2.1.

[178] This is line with the interpretation assumed by the AB in *EC-Hormones I*, para. 124.

[179] Button, *The Power to Protect*, p. 114.

[180] Scott, *The WTO Agreement on Sanitary and Phytosanitary Measures*, p. 128.

den Bossche, on the other hand, considers that the conclusions of the AB in *EC-Hormones I* and the consistent approach assumed by the Panel in *EC-Approval and Marketing of Biotech Products* in practice confines the application of the precautionary principle to Article 5.7 situations, due to the fact that the principle cannot be considered to add 'flexibility' to the scientific disciplines under the SPS Agreement.[181] However, as has been argued above, the precautionary principle can be used by Members without adding 'flexibility' to the scientific disciplines in the SPS Agreement. Since full scientific certainty is not required for the issuing of measures, and minority opinions can be relied upon, it could be argued that the 'flexibility' needed for Member States to rely on the precautionary principle is in fact already built into the SPS regime, although it does not feature as an independent ground to invoke as justification for measures.

In a letter to the parties attached to the Panel Report in *EC-Approval and Marketing of Biotech Products*, the Panel elaborated slightly on its view on scientific evidence.[182] In line with previous findings, it held that the insufficiency of scientific evidence, which triggers the use of Article 5.7, can be qualitative as well as quantitative. The Panel also repeated the position advanced in *EC-Hormones I*, that the risk assessed is 'a real world risk', and does not only have to be ascertained in a laboratory. Furthermore, the Panel recognised that scientific evidence is not necessarily static. Instead it emphasised that the emergence of new scientific evidence may contradict a previous risk assessment and lead to a situation where sufficient scientific evidence to perform a risk assessment may not be available.

Panels and the AB in cases under GATT 1994 have also emphasised the importance of a link between protective measures and scientific evidence.[183] As in the case of the SPS Agreement, minority opinions may be relied upon to substantiate such a link.[184] No explicit reference has been made to precautionary action in GATT itself or in subsequent case law. However, in relation to *EC-Asbestos* it has been argued that the AB offered some deference to Members to act quicker and on the basis of less scientific

[181] Van den Bossche, *The Law and Policy of the World Trade Organization*, 2nd edn, Section 8.4.5.3.

[182] See the letter attached to Panel Report, *EC-Approval and Marketing of Biotech Products*, Annex K.

[183] Report of the Appellate Body, European Communities – Measures Affecting Asbestos and Asbestos-Containing Products. WT/DS135/AB/R (*EC-Asbestos*), para. 115.

[184] *Ibid.*, paras. 178 and 194.

evidence in cases where health scares have resulted in a 'panicked public opinion'.[185]

The TBT Agreement has so far not been the object of interpretation by the DSU as to issues of scientific evidence, and so on.[186]

As can be deduced from the above, Panels, as well as the AB, have so far failed to identify a clear level of proof that needs to be met by a party wishing to maintain a measure. In *US-Wool Shirts and Blouses* the AB held that the level of proof required 'will necessarily vary from measure to measure, provision to provision, and case to case.'[187] Panels and the AB have explicitly stated that they will focus their attention on the establishment of a link between the scientific evidence presented and the measure at stake, and that the level required can be relatively low. Pauwelyn has criticised the case-by-case approach on the grounds that it risks leading to different levels of proof being applied in different cases.[188] However, even though it would be helpful from a legal certainty point of view to identify a fixed level of proof that applies equally in all cases, it is difficult to foresee how this would work in practice.

It should be pointed out that some authors have generally criticised the importance given to scientific evidence by Panels and the AB. It has been held that this limits the Member government's ability to choose their own levels of acceptable risk.[189] However, as has been argued here, the scope of precautionary action is significantly broader under the SPS Agreement than it is generally given credit for.

3.3.2.5 The burden of proof

The basic approach to the allocation of the burden of proof under the WTO Agreements was identified by the AB in 1997 in *US–Wool Shirts and Blouses*. The AB found that international tribunals, as well as most

[185] L. A. Ruessmann, 'Reflections on the WTO Doha Ministerial Conference: Putting the Precautionary Principle in its Place: Parameters for the Proper Application of a Precautionary Approach and the Implications for Developing Countries in Light of the Doha WTO Ministerial', *Am. U. Int'l L. Rev.*, 17 (2002), 905.

[186] This has, according to Scott, rendered it 'profoundly under-determined and intensely ambiguous' in relation to future regulatory regimes, notably the EU regulatory regime in relation to genetically modified crops and products. See Scott, 'The European Regulation of GMOs and the WTO', 231.

[187] Report of the Appellate Body. United States – Measure Affecting Imports of Woven Wool Shirts and Blouses From India. WT/DS33/AB/R (*US–Wool Shirts and Blouses*), para. 14.

[188] Pauwelyn, 'The Use of Experts in WTO Dispute Settlement', 360.

[189] See, for example, N. de Sadeleer, 'The Precautionary Principle in European Community Health and Environmental Law', *ELJ*, 12:2 (2006), 139.

domestic legal orders, positioned the burden of proof on the party asserting a fact in the affirmative. This it found to be the case regardless of whether the party assumed the complaining or defending role. The level of evidence to be reached in this initial phase is to 'raise a presumption that what is claimed is true'.[190] What is in fact required to be shown, the AB held, varies from case to case depending on the circumstances in each situation. Once the prima facie case has been made, the burden of proof shifts to the other party, which must provide 'sufficient evidence' to rebut the claim.[191] As India was claiming that the USA was in breach of one of the WTO Agreements (the Agreement on Textiles and Clothing) in *US–Wool Shirts and Blouses*, it had to provide 'evidence and legal argument sufficient to demonstrate' the non-compliance of the USA.[192] Once India successfully did this, the onus was shifted onto the USA to bring forth evidence showing its compliance.

When it comes to the SPS Agreement, the Panel, in *EC-Hormones I*, began by stating that it considered that the initial burden of proof lies with the complaining party to provide prima facie evidence that a measure is not complying with a WTO Agreement. However, when SPS measures are concerned, the Panel stated, it would be up to the party invoking an exception to support this with scientific evidence. In this case, the Panel ruled, the EC had the 'evidentiary burden' to show that it had acted in accordance with the Agreement since it chose to invoke what the Panel characterised as the exception in Article 3.3.[193] The AB did not agree with the position assumed by the Panel and called its conclusion a 'general, unqualified, interpretative ruling' which lacked support in the WTO Agreements and was inconsistent with previous AB reports.[194] Instead, it held, consistently with its position in *US–Wool Shirts and Blouses*, that the initial burden of proof should rest with the party claiming a fact. Thus, it reversed the Panel's ruling and held that the USA and Canada in *EC-Hormones I* bore the responsibility of showing prima facie non-compliance on the side of the EC. Once that had been done, the burden of proof was shifted onto the EC to disprove the claims of the USA and Canada.[195] A prima facie case the AB defined as 'one which, in the absence of effective refutation by the defending party, requires a Panel, as a matter of law, to rule in favour of

190 *US–Wool Shirts and Blouses*, p. 14.

191 *Ibid.*

192 *Ibid.*, p. 16.

193 US Panel Report. *EC Hormones I*, paras. 8.151 and 8.165 and Canada Panel Report, paras. 8.154 and 8.168.

194 AB Report, *EC-Hormones I*, paras. 99 and 102.

195 *Ibid.*, paras. 108 and 109.

the complaining party presenting the prima facie case'.[196] In the case at issue, the AB found that the USA and Canada had managed to do this, despite not being required to do so by the Panel.[197]

In *Japan-Apples* the AB held that, in relation to the SPS Agreement, the requirement by a complaining party to provide a prima facie case related to the non-compliance of the defending party of the SPS Agreement, not to the scientific issue at the heart of the matter. Thus, it was sufficient for the USA to claim that the Japanese measures were maintained without reliance on a risk assessment or a sufficient scientific basis, to shift the onus onto Japan. Japan, in turn, would then have the burden to show that it had performed a risk assessment and that this had yielded sufficient scientific evidence to be in compliance with Article 2.2.[198] The burden to provide scientific reasoning in favour of its claims thus rested firmly with Japan. More specifically, Japan had made certain statements as to the possibilities of the completion of the pathway for fire-blight in infected apples.[199] The AB found that, since the Panel had found it to be 'reasonably assumed' that these facts would not materialise in reality, Japan carried the burden of proving them before the AB.[200] The level which would have to be met was discussed above.[201]

Japan also made a secondary motion to rely on Article 5.7 of the SPS Agreement. In relation to this, the Panel held that the burden of proving compliance with this article would rest with Japan. This would appear consistent with the general approach accounted for here above, since it was Japan that asserted compliance with Article 5.7. The issue was not raised on appeal.[202]

196 *Ibid.*, para. 104.

197 *Ibid.*, footnotes 71 and 180. That the conclusion arrived at by the AB is also the most efficient in law and economics terms is argued forcefully by Grando in M. T. Grando, 'Allocating the Burden of Proof in WTO Disputes: A Critical Analysis', *Journal of International Economic Law*, 9:3 (2006), 615, 650. The issue of what constitutes sufficient scientific evidence to maintain measures was dealt with in the section on scientific evidence and the level of proof.

198 AB Report, *Japan-Apples*, para. 160.

199 See, for example, Panel Report, *Japan Apples*, paras 4.82 and 4.83.

200 AB Report, *Japan-Apples*, para. 228.

201 See previous Section.

202 AB Report, *Japan-Apples*, para. 175, and note 316. Bernasconi-Osterwalder, on the other hand, holds the approach by the Panel to be inconsistent with the previous practice. It appears that this is based on the assumption that it is always the party asserting a breach of the agreements that has the initial burden of proof. However, as was accounted for above, it is clear from *US-Wool Shirts and Blouses* that it is the party asserting a fact, whether it be the complainant or the defendant, that has this burden. See Bernasconi-Osterwalder *et al.*, *Environment and Trade*, p. 290.

In *EC-Approval and Marketing of Biotech Products* the Panel took a different view on the burden of proof under Article 5.7 compared to the approach assumed by the Panel in *Japan-Apples*. In contrast with the Panel in *Japan-Apples*, the Panel in *EC-Approval and Marketing of Biotech Products* stated that it is for the complainant to show that an SPS measure is in breach of one of the requirements in Article 5.7.[203] If this approach is upheld in future practice, it will lead to a situation where Article 5.7 becomes an exemption from the general rules on the burden of proof under the WTO Agreements. It would possibly be more prudent to view the approach assumed by the Panel in *Japan-Apples* as the correct one. Thus Article 5.7 should be considered an exception from the general rule in Article 2.2, and the party wishing to invoke the exception would carry the burden of proving that the requirements for the exception are fulfilled. That Article 5.7 is to be construed as an exception to Article 2.2 is also supported by the wording of Article 2.2 which states that the requirements on scientific evidence need to be met 'except as provided for in paragraph 7 of Article 5'.[204] This also appears to be the opinion of the AB following the *EC-Hormones II* ruling.[205]

The allocation of the burden of proof with regard to precautionary measures thus depends on which provision is invoked by the Member issuing the measures. A Member challenging a measure has to show prima facie that the measures are introduced or maintained without basis in a risk assessment which has yielded 'sufficient scientific evidence' in support of the measures. This burden does not necessarily extend to the scientific issue at hand. Instead it is enough that the challenging party makes a prima facie case that the defending party is in violation of the WTO Agreements. The burden is then shifted onto the defendant to show that sufficient scientific evidence does exist in support of the measure. Alternatively, the defendant can invoke Article 5.7, in which case it has the burden of proving that the requirements of the article are met.

[203] Panel Report, *EC-Approval and Marketing of Biotech Products*, paras. 7.2969 and 7.2976.

[204] This issue is also discussed briefly in Prevost, 'Opening Pandora's Box', pp. 84–5, footnote 81. See also, the similar opinion in Scott, *The WTO Agreement on Sanitary and Phytosanitary Measures*, pp. 114–15. See also, a general discussion on the burden of proof under the WTO Agreements in Grando, 'Allocating the Burden of Proof in WTO Disputes', 615–58.

[205] AB Report, *EC-Hormones II*, para. 674.

3.3.2.6 The targeting of risks, proportionality and cost–benefit analysis

Articles 5.5 and 5.6 of the SPS Agreement lay down a number of qualitative requirements, which SPS measures applied in different situations must fulfil. Article 5.5 states that when Members set their chosen level of protection, they must avoid arbitrary or unjustifiable distinctions or distinctions that result in discrimination or a disguised restriction on international trade. Therefore, measures that result in the setting of a level of protection breaching the article shall be avoided. Furthermore, Article 5.6 holds that measures introduced or maintained should not be more trade-restrictive than required to achieve their intended aim, taking into account technical and economic feasibility.

The AB in *EC-Hormones I* stated that Article 5.5 must be read in the context of Article 2.3, which holds that measures issued by a Member shall not arbitrarily or unjustifiably discriminate between Members where identical or similar conditions prevail. The two provisions aim together to ensure a 'consistent' application of measures enacted to ensure the chosen level of protection.[206] On close inspection, the AB found that Article 5.5 prescribes that in order for the Article to be breached, three requirements must be fulfilled.[207] First, the Member imposing the SPS measures must have adopted different levels of protection in several comparable situations. Second, the contested measures must be arbitrary or unjustifiable in comparison with other comparable measures. Third, these arbitrary or unjustifiable differences must result in discrimination or a disguised restriction on trade. The requirements are, in principle, considered cumulative, but the AB has specifically emphasised the existence of the last two.[208] In *EC-Hormones I* the AB identified five different situations where different levels of protection against hormones in food were applied by the EC.[209] For example, it chose to ban hormones used as growth promoters, but it did not take any action at all against hormones occurring naturally in, for instance, broccoli and eggs. The Panel found that this was enough of a discrepancy to consider the arbitrary or unjustifiable criteria fulfilled.[210] The AB, however, found it impossible to equate

[206] AB Report, *EC-Hormones I*, paras. 212 and 213.

[207] For a discussion on this, see Ruessmann, 'Reflections on the WTO Doha Ministerial Conference', 933–5.

[208] AB Report, *EC-Hormones I*, paras. 214 and 215. [209] *Ibid.*, para. 218.

[210] US Panel Report. *EC-Hormones I*, paras. 8.213, 8.264 and 8.265; Canada Panel Report. *EC-Hormones I*, paras. 8.216, 8.267 and 8.268.

the regulation of naturally occurring hormones with added hormones. Thus, it could not be considered inconsistent that the EC did not regulate naturally occurring hormones in the same way as it regulated added hormones.[211]

The AB found the fact that the EC chose to regulate the use of hormones as growth promoters in beef, but allowed for the use of antibiotics as growth promoters in pork, to be more problematical. This was especially the case since the carcinogenic risk to humans appeared to be greater in the latter situation. The Panel found that this was enough to consider that the measures were in violation of Article 5.5.[212] This the AB agreed with, but it found that this did not necessarily mean that the third requirement – that the measures constituted a disguised restriction on trade – had been fulfilled. Instead, it investigated the reasons behind the EC approach and found no justification for assumptions made by the Panel that the EC had acted out of protectionist concerns. This was further highlighted by the fact that the measures applied equally to European and foreign produce. The AB concluded that it found no other reasons behind the EC ban than a genuine wish to protect its population from cancer.[213] Thus, the AB found that the measures could not be considered to 'result in a disguised restriction on international trade'.[214]

The AB furthermore stated that a measure may be allowed, even if it is arbitrary or unjustified, as long as it does not result in a disguised restriction on international trade. To investigate if a measure constitutes a disguised restriction on international trade, the AB appears to investigate not only the effects of the measures, but also the reasons behind them.[215] In *EC-Hormones I* the AB put weight on the fact that when investigating the internal legal acts constituting the measures, it was clear that the intention behind the measures was to protect citizens against cancer rather than protecting the EC market. Furthermore, the AB investigated the objective criteria of whether the disputed measures led to a more favourable treatment of domestic goods than of imports. It found this not to be the case.[216]

211 AB Report, *EC-Hormones I*, para. 221.

212 US Panel Report, *EC-Hormones I*, para. 8.182; Canada Panel Report, *EC-Hormones I* para. 8.240.

213 AB Report, *EC-Hormones I*, paras. 244 and 245.

214 *Ibid.*, para. 246.

215 *Ibid.*, paras. 244–5.

216 For a brief discussion on this, see J. Pauwelyn, 'The WTO Agreement on Sanitary and Phytosanitary (SPS) Measures As Applied in the First Three SPS Disputes', *JIEL*, 2:4 (1999), 641, 654.

Thus, the conclusion is that Members are, in principle, allowed to target risks provided that these measures do not lead to the setting of a level of protection that is unjustified or arbitrary. However, even when unjustified and arbitrary, measures can be maintained as long as they do not constitute a disguised restriction on trade. The Panel in *EC-Hormones I* had wished to go further in setting stricter efficiency requirements on Members. The AB, on the other hand, did not only stay clear of establishing efficiency or cost-effectiveness requirements for measures, but it limited itself to establishing a minimum test to ward off outright protectionism.

3.4 Conclusion

The precautionary principle has emerged rapidly in international environmental law over the last two decades. In international agreements, reference to the principle is often included in sector-specific agreements, but more rarely elaborated on in any detail. The non-binding Principle 15 of the Rio Declaration is the most cited and conclusive definition of the principle in effect at the international level. However, international tribunals have proven reluctant to base rulings on the precautionary principle as a fully-fledged part of international custom or as a general principle of international law.

In general, relatively widespread international agreement seems to exist on the importance of acting in a precautionary fashion when dealing with the environment. Thus, a broad definition of the precautionary principle, without direct legal effects, seems to enjoy wider consensus.

However, at present international consensus does not appear to exist on a narrower, prescriptive definition of the precautionary principle, including negative measures and a reversal of the burden of proof. In all cases from international tribunals discussed above, parties have agreed to the importance of precaution in the face of uncertainty, but have disagreed as to the modalities, means and goals of that precaution.

Potentially, the weak, *de minimis* version of the precautionary principle, which today appears to enjoy international acceptance, is more reminiscent of a policy goal than a legal principle. A policy, according to Dworkin, sets out a standard to be reached in some political, social or economic area.[217] It does not enjoy the same legal effects that are associated with legal principles. In sum, at present it must be concluded that the

[217] See the discussion above in Section 2.5.

precautionary principle does not constitute a generally binding norm of international law.

In WTO law it is obvious that scientific evidence is of great significance when Members issue or maintain protective measures. The precautionary principle as such is not mentioned or defined in the WTO Agreements. However, it has widely been understood, by Panels and the AB, as well as by commentators, that it finds its clearest embodiment in situations covered by Article 5.7 of the SPS Agreement. Precautionary measures under Article 5.7 are only allowed in situations where so little scientific evidence exists that a risk assessment is impossible to carry out. If the precautionary principle was deemed to be limited to this article, precautionary measures could only be taken under one type of scientific uncertainty, namely where scientific evidence is lacking. In instances where other types of uncertainty prevail, such as scientific controversy, it would not be possible to apply more precautionary measures than have been prescribed in international standards.

It is important to note that this, however, is not the only place the precautionary principle assumes in the WTO regime. The situation in Article 5.7 only aims at describing situations where a risk assessment does not need to precede the enactment of precautionary measures. However, the AB has made it clear that a risk assessment must also include an assessment of minority scientific opinion, and it is not excluded that such opinion may fulfil the requirement on sufficient scientific evidence, which may justify precautionary measures. As was discussed above, in Chapter 2, situations where minority scientific evidence exists are by definition situations surrounded by uncertainty due to conflicting scientific evidence. In allowing for either majority or minority opinion to be the basis for regulation, there is a role to be played for the precautionary principle under scientific uncertainty due to conflicting evidence.

Moreover, the scientific evidence relied upon by a Member to justify precautionary measures must establish a clear link between the activity and the risk, but the risk established must not be scientifically certain. This also leaves room for Members to act precautionary. However, so far, in all cases where the precautionary principle has been at issue, the parties wishing to maintain precautionary measures have failed to establish such a link. Furthermore, the evidence must be specific enough to the case in hand. More general evidence, for example, the risks associated with a general category of substances, will be considered but is not enough on its own. Measures may not aim to establish a zero-risk level on the market of the Member issuing the measures.

A number of issues influence the issuing of precautionary measures outside of Article 5.7. First, a risk assessment must be carried out. Second, the risk assessment must establish some sort of link between the scientific evidence and the measures that are being introduced. Third, the link established does not have to be based on the majority of scientific studies. A minority scientific opinion, provided that it is specific to the case in question and comes from a reputable source, can be relied upon. Finally, the measures need to fulfil a few qualitative requirements. For example, they must be the proportional, the least trade-restrictive, and so on. Panels and the AB will investigate measures as well as scientific studies on which they are based very closely. The review in the WTO regime is deep and wide compared to most other legal systems discussed below.

The burden of proof under uncertainty follows the general burden of proof under the WTO Agreements. The party asserting a fact has the burden of proving that fact. Since the presumption under the SPS Agreement is that SPS measures are allowed, the initial burden of proof rests with the challenging party to show that the contested measures are not in conformity with the agreement. Once a prima facie case showing this has been made, it is up to the party wishing to maintain the measures to show that they were issued and upheld in conformity with the agreement. With regard to the invoking of Article 5.7, Panels have reached different conclusions concerning the burden of proof. However, it appears reasonable to consider Article 5.7 to be an exception to the requirement of scientific evidence in Article 2.2, and thus the party invoking the exception would have the burden of proving that the requirements of the exception are fulfilled. This also appears to be the position of the AB following *EC-Hormones II*. Thus, a Member that claims that the scientific data is so scarce that a risk assessment cannot be carried out has the burden of proving that this indeed is the case.

Even if SPS measures are found to be allowed in principle, they must not lead to the imposing of a level of protection which would be in violation of Article 5.5 of the SPS Agreement, as was mentioned above. In order for Article 5.5 to be breached, comparable levels of protection must have been adopted in several situations. The chosen level of protection cannot be found to be arbitrary or unjustifiable or result in discrimination or a disguised restriction on international trade when compared to other situations. Furthermore, and most importantly, the chosen level of protection must not result in a disguised restriction on international trade. By this the AB appears to mean that the rationale behind the chosen level of protection may not be a wish to protect a Member's own market. In investigating this, the AB will go to great lengths to investigate the intent behind measures imposed by a Member. However, the AB does not make any cost–benefit or other efficiency demands on measures, and Members are, in practice, free to target risks at will.

In sum, precautionary measures under the WTO regime must be based in science and conform to the basic SPS obligations of Articles 2.2 and 5.1. The only exemption from this rule is when the scientific evidence is so scarce that it is not possible to carry out a meaningful assessment of it. However, the firm reliance on science does not exclude the use of the precautionary principle. The AB has made it clear that it does not require conclusive or even majority scientific evidence to allow measures under the SPS Agreement. This indicates that the precautionary principle can also come into effect in cases of scientific controversy. However, the party wishing to maintain its measures must show that scientific evidence providing a clear link between the action and the risk exists. It is not enough to justify precautionary measures with a wish to maintain scientifically unsubstantiated higher levels of protection or the opinion of the public. So far no Member has successfully relied on minority opinion before the WTO DSB. Furthermore, the bulk of the case law has been based on the SPS Agreement; therefore, it is unclear if what has been established by Panels and the AB is directly applicable on measures taken under GATT 1994 and the TBT Agreement. However, it would appear that this would be the case since the WTO strives to ensure consistency in the application of the WTO agreements.

4

The precautionary principle in EU law

4.1 Introduction

That the European Union is an outspoken champion of the application of a precautionary principle with clear legal implications in the international arena was shown in the previous chapter.[1] Originally developed in Germany, the principle has matured at European level only in the last ten to fifteen years. The original German *vorsorgeprinzip* has been interpreted as a highly ideological policy instrument with a wider and more complex scope than the precautionary principle generally discussed in today's literature and international trade conflicts.[2] The purpose of the original concept answered to unease in German society with regard to the risks inherent in modern society,[3] and was part of a wider ideology that was sometimes called 'ecological modernisation'.[4] Some commentators hold that Germany, fearing that its own industries would be disadvantaged by its strict environmental policies, pushed through the precautionary principle at the European level in order to create a least common denominator of environmental protection in its most important export market.[5] Others see the development as a natural response to the

[1] See, for example, the Report of the Appellate Body, US – Continued Suspension of Oblgiations in the EC-Hormones Dispute, WT/DS320/AB/R (*EC-Hormones II*); the Report of the Appellate Body, EC Measures Concerning Meat and Meat Products (*EC-Hormones I*). WT/DS26/AB/R and Panel Reports, European Communities - Measures Affecting the Approval and Marketing of Biotech Products, WT/DS291/R, WT/DS292/R WT/DS293/R (*EC-Approval and Marketing of Biotech Products*). See also, Scott, *The WTO Agreement on Sanitary and Phytosanitary Measures*, p. 128.

[2] See, for example, K. von Moltke, 'The Vorsorgeprinzip in West German Environmental Policy', in *Royal Commission On Environmental Pollution*, Twelfth Report, HMSO 1988.

[3] See Beck, *Risk Society*, and Beck, *World Risk Society*.

[4] M. Hajer, *Environmental Discourse* (Clarendon Press, 1995); A. Jordan, 'The Precautionary Principle in the European Union', in T. O'Riordan, J. Cameron and A. Jordan (eds.), *Reinterpreting the Precautionary Principle* (Cameron May, 2001); A, Weale, 'Ecological Modernisation and the Integration of European Environmental Policy', in: J. D. Lifferink, *European Integration and Environmental Policy* (Belhaven Press, 1993).

[5] Jordan, 'The Precautionary Principle in the European Union'.

growing awareness of the risks inherent in modernisation.[6] During the 1990s the appeal of the precautionary principle increased dramatically among European citizens as well as among their elected politicians and policy-makers. Especially tangible is the rise of the use of the principle in health and consumer protection. Several explanations have been proposed to explain this newfound fondness for the precautionary principle in Europe, ranging from theories of political cycles in regulation,[7] to the assertion that European legal systems are more suited to the principle as such.[8]

The most readily available explanation as to why the EU has become more 'precautionary' and risk-averse since the mid-1990s lies in the food scares experienced in Europe in the late 1990s. BSE in cattle, foot-and-mouth disease and AIDS-contaminated blood are examples of unfamiliar, unnatural and involuntary risks, which tend to be deemed to be less acceptable by the public than familiar, natural and voluntary risks.[9] These recent risks were highly publicised in the media and thus firmly planted in the public conscience. It is possible that the perception of risks in relation to, for example, BSE, might have 'spilled over' to, for example, uncertain risks in relation to genetically modified organisms (GMOs); the perceived link between the two causally unrelated areas being that both were creating involuntary risks to the public by 'unnatural' behaviour, such as the feeding of bone meal to cattle, and altering DNA codes to create new kinds of crops. This seems to have led to an increased pressure on European legislators to regulate uncertainty by erring on the side of caution. The instrument chosen for this has, in many cases, been a strict interpretation and application of the precautionary principle.

This chapter begins with an outline of the development of the precautionary principle in EU environmental law, and its subsequent emergence in the field of public health and consumer protection.

The precautionary principle in the EU is a double-edged sword.[10] First, it can be used by the European institutions as a basis for legislation in harmonised areas. This is the most widely discussed use of the precautionary principle in the EU, and the one that will receive the most attention

[6] See, for example, Beck, *Risk Society*.

[7] R. Löfstedt and D. Vogel, 'The Changing Character of Consumer and Environmental Regulation: A Comparison of Europe and the United States', *Risk Analysis*, 21:3 (2001), 399.

[8] Wiener and Rogers, 'Comparing Precaution in the US and Europe', 340–2.

[9] P. Slovic, 'Perceptions of Risk', *Science*, 236:4799 (1987), 280–5.

[10] For a similar analogy, see Scott and Vos, 'The Juridification of Uncertainty', pp. 253–86.

below. Second, the precautionary principle can be used by individual Member States as a shield against the free movement of products in non-harmonised areas under Article 36 of the Treaty on the Functioning of the European Union (TFEU). It may also be possible for Member States to make use of the principle in harmonised areas by invoking the exemptions to harmonisation measures included in Article 114 (4) and (5) TFEU.

Owing to the restraints on the judicial review permitted by the European Court of Justice (ECJ), the margin of discretion afforded to the European institutions, primarily the European Commission, in harmonised areas is considerable.[11] The same is true with regard to Member State measures in non-harmonised areas. This raises some concern in relation to the implications and actual content of the precautionary principle in the EU. Is there a workable definition of the principle, and what are the thresholds to be met in order to invoke it in the EU? What are the implications of the limits of the Court's judicial review with regard to the precautionary principle? If the Court does not evaluate the scientific evidence put before it, how can a claimant successfully argue that the European institutions have acted erroneously? This raises questions as to how the European institutions and the Court deal with issues of scientific uncertainty and scientific evidence. Furthermore, what are the implications for the completion of the internal market if Member States are allowed to employ their own precautionary reasoning in the face of harmonising measures? Finally, and perhaps most importantly, does the application of the precautionary principle in the EU today lead to a situation in which the risks are targeted in an efficient and comprehensive way? This chapter aims to address these questions by analysing the constituent parts of the precautionary principle. In this regard, it follows the outline of the previous chapter on the precautionary principle in WTO law.

4.2 Treaties, legislation and soft law

4.2.1 Early developments

The wish of responsible governments to protect public health and sometimes err on the side of caution is not something that was born with the concept of the precautionary principle. Instead, it is a fundamental part

[11] See, for example, the rulings of the ECJ in Case T-13/99 *Pfizer Animal Health SA* v. *Council of the European Union* [2002] ECR II-03305 and Case C-3/00 *Denmark* v. *Commission* [2003] ECR I-02643.

of most legal systems.[12] However, what is investigated here is the origin of the principle of precaution as applied in the EU today.

It is widely assumed that the precautionary principle stems from environmental law as it developed from the 1970s onwards. Some authors claim that this development did, in fact, borrow precautionary thinking from the field of health protection, which, in many legal systems, already employed pre-market authorisation requirements for certain products, and so on.[13] In the EU context, the first time that the embryos of the precautionary principle were considered in policy documents and legislation was in the environmental context. In the First Environmental Action Programme (1973–1976) (EAP I),[14] precaution was discussed as something to be considered for emergency measures, as it was pointed out that available knowledge might, in some cases, be inadequate. Consequently, technological knowledge in the European Community was to be improved and research was to be encouraged:

> [This] ought not, however, to have any delaying effect on the application of emergency measures in cases where there is a real or potential danger to man or his environment, since these measures can be reviewed and modified subsequently in the light of the results of research carried out.[15]

The reference here to 'potential danger' suggests that measures taken without full scientific knowledge would also be covered. It is also interesting to note that no distinction was made between health and environment, and the wording seems to apply to both. However, more emphasis was placed on potentially precautionary measures later on in the same document:

> [In certain cases] urgent reasons based on the protection of human health or protection of the environment may require that provisional common standards be laid down before criteria and quality objectives have already been established at Community level, where the relevant criteria

[12] D. P. Fidler, *International Law and Public Health* (Transnational Publishers Inc., 2000); T. Christoforou, 'Science, Law and Precaution in Dispute Resolutions on Health and Environmental Protection: What Role for Scientific Experts?', in J. Bourrinet and S. Maljean-Dubois, *Le commerce international des organismes génétiquement modifies* (Aix-Marseilles/Paris, 2002).

[13] Christoforou, 'Science, Law and Precaution in Dispute Resolutions on Health and Environmental Protection', pp. 217–18.

[14] Declaration of the Council of the European Communities and of the Representatives of the Governments of the Member States, meeting in the Council on the Programme of action of the European Communities on the Environment, OJ 1973, C112/1.

[15] *Ibid.*, p. 33.

> already existing on an international or national level are considered to be adequate.[16]

The measures referred to are to be of provisional character and are to be employed pending further research. They are to be taken up on the basis of urgent reasons in response to a threat to health or the environment. It is not spelled out that the measures are of a precautionary nature, but, in holding that they should be issued before 'criteria and quality objectives' have been established, it would seem that emerging uncertain risks might be referred to. In this regard, it should be pointed out that the provision bears resemblance to the wording of the SPS Agreement.[17]

It is curious to notice that the EAPs that followed during the 1970s and the first half of the 1980s no longer made explicit references to precaution. EAP IV[18] was issued in the same year as the Brundtland Report, but did not mention sustainability or precaution by name. However, it prescribed precautionary approaches in a number of identified sectors, such as gene technology and nuclear power.[19] Other policy documents elaborated on different environmental principles such as the polluter pays and prevention, but they all stayed clear of precaution.[20] However, in the Dublin Declaration on the Environmental Imperative of 1990, the EU Heads of State for the first time expressed their intention that action by the European Community as well as by the Member States 'will be developed on a co-ordinated basis and on the principles of sustainable development and preventive and precautionary action'.[21] Although this concept was not developed further, it shows the willingness of the EU to embrace the precautionary principle on a horizontal level to function as an underlying principle for all action in the EU. Two years later the principle was to find its way into the EC Treaty.[22]

Precautionary legislation predating the Maastricht Treaty tends to be sectoral, applying a precautionary rationale vertically to a specific area.

16 *Ibid.*, p. 49.

17 For a discussion on the wording of the SPS Agreement see, above, Section 3.3.1.

18 Resolution of the Council and Representatives of the Governments of the Member States, meeting within the Council on the continuation and implementation of an EC policy and action programme on the environment (1987–1992), OJ 1987, C328/1.

19 For example, it is established that the ALARA (as low as reasonably possible) principle should be guiding nuclear safety (Point 4.6.3). In relation to biotechnology it was held that 'special precautions may be necessary' (Point 4.4.8).

20 Douma, 'The Precautionary Principle', pp. 202–16.

21 Bulletin EC 6-1990, Conclusions of the Presidency, Point 1.14 and Annex II.

22 For a more in-depth account of early soft law dealing with sustainable development and precaution in the EC, see Douma, 'The Precautionary Principle', pp. 191–254.

The actual word precaution is not mentioned until the Dublin Declaration of 1990 as discussed above. However, even though the word 'precaution' was not mentioned or formulated as an overarching, horizontal policy or principle, a general wish to be precautionary heavily influenced certain areas.

In 1965 Directive 65/65 established a prior approval mechanism for the placing of medicinal products on the market.[23] The same is true with regard to the Directive on Dangerous Substances, which also established a prior approval system (positive or reverse listing) for substances such as pesticides. This meant that substances were not allowed onto the markets unless proven safe.[24] Establishing a system for the testing of products before they are allowed onto the market is essentially precautionary in that it foresees certain possible risks and aims to analyse these risks before the products in question are allowed. Thus, these proved to be early examples of a precautionary approach common to many legal systems. The same is true, for example, for Directive 70/534 concerning Additives in Feeding Stuff,[25] Directive 75/319 on Medicinal Products,[26] Directive 75/442 on the Reduction or Avoidance of Risks relating to Waste[27] and Directive 79/831 relating to the Testing of Chemicals.[28]

Other examples of early precautionary thinking include the Birds Directive of 1979,[29] especially since this directive includes the protection of birds in no imminent danger of extinction. Council Decision 80/732 Concerning CFCs in the Environment even mentions precaution

[23] Directive 65/65/EEC of 26 January 1965 on the approximation of provisions laid down by Law, Regulation or Administrative Action relating to proprietary medicinal products, OJ 1965 L 22 p. 369. See Christoforou, 'Science, Law and Precaution in Dispute Resolutions on Health and Environmental Protection', p. 232.

[24] Directive 67/548/EEC of 27 June 1967 on the Approximation of the Laws, Regulations, and Administrative Provisions of the Member States Relating to the Classification, Packaging, and Labelling of Dangerous Substances, OJ 1967 L 196 p. 1.

[25] Directive 70/534/EEC of 23 November 1970 concerning additives in feeding-stuffs, OJ 1970 L 270, p. 1.

[26] Directive 75/319/EEC of 20 May 1975 on the approximation of provisions laid down by Law, Regulation or Administrative Action relating to proprietary medicinal products, OJ 1975 L 147, p. 13.

[27] Directive 75/442/EEC of 15 July 1975 on waste, OJ 1975 L 194, p. 39.

[28] Directive 79/831/EEC of 18 September 1979, amending for the sixth time Directive 67/548/EEC on the Approximation of the Laws, Regulations and Administrative Provisions of the Member States Relating to the Classification, Packaging and Labelling of Dangerous Substances, OJ 1979 L 259, p. 10.

[29] Directive 79/409/EEC of 2 April 1979 on the conservation of wild birds, OJ 1979 L 103, p. 1.

explicitly, and is thus 'the earliest example of an explicit reference in EC legislation to a precautionary measure'.[30]

The ECJ contributed greatly to affording weight and justification to precautionary measures during the 1980s. In the landmark *Cassis de Dijon* case, the ECJ had already established that it would be possible for Member States to restrict the free circulation of goods in the absence of harmonisation measures. These restrictions, which would have to be based upon Article 3 TFEU (formerly Article 30 EC)., had to be necessary to satisfy mandatory requirements such as fiscal controls, fair trading practices and consumer protection. As the Court held:

> Obstacles to movement within the Community resulting from disparities between the national law relating to the marketing of the products in question must be accepted in so far as those provisions may be recognised as being necessary in order to satisfy mandatory requirements relating in particular to the effectiveness of fiscal supervision, the protection of public health, the fairness of commercial transactions and the defence of the consumer.[31]

The list provided in the case was not intended to be exhaustive.[32] In the early 1980s the Court decided to give the Member States the freedom to set the level of protection also in complex cases surrounded by scientific uncertainty as to the effects and causality of products. In the *Melkunie* case a Dutch ban on certain milk products was upheld by the Court. The ban was motivated by a wish by the Dutch authorities to maintain a high level of protection for consumers who might be particularly sensitive towards non-pathogenic micro-organisms in pasteurised milk. That the exact effects of the consumption of the micro-organisms were uncertain was not contested by the parties. The Court stated that the Dutch prohibition was a measure having equivalent effect to a quantitative restriction under Article 34 TFEU (formerly Article 30 EEC). However, this was justified under the exemptions stated in Article 36 TFEU (formerly Article 36 EEC). In the words of the Court:

> the data available at the present stage of scientific research do not make it possible to determine with certainty the precise number of non-pathogenic micro-organisms above which a pasteurised milk product

[30] Nigel Haigh as quoted in Trouwborst, *Evolution and Status of the Precautionary Principle in International Law*, p. 143.

[31] Case 120/78 *Rewe* v. *Bundesmonopolverwaltung* ('Cassis de Dijon') [1979] ECR 649, para. 8.

[32] L. Krämer, *EC Environmental Law* (Sweet and Maxwell, 2000), p. 75.

> becomes a source of danger to human health. In the absence of harmonization in this field, it is for the Member States to determine, with due regard to the requirements of the free movement of goods, the level at which they wish to ensure that human life and health are protected.[33]

Thus, the conclusion of the Court was that, in non-harmonised areas, when the present state of scientific research does not allow for a certain estimation, it is up to the individual Member State to set the level at which it considers life and health to be protected. However, the Member State is still bound to respect the principle of proportionality in enacting measures to reach these levels.[34] This has been called by some authors the 'reversed "Cassis de Dijon" principle', and signifies the limits put upon the principle of mutual recognition.[35] The *Melkunie* judgment significantly demonstrates how the ECJ had come to regard both risk regulation and its own abilities to review measures taken under uncertainty. Here, trade restrictions based upon uncertain risks were allowed to override the 'Cassis de Dijon' principle of mutual recognition. In allowing for this, the Court recognised how sensitive the issue of risk regulation is for the Member States in Europe. By approving restrictive measures based upon scientific uncertainty, the Court essentially allowed the Member States to enact precautionary measures in non-harmonised areas.

Furthermore, the Court effectively relieved itself of the responsibility to base its judgment on the merits of the scientific evidence put before it in such cases. The level of proof needed to be met by the derogating party is low. It would seem to be clear that the latter does not need to show that the risk, in this case the presence of non-pathogenic micro-organisms in milk, is likely. Instead, the Court seemed satisfied that it was unclear at what moment the risk might appear.[36] In the light of this approach by the Court, it is hardly surprising that it would impose limits on its judicial review in order to avoid the sensitive issue of scientific evidence in harmonised areas as well. This is dealt with in more detail below.

[33] Case 97/83 *Melkunie* [1984] ECR 2367, para. 18.

[34] Jans, 'Proportionality Revisited', 252.

[35] Krämer, *EC Environmental Law*, pp. 75 *et seq.*

[36] Such a level would be found somewhere between 0 and presumable/acceptable in the chart on burden of proof introduced in the previous chapter. It appears that a challenging party would have to reach at least proven/verified in the same chart to be successful. However, the Court does not touch on questions of uncertainty or burden of proof in detail.

In the *Heijn*[37] and *Mirepoix*[38] cases the ECJ considered pesticides to constitute a major risk to human and animal health since 'the quantities absorbed by the consumer can neither be predicted nor controlled'. It is interesting to note that the Court took the notion of uncertainty into account not only with regard to the product, but also with regard to the (involuntary) consumption of the substance, as a justification of precautionary measures. Even more notable is the fact that the Court clearly stated that, where pesticides are concerned, a risk does not need to be established for the individual product since it is not disputed that pesticides per se constitute a major risk to human and animal health.[39]

4.2.2 The Single European Act

As the Single European Act (SEA) came into force 1 June 1987 the term 'the internal market' was introduced into the Treaties of the (then) European Economic Community (EEC). This came to be the central notion around which the European Communities (and later the EU) would gravitate. The fundamental idea of a mutually beneficial common market for the use of the Member States has proven to be the incentive needed for increased integration. It is, therefore, understandable that derogations from what has been commonly decided in relation to the internal market, and regarding the free movement of goods in particular, has been viewed by the European Commission as potentially threatening to the completion of the internal market.

Not only did the SEA introduce the concept of the internal market, it also abolished the unanimity requirement for decisions relating to its completion. Instead, a system of qualified-majority voting was established in Article 100a of the SEA. However, some Member States feared that they would be overruled in the Council under the new provisions and were reluctant to agree on a qualified-majority regime. In order to meet these fears, Article 100a(4) was included in the SEA.[40] This article provides that:

> If after the adoption of a harmonization measure by the Council acting by a qualified majority, a Member State deems it necessary to apply national

[37] Case 94/83 *Heijn* [1984] ECR 3263.

[38] Case 54/85 *Mirepoix* [1986] ECR 1067.

[39] Case 94/83 *Heijn*, para. 13 and Case 54/85 *Mirepoix*, para. 13.

[40] C.-D. Ehlerman, 'The Internal Market Following the Single European Act', *CMLRev.*, 24 (1987), 9.

> provisions on grounds of major need referred to in Article 36, or relating to the protection of the environment or the working environment, it shall notify the Commission of these provisions.
> The Commission shall confirm the provisions involved after having verified that they are not a means of arbitrary discrimination or a disguised restriction on trade between Member States.

The article proved controversial and received a fair share of criticism from commentators who feared that it would erode the application of internal market legislation.[41] These fears have, at least until recently, appeared to be relatively unfounded, since only few derogations have been notified and allowed under the provision.[42]

The second major development in the SEA was the inclusion of environmental protection as a major aim of European co-operation.[43] It was not included as one of the tasks of the EEC as enumerated in Article 2, but Article 130R, S and T established that the objectives in relation to the environment should be:

(i) to preserve, protect and improve the quality of the environment;
(ii) to contribute towards protecting human health;
(iii) to ensure a prudent and rational utilisation of natural resources[44]

The SEA continued to establish that three environmental principles should guide environmental protection in the EEC: the principle of preventative action, the rectification-at-source principle and the polluter-pays principle.[45] The precautionary principle was not yet mentioned, due to the fact that it had not yet received sufficient support among the Member States.[46] Five years later, at the time of the signing of the Treaty of Maastricht, this had changed.

The inclusion of environment and health protection as important objectives in the SEA signified competition for the traditional free movement objectives. A balance had to be struck between the protection of

[41] See, for example, P. Pescatore, 'Some critical remarks on the "'Single European Act"' *CMLRev.*, 24 (1987), 9.

[42] However, as of late, this may be changing. See, for example, Scott and Vos, 'The juridification of Uncertainty', and J. Zander, 'The Green Guarantee in Article 95 of the EC Treaty: Two Recent Cases', *Journal of Environmental Law*, 16:1 (2004).

[43] In Case 240–83 *Waste Oils* [1985] ECR 531 the ECJ had already made it clear that it considered environmental protection to be an essential objective of the Community and that this may justify certain limitations as to the free movement provisions.

[44] Article 130R(1) EEC.

[45] *Ibid.*, Article 130R(2) EEC.

[46] N. Haigh, 'The Introduction of the Precautionary Principle into the UK', in T. O'Riordan and J. Cameron (eds.), *Interpreting the Precautionary Principle* (Earthscan, 1994).

health and the environment on the one hand, and the completion of the internal market on the other. It fell to the Court to find this equilibrium. In this, the Court was helped by Article 130R(3), in which it was established that scientific and technical data was to be the basis for legislation in this area, and that all measures should be subjected to a cost–benefit analysis. This made the environment a unique policy area since no other field was subjected to a cost–benefit analysis as a formal requirement.[47] However, it could be argued that cost–benefit analysis was nothing new to the EEC legal order, or specific to environmental and health protection for that matter. The exemptions from the free movement provisions included in Article 36 EEC (now 36 TFEU), with its emphasis on necessity, proportionality and least trade-restrictiveness, entail elements of cost–benefit reasoning. This would also be supported by the reasoning in the 'Cassis de Dijon' case.[48] Following this assumption, not only is environment and health protection subject to some cost–benefit balancing, but so are all derogations from basic free movement provisions under Article 36 TFEU. However, it should be pointed out that a cost–benefit analysis within the framework of the proportionality principle in Article 36 TFEU refers primarily to the weighing of the proportionality of the measures in relation to their possible hindrance of intra-EU trade. Cost–benefit analysis *strictu sensu*, as it is generally understood, has a broader scope, which aims to compare the general cost-effectiveness of a measure with other policy choices.[49]

The best examples of precautionary legislation enacted by the EC legislature in the time between the Treaty changes in 1987 and 1992 are the directives on genetically modified organisms of 1990.[50] The precautionary principle is not mentioned explicitly, but the directives, nevertheless, establish a highly precautionary approach to GMOs. Directive 90/220/EC provides for a comprehensive system for the authorisation of GMOs to be released into the environment in the EU. Anyone wishing to release GMOs into the environment needed to produce an assessment of the risks associated with the organism in question to the competent authority.[51]

[47] Douma, 'The Precautionary Principle', pp. 258–9.

[48] See, for example, the reasoning of Jans in 'Proportionality Revisited'.

[49] For a detailed discussion of the scope and purpose of cost-benefit analysis see, above, Section 2.3.2.

[50] Directive 90/219/EEC of 23 April 1990 on the contained use of genetically modified micro-organisms, OJ 1990, L 117, p. 1 and Directive 90/220 of 23 April 1990 on the deliberate release into the environment of genetically modified organisms, OJ 1990, L 117, p. 15.

[51] Directive 90/220/EEC, Article 11(1).

Before the product could be placed on the market, all other Member States were to be consulted.[52] Such a system, including risk assessment requirements and wide consultation among Member States in the risk management phase, has become the model of European precautionary regulation in relation to GMOs.

4.2.3 *The Treaty of Maastricht*

As the Treaty of the European Community came into force on 1 November 1993, the precautionary principle assumed its name and official place in the European legal order. Included in the modified Article 130R, it was clear that it was now an environmental principle considered to be on an equal footing with the principles of prevention, the rectification-at-source and polluter-pays.[53]

It is interesting to observe that, in the first paragraph of the article, the wording on human health has been strengthened. Where the article inserted into the EEC Treaty by the SEA stated that action by the European Community should aim to 'contribute towards protecting human health', the new version states that it should aim at 'protecting human health'. The use of the more direct language introduced by the SEA signifies an increased emphasis on health protection in the EU.

At the time of its inclusion in the Treaty, the precautionary principle drew limited attention from authors active in the field of European environmental law. Kiss and Shelton elaborated somewhat on the concept, and argued that a distinction should be made between a strong version of the principle and a weaker version. The former would have the capacity to be used as a legal justification for regulating uncertain risks and reversing the burden of proof.[54] However, the precautionary principle included in

[52] *Ibid.*, Article 13.

[53] The relevant parts of the Article reads:'Community policy on the environment shall contribute to pursuit of the following objectives:– preserving, protecting and improving the quality of the environment;– protecting human health– prudent rational utilization of natural resources;– promoting measures at international level to deal with regional or worldwide environmental problems.Community policy on the environment shall aim at a high level of protection taking into account the diversity of situations on the various regions of the Community. It shall be based on the precautionary principle and on the principles that preventive action should be taken, that environmental damage should as a priority be rectified at source and that the polluter should pay. Environmental protection requirements must be integrated into the definition and implementation of other Community policies.'

[54] A. C. Kiss and D. Shelton, *International Environmental Law* (Transnational Publishers, 1994).

Article 130R lacked any concrete definition, and case law explicitly referring to the principle was not yet present, so the reluctance of commentators to speculate is understandable.

It took until the late 1990s for the discussion to take off in the wake of the BSE crisis, the Treaty of Amsterdam and the subsequent publication of the Commission Communication on the Precautionary Principle.[55] In the meantime, the European institutions, primarily the ECJ and the European Commission, tentatively started to make use of the principle. In addition, in European legislation, the precautionary principle began to appear in the time between the Treaties of Maastricht and Amsterdam, even though it was still not made explicit in the legislative acts.[56]

The real turning point, which propelled the precautionary principle into a force to be reckoned with in Europe, came with the BSE crisis in 1996. Until the crisis was a reality, European Community regulation on food safety was established ad hoc and mainly guided by pragmatic considerations.[57] The EU system of independent scientific committees was widely regarded as a functioning solution in a community which was not primarily concerned with risk regulation. This image was severely damaged by the failure to act adequately at an early stage against the threat posed by BSE.[58]

Bovine spongiform encephalopathy was discovered as early as 1985, but it was not until March 1996 that the UK authorities went public with their finding that BSE could (possibly) be transferred to humans and appear as Creutzfeldt-Jakob disease. Following the UK report, the Commission acted immediately and issued a decision banning the importation of British beef to the rest of the EU.[59] The decision bore all the hallmarks of a precautionary measure, even if the word 'precaution' was not employed as such. However, in the preamble the Commission made it clear that a

[55] Communication from the Commission on the precautionary principle, COM 2000/1/ Final.

[56] See, for example, Directive 94/67 of 16 December 1994 on the incineration of hazardous waste, OJ 1994 L 365/34, which aimed, among other things, to regulate dioxin emissions, even though scientific evidence regarding the effects of dioxin was not yet conclusive at the time of adoption, and Directive 96/61 of 24 September 1996 on integrated pollution prevention and control, OJ 1996 C 137/20, in which a loose reference to the precautionary principle is made in Annex IV.

[57] Vos, 'EU Food Safety Regulation in the Aftermath of the BSE Crisis', *Journal of Consumer Policy*, 23 (2000), 231.

[58] *Ibid.*

[59] Council Decision 96/239 of 27 March 1996 on emergency measures to protect against bovine spongiform encephalopathy, OJ L 078.

definitive stance on the possibility of the transmission of BSE was not possible at that that time, but that it could not be excluded that BSE could transfer to humans. Furthermore, the preamble refers to the serious concern that this uncertainty had caused among European consumers. For these reasons, a temporary ban on the importation of British beef was deemed warranted.[60] But when the UK report was published and the ban had been issued by the Commission, it was already too late: the confidence crisis in not just British, but also European, food safety regulation was already a fact.[61] Inquiries set up to investigate how the BSE crisis had been allowed to go as far as it had revealed that the European scientific committees were very strongly influenced by 'British thinking', and that the UK policy-makers had been instructed to defend the economic interests of farmers at the same time as they were to take health measures restricting the trade of the same farmers.[62] This inevitably confused the roles and delayed any decisive form of action from the British, as well as the European authorities. A lack of transparency in the decision-making process delayed the reaction of the public.

As the ban on British beef was challenged before the ECJ, it became clear that the Court would allow outright precautionary measures by the European institutions taken in the face of uncertainty.[63] The case was to become the foundation of a new, outspoken precautionary principle in the EU. It also became clear that the Court did not see the principle to be confined to environmental protection alone but also to the protection of human health.

Faced with the factual findings of the Inquiry Committee on BSE, which had been set up by the European Parliament to investigate the scandal, as well as severe criticism from the media of how it handled the crisis, the Commission quickly developed a new approach to food safety. One part of the approach consisted of systematising the scientific committees and placing them under the joint heading of DG XXIV (now DG SANCO) in order to be able to control them and protect them from 'external' and 'non-scientific' influence.[64] A second part of the Commission's

60 *Ibid.*, recital 9.

61 Vos, 'EU Food Safety Regulation in the Aftermath of the BSE Crisis'.

62 *Ibid.*, and P. Zwanenberg and E. Millstone, 'Mad Cow Disease 1980s–2000: How Reassurances Undermined Precaution', in *Late Lessons From Early Warnings: The Precautionary Principle 1896–2000*, Environmental Issue Report No. 22, 2002. Available at: http://reports.eea.europa.eu/environmental_issue_report_2001_22/en/ (accessed 1 March 2010).

63 The *BSE* case before the European Court of Justice is dealt with in detail below.

64 Vos, 'EU Food Safety Regulation in the Aftermath of the BSE Crisis', 233–4.

response was the issuing of a communication outlining a new approach to food safety in the EU.[65] In this new approach the Commission moved away from an approach centred on food security as previously considered in relation to agricultural policy, and moved towards the protection of the consumer. According to the Commission, there were two distinct aspects of the precautionary principle: the political decision to act or not to act, and the decision on which measures to use when action has been decided upon. It stressed that three instruments were essential for an effective consumer health policy: namely, scientific advice, risk analysis and control. The Commission also defined the three stages of dealing with risks: risk assessment, risk management and risk communication.[66] Furthermore, it made it clear that scientific advice is the basis of all regulatory initiatives in the area of food safety and consumer protection, and, to ensure this, it emphasised the importance of the scientific committees.

4.2.4 *The Treaty of Amsterdam and the Lisbon Treaty*

The focus on science-based decision-making in environmental and health matters was also highlighted in the Treaty of Amsterdam:

> The commission, in its proposals envisaged in paragraph 1 concerning health, safety, environmental protection and consumer protection, will take as a base a high level of protection, taking account in particular of any new development based on scientific facts. Within their respective powers, the European Parliament and the Council will also seek to achieve this objective.[67]

This was not the only change to the former Article 100(a) that was to be undertaken when the article was transformed into Article 95 by the Treaty of Amsterdam. The treaty also clarified the legal basis and requirements for Member States to derogate from harmonisation measures taken under Article 95. As mentioned briefly above, the derogation possibilities in Article 100(a)(4) were perceived (according to its wording) as being limited to the maintenance of stricter national legislation, and did not appear to stretch as far as authorising Member States to introduce new, stricter legislation in the face of harmonisation. However, the exact state of the law was unclear at this point. When Article 95 clearly separated the maintenance of the existing measures from the introduction of new ones,

[65] Communication of the Commission on Consumer Health and Food Safety, COM/1997/83/Final.

[66] COM/2000/1, pp. 12 *et seq.* [67] EC Treaty (after Amsterdam) Article 95(3).

it became clear that both situations were allowed provided that somewhat different requirements were fulfilled.[68]

One of the main aims of the new provisions was to establish in which situations derogation would be allowed.[69] That the possibilities for Member States to derogate under the new provisions seem greater than under Article 100(a)(4) has been motivated by the fact that the introduction of the protection of health and environment as fundamental objectives of the EU shifted the focus of the EU away from strictly economic considerations. Allowing Member States to maintain or introduce stricter regulation than the harmonisation measure gave them the opportunity to be more protective than the rest of the EU.[70] However, an excessively lenient view of such derogation might also lead to greater fragmentation of the internal market, since the harmonisation rules might be set aside if the relevant requirements were fulfilled.[71]

Once again, it was up to the Court and the Commission to decide on how to distinguish between the two paragraphs, how to define key concepts such as 'new scientific evidence' and where the outer boundaries for the application of what has been called the 'green guarantee' should be drawn. The issue of the application of the Article 95 exemptions will be returned to shortly, in the section on case law.[72]

[68] The Treaty of Amsterdam thereby ended the primarily academic discussion during the 1990s on whether Article 100(a) also could make room for the introduction of new measures, that is a national measure introduced after a harmonisation measure had been enacted. The issue remained open until the Treaty of Amsterdam came into force, since the Commission had not had the opportunity to issue any decisions in reply to such derogation requests (Krämer, *EC Environmental Law*, pp. 98 and 103).

[69] G. van Calster, *International and EU Trade Law: The Environmental Challenge* (Cameron May, 2000), p. 398.

[70] N. de Sadeleer, 'Procedures for Derogations from the Principle of Approximation of Laws under Article 95 EC', *CMLRev.*, 40 (2003), 889.

[71] The relevant parts of Article 95 EC read:'4. If, after the adoption by the Council or by the Commission of a harmonisation measure, a Member State deems it necessary to maintain national provisions on grounds of major needs referred to in Article 30, or relating to the protection of the environment or the working environment, it shall notify the Commission of these provisions as well as the grounds for maintaining them. 5. Moreover, without prejudice to paragraph 4, if, after the adoption by the Council or by the Commission of a harmonisation measure, a Member State deems it necessary to introduce national provisions based on new scientific evidence relating to the protection of the environment or the working environment on grounds of a problem specific to that Member State arising after the adoption of the harmonisation measure, it shall notify the Commission of the envisaged provisions as well as the grounds for introducing them.'

[72] The issue of the application of the Article 95 EC exemptions is returned to below.

The entry into force of the TFEU on 1 December 2009 did not substantively change the situation with regard to the precautionary principle. The relevant provisions were taken over verbatim from the EC Treaty and only the numbering of the articles was modified.

4.2.5 *The Commission Communication on the Precautionary Principle*

4.2.5.1 The purpose of the Communication

In answer to an appeal by the Council, the Commission issued a Communication in 2000 which aimed to establish guidelines for the application of the precautionary principle in the EU.[73] The addressees of the Communication are clearly the EU institutions and the Member States, but are also the trading partners of the EC, who had started to fear that the precautionary principle might be used to protect the European market against foreign competition.[74] The Communication aimed to outline a systematic, science-based approach to precaution, and thus ease some of the concerns of the critics of the precautionary principle. It also attempted to link the European approach to Article 5.7 of the SPS Agreement.[75] This indicates that the EU sees a broader scope for precautionary action under this article than had been considered by WTO Panels and the Appellate Body. As discussed in the previous chapter, Article 5.7 SPS has consistently been considered by Panels and the Appellate Body to apply in situations of insufficient scientific data.[76] In situations in which the data available has been sufficient to allow for differing scientific conclusions, the article has not been considered to apply. This indicates, as was concluded in the previous chapter, that Article 5.7 SPS only applies in situations in which the available scientific situations are not sufficient to allow for a risk assessment to be concluded. If a risk assessment can be performed, the other articles of the SPS agreement apply, notably Articles 2 and 5.1. Thus, it appears that precautionary action under Article 5.7 SPS is limited to allowing for measures to be enacted without recourse to a risk assessment in certain well-circumscribed cases. However, this does not limit the role that

[73] COM/2000/1.

[74] This is highlighted by the fact that the Communication was submitted to the WTO Trade and Environment and Technical Trade Barriers Committees. See WTO documents WT/CTE/W147 and G/TBT/W137 respectively.

[75] COM/2000/1, pp. 11–12.

[76] Compare with the discussion outlined above in Section 3.3.

the risk assessment plays under the WTO Agreements.[77] Precautionary action can be taken if a scientific link can be shown to exist between the risk at hand and the measures enacted. This risk does not have to be conclusively shown, and it can be based upon minority opinions. It would have been wise for the European Commission also to address this issue, and to stress in the Communication that precautionary action under the WTO regime is not confined to Article 5.7 of the SPS Agreement.

4.2.5.2 The definition of the precautionary principle

The Communication, like the TFEU, does not provide a definition of the precautionary principle. It merely indicates the circumstances in which it should be employed by stating that it:

> covers those specific circumstances where scientific evidence is insufficient, inconclusive or uncertain and there are indications through preliminary objective scientific evaluation that there are reasonable grounds for concern that the potentially dangerous effects on the environment, human, animal or plant health may be inconsistent with the chosen level of protection.[78]

Furthermore, the application of precautionary measures may not be of an arbitrary nature.[79]

It has been held by proponents of a strict application of the precautionary principle that a universally acceptable definition of the principle is utopian and unnecessary due to the case-by-case based approach that a state needs to employ in order to achieve its desired level of protection. Moreover, the level of acceptable risk may vary over time due to public acceptance and scientific advances; therefore, the exact definition of the instrument used to achieve the level may also vary.[80] Critics of the approach claim that the lack of a definition runs the risk of turning the principle into a self-serving instrument which can be used to achieve any desired result.[81]

4.2.5.3 Judicial review

The Commission emphasises that it is ultimately the responsibility of the Court to decide upon the legality of EU measures based upon the

[77] *Ibid.* [78] COM/2000/1, p. 10.

[79] *Ibid.*, pp. 13, 16, 19 and 22.

[80] Christoforou, 'Science, Law and Precaution in Dispute Resolutions on Health and Environmental Protection', p. 218.

[81] Bergkamp, 'Understanding the Precautionary Principle (Part I)', 21.

precautionary principle. However, the Commission points out that, in so doing, the ECJ is restricted with regard to the nature and scope of its review. Consistent case law holds that, in cases where the EU institutions enjoy broad discretionary powers, the review of the Court is limited to examining whether manifest errors have been committed, or whether the relevant institution has misused its powers or otherwise acted ultra vires.[82] This has led to a situation in which the judicial review of the Court is restricted and limited in nature.

4.2.5.4 Risk assessment and risk management

The Communication establishes that the approach to risk analyses should be based upon a structured approach which separates risk assessment, risk management and risk communication.[83] The content of the risk assessment is further elaborated on in Annex III, where its four components are listed: hazard identification, hazard characterisation, appraisal of exposure and risk characterisation.

Three factors need to be present initially in order to trigger the use of the precautionary principle: (1) Potentially negative effects; (2) A scientific evaluation; (3) Scientific uncertainty.[84]

After potentially negative effects of a certain activity have been identified, these have to be evaluated scientifically in order to ascertain (or at least estimate) their gravity. This is done by conducting a scientific evaluation (or 'risk assessment', as it may be more commonly known). Interestingly enough, the Commission does not consider this obligation to be absolutely binding. It states that a risk assessment 'should be considered where feasible', which would seem to allow for a risk assessment not to be carried out where it is not 'feasible'.[85] How, and by whom, this is to be determined is not clear. It would appear that, even in cases of deep uncertainty, a risk assessment could be carried out, even if the result would turn out to be inconclusive. As discussed in Chapter 2, a risk assessment cannot always be expected to put an end to the scientific uncertainty surrounding a risk.[86] However, a risk assessment can identify the nature, scope and degree of the scientific uncertainty in question.[87] This will assist the risk manager when deciding on the appropriate measures to take, and the assignment of the burden of proof. It must thus be

[82] See the discussion on judicial review below in Section 4.3.3.

[83] COM/2000/1, p. 8. [84] *Ibid.*, p. 13.

[85] *Ibid.* [86] See, above, Sections 2.3.1 and 2.3.2.

[87] See, for example, van Asselt, *Perspectives on Uncertainty and Risk*, pp. 82–92.

considered that, also under uncertainty, a risk assessment provides essential information. This is also indicated in the Communication. However, the wording of the Communication seems to imply that the 'possibility' of conducting a risk assessment could be a policy decision, thus leaving it up to the policy-maker to decide whether it is feasible to conduct a risk assessment or not.[88] This could be interpreted as undermining the strictly science-based approach otherwise advocated in the Communication and stressed in Article 114 (3) TFEU. A strict requirement to conduct a risk assessment, which makes allowance for the possibility that it may not provide a final scientific answer, would have been preferable in order to avoid any confusion. This becomes particularly significant in the WTO context, where, as has been shown previously, the requirement to perform a risk assessment before taking precautionary action is strict.[89] The only exemption under the SPS Agreement is for a Member to seek recourse in Article 5.7, but this can only be done in the clearly defined situation in which the scientific data upon which to base a risk assessment is insufficient.[90] Panels will thoroughly investigate whether this is the case. In comparison to the science-oriented approach in the WTO regime, the Commission's approach in the Communication appears to be significantly weaker.

Finally, the activity to be regulated needs to be scientifically uncertain. The Commission affords a broad interpretation to the term 'scientific uncertainty', which generally corresponds to the interpretation referred to in Chapter 2. In addition, the risk assessment should show that the desired level of protection for the environment or a population group could be jeopardised. Furthermore, the conclusions should include an assessment of the scientific uncertainty, and an assessment of the consequences of inaction may be used by the decision-maker to trigger action.[91]

4.2.5.5 Scientific evidence and the introduction of precautionary measures

That the Commission wants to emphasise its view that the application of the precautionary principle is fundamentally science-based is clear. The term 'scientific evidence' is mentioned twelve times in the twenty-two-

[88] COM/2000/1, p. 13.

[89] At least under the SPS Agreement, see, above, Section 3.3.2.3.

[90] *Ibid.*

[91] COM/2000/1, p. 17. In particular, the inclusion of a requirement to assess the existing scientific uncertainty should be welcomed. As discussed in Chapter 2, Sections 2.3.1 and 2.3.2, scientific uncertainty can be due to a magnitude of factors and is far from being a monolithic concept. Different types of uncertainty warrant different approaches and this should be identified and analysed at as early a stage as is possible.

page document. This can be seen as a wish on the part of the Commission to assure its trading partners that it, too, wishes to comply with the science-oriented approach to risk regulation that has crystallised within the WTO regime. However, the Commission's requirements on what is needed to trigger precautionary action are somewhat different than those established in the WTO, even if obvious similarities in the approaches also exist.

First of all, as in the regime under the SPS Agreement, the Commission does not require that the scientific evidence relied upon represents the majority opinion of the scientists active in the field. The Communication explicitly allows for reference to minority opinion provided that it is credible and comes from a reputable source.[92] Furthermore, the Commission holds that all evaluation of scientific evidence requires 'reliable scientific data and logical reasoning'.[93] However, the manner in which this is to be determined is not clear. In the following sentence, it is established that the desired level of protection may not aim at zero risk.[94] This requirement, which is consistent with statements made by the WTO Panels and the AB, seems to indicate that some level of scientific evidence must be met before precautionary measures can be taken. The emphasis on this may have been included as a response to critics who claimed that the EU was aiming to use the precautionary principle to justify policy choices without having to provide appropriate justification.[95] However, the Commission seems to have undermined this by enigmatically stating:

> However, in certain cases, an incomplete assessment of the risk may considerably limit the number of options available to the risk managers.[96]

The meaning of this sentence is ambiguous, and it implies that the prohibition of zero-risk levels may be negotiable. Like the hollow requirement to perform risk assessments dealt with above, the Commission here does not seem to want to create strict recommendations as to what risk levels the decision-maker is to employ. Not only does this confuse the issue of the meaning of zero risk, but it also concludes the Commission's opinion on risk levels and scientific evidence. Undoubtedly, it is impossible to establish a general formula that would be generally applicable for how scientific

[92] COM/2000/1, p. 17. As shown above in Sections 3.3.2.4 and 3.3.2.5, this is also the opinion held consistently by Panels and the AB under the SPS Agreement.

[93] *Ibid.*, p. 14. [94] *Ibid.*, p. 17.

[95] See, for example, 'The Battle Over Caution', *Newsweek*, 13 December 2003. Available at: www.newsweek.com/id/60863 (accessed 1 March 2010).

[96] COM/2000/1, p. 17.

evidence is to be examined and what levels of risk are to be met. However, in avoiding this topic in its entirety, the Commission has missed an important opportunity to strengthen the legal certainty and consistency in the application of the principle. More elaborate formal requirements on scientific evidence would have provided the EU institutions (including the ECJ) with information against which the application of the principle could be tested. This reluctance to be specific does not help to clarify the content of the precautionary principle and seriously undermines the practical use of the Communication.

The Commission finally stresses that, when precautionary measures are taken, further scientific research should be conducted with the aim of ending the scientific uncertainty. As with the above-mentioned requirements, a strict application of this would provide the Court with a formal test of the actions taken by the decision-maker when enacting precautionary measures. It would be possible for the Court to carry out such a formal test, even under the limited review it currently employs.[97]

4.2.5.6 The burden of proof

With regard to the legally controversial issue of the reversal of the burden of proof in cases of scientific uncertainty, the Commission begins by stating that one way that this instrument can be used is through prior authorisation schemes, or so-called positive lists.[98] This approach has a long tradition in EU Member States, at EU level and in third countries.[99] Even though they might be burdensome for some actors, such systems are relatively accepted and uncontroversial since they are generally in place before a product is developed and manufactured. Thus, the actor wishing to market a product knows what is required and can decide whether it wishes to pursue the activity or not. Therefore, the legal situation is consistent and foreseeable.

The situation in which no such authorisation scheme exists and the burden of proof is reversed ad hoc is more contentious, though. Here, the Commission, once again, omits to give any guidance. It is satisfied to state that action taken under the heading of the precautionary principle 'must, in certain cases, include'[100] a reversal of the burden of proof, and that this is to be determined on a 'case-by-case basis'.[101] It is clear

[97] COM/2000/1, pp. 20–1. [98] *Ibid.*, p. 21.

[99] For a brief discussion of prior authorisation schemes at the Community level see above, Section 4.2.

[100] COM/2000/1, 21. [101] *Ibid.*

that the entity wishing to rely on the precautionary principle has to provide some form of proof of the potential occurrence of a risk in order to trigger the use of the precautionary principle.[102] What is required to fulfil this is not clear, but it indicates that the initial burden of proof lies with the party invoking the principle, and, once an uncertain level of proof has been met, the burden shifts to the other party. In practice, this would mean that, if a Member State wished to restrict the free movement of goods on the internal market by invoking an exemption in Article 36 TFEU, or Article 114 (4) or (5) in a harmonised area, it would have to show that some kind of risk was present. Similarly, the European Commission or Council would have to make a case in favour of taking precautionary measures when they were mandated to do so. However, as in the case of the WTO, a party challenging a precautionary measure would have to show prima facie that a legislative provision had been breached before the burden could be switched.[103]

4.2.5.7 Cost–benefit analysis and proportionality

The cost–benefit analysis envisaged in the Communication appears, at first sight, to be a strict requirement, and seems to be operating outside of the proportionality test.[104] As discussed earlier in this chapter, the proportionality test in EC law is aimed at weighing the restrictions that measures impose on trade in the internal market. Cost–benefit analysis, on the other hand, is generally understood as a procedure which aims to analyse the general cost-effectiveness of a measure.[105] This may include effects on trade, but also includes comparisons with how the same resources could have been put to use in other areas of policy. However, such a wide analysis seems to be precluded from the concept of proportionality.[106]

The Commission rightly states that a cost–benefit analysis cannot be reduced to taking only economic considerations into account, but must also include non-economic considerations. This is essential if a

[102] *Ibid.*, p. 13: 'The precautionary principle is relevant only in the event of a potential risk.'

[103] Thus, it would be for a challenging party to show, for example, that a particular measure or set of measures infringes the basic free movement provision in Article 28 EC. Once this burden has been fulfilled, it is for the party which has invoked the measures to justify these under one of the exemptions in Article 30 EC. For a discussion on the general burden of proof regime in the WTO context see, above, Section 3.3.2.5.

[104] COM/2000/1, p. 18: 'A comparison must be made between the most likely positive or negative consequences of the envisaged action and those of inaction.'

[105] See, above, Section 2.3.2.

[106] For an authoritative discussion on the proportionality assessment in EC law, see Jans, 'Proportionality Revisited'.

cost–benefit analysis is going to be performed when applying the precautionary principle, since the principle aims primarily to protect non-economic interests. If only economic interests were taken into account, the precautionary principle would rarely be applied. It must also be considered appropriate that the Commission considers public health concerns to be of greater weight than economic concerns. Any other approach would seem callous.[107] However, the Commission goes on to state that an economic cost–benefit analysis should only be carried out where this is 'appropriate and possible'. Once again, it seems that a requirement which, prima facie, appears to be binding is in fact hollow, and it is, thus, left to the discretion of the decision-maker to take into account or not.

Once again, one can suspect that the treatment of cost–benefit analyses in the Communication has more to do with formally aligning the EU policy towards the precautionary principle and with the SPS Agreement, than with a heartfelt desire to rationalise the targeting of risks. In Article 5.3 of the SPS Agreement, cost–benefit analysis is established as one of the requirements made on all measures taken with the SPS Agreement as a basis. However, Panels and the AB have stopped well short of turning this requirement into a binding qualitative norm, and they have, instead, chosen to investigate subjectively whether contested measures were taken for protectionist reasons or not.[108] Thus, the WTO jurisprudence, like the Commission and the ECJ, is also closer to the performance of a proportionality test which centres on the effects that measures will have on trade, rather than on setting efficiency requirements for the Members.

4.2.5.8 The targeting of risks and risk trade-offs

As was touched upon above in the context of judicial review, the Communication echoes the opinion of the ECJ with regard to limited judicial review in cases where broad discretionary powers have been vested in the EU institutions.[109] This means that the Court's powers are circumscribed to the consideration of formal complaints and manifest error in the actions of the institutions. The reason for this is that it is not for the Court to determine the appropriateness of the content of certain measures, but only to consider whether they have been issued in the appropriate fashion.

[107] See, below, how the ECJ has used this requirement, worded as 'public health concerns always take precedence over economic concerns', in the *BSE* case and more recent case law. This is a stricter and much more problematical formulation.

[108] See, above, Section 3.3.2.6. [109] See, below, under Section 4.3.3.

The rationale behind this is established in the Communication, in which the Commission holds that '[j]udging what is an "acceptable" level of risk for society is an eminently political responsibility'.[110] Therefore, it is not for the Commission or the Court to second-guess what has been agreed upon by elected politicians. Nothing in the Communication indicates that risks need to be dealt with in an effective or comprehensive manner. This leads to the situation in which politicians may decide to target certain risks while leaving other risks unregulated.

It would have been possible for the Commission to state that the institutions would have to consider, for example, risk trade-offs when preparing legislation, and that the Council and European Parliament would have to take this into account when deciding on measures. As shown in the previous chapter, allowing such concerns to play a role in risk regulation leads to the consideration of the whole risk spectrum, rather than targeting individual risks in a seemingly haphazard fashion. The deferential approach advocated in the Communication is in line with how the issue has been dealt with at WTO level.[111]

4.2.5.9 Concluding remarks on Communication (2000)1

The Commission Communication on the Precautionary Principle is of non-binding character. Nonetheless, it is of great significance when discussing the precautionary principle in the EU. In particular, this is due to the fact that, when it was issued, it aimed both to codify the precautionary principle as it was being interpreted and used at the time, and to function as guidance for the future. The significance of the document was further underlined by the fact that it was formally supported by the Council,[112] the European Parliament[113] and, in principle, upheld in ECJ case law.[114] Thus, it is prudent to view the Communication as a formal manifestation of the official EU view on the precautionary principle. As such, the Communication is used below as a benchmark against which the application of the precautionary principle can be measured *in casu* at EU level as well as at Member State level.

[110] COM/2000/1, p. 4. [111] See, above, Sections 3.3.2.2 and 3.3.2.6.

[112] Conclusions of the Presidency, European Council at Nice 7–10 December 2000, Annex III, Point 1. Available at: www.europarl.europa.eu/summits/nice1_En.htm (accessed 1 March 2010).

[113] EU Bulletin 12–2000, Chapter 10 Community institutions, point 1.4.75.

[114] See, below, Section 4.3.

However, the Communication provides little in the way of concrete requirements on the parties to which it is addressed. Instead, it describes a number of issues that are to be viewed as components of the principle. However, in many cases, it also allows for the party that is applying the principle to derogate from a strict interpretation of these guidelines. A number of such cases have been dealt with above. This is particularly the case in relation to the requirements on risk assessment, cost–benefit analysis and risk trade-offs. In not elaborating on these issues, the Communication fails to make any qualitative requirements on the application of the precautionary principle.

A number of other weaknesses of the Communication have been pointed out.[115] First, it does not make it clear who is to use the precautionary principle. It seems that its primary addressees are the European institutions and the trading partners of the EU. Recently, however, it has seemed that there is increasing pressure from the Member States to make use of the principle as a shield against EU legislation under Articles 36 and 114 (4) and (5) TFEU.[116] A widespread use of the principle to derogate from the basic free movement provisions and the harmonisation measures on the internal market could pose a serious threat to the completion of the internal market. However, this issue is not dealt with in the Communication.

Second, and related to the first, the Communication remains silent on the institutional framework for the application of the principle. This risks creating ambiguity and disagreement with regard to at which level the principle is to be applied within the EU structure.

Third, in not clarifying the usefulness of cost–benefit analysis or considering risk trade-offs, the Commission fails to make any qualitative requirements for the application of precautionary measures.

Despite this lack in rigour and clarity, the Communication is the only document at European level that provides some form of *de minimis* requirements on the precautionary principle. Because of this, it remains a useful blueprint for the structure around which to build a discussion on the precautionary principle in the EU.

115 Vos, 'EU Food Safety Regulation in the Aftermath of the BSE Crisis', 248–9.

116 See, for example, Case C-3/00 *Denmark* v. *Commission* (discussed below) and Joined Cases T-366/03 and T-235/04, *Austria* v. *Commission* [2005] ECR II-4005. For a discussion on earlier case law in this regard, see Scott and Vos, 'The Juridification of Uncertainty'.

4.2.6 *Recent examples of the influence of the precautionary principle in secondary EU legislation*

Since the publication of the Commission Communication, the precautionary principle has been a fundamental element of EU harmonising measures in areas affected by health and environmental concerns.

The rules regulating the use of GMOs in the EU do not define the use of the precautionary principle, or even explicitly mention it. However, the establishment of an authorisation procedure in Directive 2001/18/EC is essentially precautionary in nature.[117] The rules regarding labelling and traceability are also intended to address the uncertainties regarding GMOs.[118] The labelling requirements allow consumers to avoid genetically modified organisms if they wish, and the traceability requirements allow for the possibility of tracing individual GMOs should risks materialise.

In the EU General Food Law, a comprehensive definition of the precautionary principle is included.[119] The definition is nuanced and stresses that provisional precautionary measures can be introduced in order to safeguard a 'high level of health protection'. Such measures must be preceded by a risk assessment and should be revisited in the light of emerging scientific data.[120]

Most recently, the REACH Regulation concerning the EU's legislative framework for registration, evaluation, authorisation and restriction of

117 Directive 2001/18/EC of the European Parliament and of the Council of 12 March 2001 on the deliberate release into the environment of genetically modified organisms and repealing Council Directive 90/220/EEC, OJ L 106, p. 1.

118 Regulation (EC) No 1829/2003 of the European Parliament and of the Council of 22 September 2003 on genetically modified food and feed, OJ L 268, pp. 1, 18.10.2003.

119 Article 7 of Regulation (EC) No 178/2002 of the European Parliament and of the Council of 28 January 2002 laying down the general principles and requirements of food law, establishing the European Food Safety Authority and laying down procedures in matters of food safety, OJ L 31, p. 1, 1.2.2002.

120 Article 7 of Regulation (EC) No 178/2002 holds:'1. In specific circumstances where, following an assessment of available information, the possibility of harmful effects on health is identified but scientific uncertainty persists, provisional risk management measures necessary to ensure the high level of health protection chosen in the Community may be adopted, pending further scientific information for a more comprehensive risk assessment.2. Measures adopted on the basis of paragraph 1 shall be proportionate and no more restrictive of trade than is required to achieve the high level of health protection chosen in the Community, regard being had to technical and economic feasibility and other factors regarded as legitimate in the matter under consideration. The measures shall be reviewed within a reasonable period of time, depending on the nature of the risk to life or health identified and the type of scientific information needed to clarify the scientific uncertainty and to conduct a more comprehensive risk assessment.'

chemicals is based upon the precautionary principle.[121] As a consequence of this, a registration system has been established for chemicals, and the burden of providing the scientific evidence when applying for registration has been placed on the applicant.[122]

4.3 EU case law

4.3.1 *Introduction*

One of the criticisms of the Commission's Communication on the Precautionary Principle is that it does not treat the issue of who is to apply the precautionary principle in any detail. The inclusion of the precautionary principle in the EU indicates that it should function as an instrument to be used by the European institutions. However, the Member States also rely on precautionary considerations when derogating from the basic free movement provisions under Article 36 TFEU,[123] and, perhaps increasingly, when derogating from harmonisation measures using exemptions included in safeguard clauses and in Article 114 (4) and (5) TFEU.[124] The ECJ has generally allowed for precautionary considerations in its case law by refraining from assessing the scientific evidence put before it in any detail, and, instead, allowing the Member States or the European institutions to choose the level of protection that they find appropriate.[125]

Below, a number of cases are dealt with in detail, all of which address different aspects of how the Court approaches precaution and the

121 See, especially, Article 1.3 of Regulation (EC) No. 1907/2006 of the European Parliament and of the Council of 18 December 2006 concerning the Registration, Evaluation, Authorisation and Restriction of Chemicals (REACH), establishing a European Chemicals Agency, amending Directive 1999/45/EC and repealing Council Regulation (EEC) No 793/93 and Commission Regulation (EC) No 1488/94 as well as Council Directive 76/769/EEC and Commission Directives 91/155/EEC, 92/67/EEC, 93/105/EC and 2000/21/EC, OJ L 396, p. 1, 30.12.2006.

122 See, for example, Article 7 and 19 of Regulation (EC) No. 1907/2006.

123 See, for example, Case 94/83 *Heijn* [1984] ECR 3263 and Case 54/85 *Mirepoix* [1986] ECR 1067.

124 In relation to safeguard clauses see Case C-236/01 *Monsanto Agricoltura* [2003] ECR I-8105 and Joined Cases T-366/03 and T-235/04 *Austria* v. *Commission*. In relation to the general provisions in Article 95 (4) and (5), see Case 512/99 *Germany* v. *Commission* [2003] ECR I-845 and Case C-3/00 *Denmark* v. *Commission*. For a discussion on the reliance by MS on the precautionary principle see Scott and Vos, 'The Juridification of Uncertainty', pp. 260–5.

125 See, for example, Case 227/82 *Van Bennekom* [1983] ECR 3883, Case 174/82 *Sandoz BV* [1983] ECR 2445 and Case 331/88 *Fedesa and others* [1990] ECR I-4023.

precautionary principle.[126] In early cases, such as the *BSE* cases[127] and the *Greenpeace* case,[128] the Court explicitly endorsed precautionary measures and used the term 'precautionary principle' for the first time. More recently, in the *Pfizer* case[129] the Court of First Instance made use of the Commission Communication on the Precautionary Principle to flesh out the content of the principle. Some authors have done important work in comparing how Member States on the one hand, and the European institutions on the other, have made use of the precautionary principle.[130] Others have treated the precautionary principle as applied by the Member States and the European institutions interchangeably in order to identify common properties in its application.[131] In the case law of the ECJ, it appears that the Court often refers to how the precautionary principle has been applied under other provisions than those at issue in the case in order to reach its decisions. This indicates that the Court views the precautionary principle as a concept with inherent properties which should be applied similarly, regardless of whether it is being invoked by a Member State or an EU institution.[132] This approach is followed below. Whenever discrepancies between the use of the precautionary principle by a Member State and by an EU institution appear, this is highlighted.

[126] For other recent studies of EC case law in the area see, N. de Sadeleer, 'The Precautionary Principle in European Community Health and Environment Law: Sword or Shield for the Nordic Countries?', in N. de Sadeleer, *Implementing the Precautionary Principle: Approaches from the Nordic Countries, EU and USA* (Earthscan, 2007), pp. 10–59; de Sadeleer, 'The Precautionary Principle in European Community Health and Environmental Law', 139–72; and E. Fisher, *Risk Regulation and Administrative Constitutionalism* (Hart Publishing, 2007), pp. 218–43.

[127] Cases C-157/96 *The Queen* v. *Ministry of Agriculture, Fisheries and Food, ex parte National Farmers' Union et al.* [1998] ECR I-2211 and C-180/96 *United Kingdom* v. *Commission* [1998] ECR I-2265 respectively. For the purposes of this discussion the two cases are identical and for the sake of clarity footnotes below refer to Case C-180/96 (*BSE*).

[128] Case C-6/99 *Association Greenpeace France and Others* v. *Ministere de l'Agriculture et de la Pêche and Others* [2000] ECR I-1651.

[129] Case T-13/99 *Pfizer Animal Health SA* v. *Council of the European Union* [2002] ECR II-03305.

[130] Scott and Vos, 'The Juridification of Uncertainty'.

[131] de Sadeleer, 'The Precautionary Principle in EC Health and Environmental Law', 139.

[132] See, for example, the approach assumed by the ECJ in Case C-192/01 *Commission* v. *Denmark* [2003] ECR I-9693, para. 49 and C-41/02 *Commission* v. *Netherlands* [2004] ECR I-11375, para. 45 which both dealt with measures imposed by Member States under Articles 28 and 30, but where the ECJ relied on the definition of the precautionary principle elaborated on in Case 157/96 *The Queen* v. *Ministry of Agriculture, Fisheries and Food, ex parte National Farmers' Union and Others* [1998] ECR I-2211, which dealt with measures imposed by the EU institutions to ban the import of British beef.

This section, like other sections in this book, is divided into a number of sub-sections, each dealing with different (but often related) aspects of the precautionary principle as interpreted in EU case law. As mentioned above, the Commission Communication is used as a blueprint for how the EU institutions intended to make use of the principle. The topics dealt with here are loosely based upon the Communication, and the document is used to compare and contrast how the ECJ and the Court of First Instance (CFI) have chosen to interpret the principle in their case law.

4.3.2 *The definition of the precautionary principle*

Even though precautionary approaches had been applied in specific cases earlier in EU case law,[133] it was not until the *BSE* case[134] that the Court made it clear that it intended to treat the precautionary principle as a general principle.[135]

The Court stated in paragraph 99 of the judgment that:

> Where there is uncertainty as to the existence or extent of risks to human health, the institutions may take protective measures without having to wait until the reality and seriousness of those risks become fully apparent.

This has become a frequently cited definition of the precautionary principle in the EU.[136] Some commentators claim that it is the only definition needed in EU law.[137] Others have pointed out that the *BSE* case related

[133] See, for example, Case 331/88 *Fedesa and others* [1990] ECR I-4023 and Case C-405/92 *Etablissement Armand Mondiet SA* v. *Amement Ijslais SARL* [1993] I-6133.

[134] Case C-180/96 *United Kingdom* v. *Commission* [1998] ECR I-2265. For a discussion of the background to the ban see, for example, L. Antonopolou and P. van Meurs, 'The Precautionary Principle within European Union Public Health Policy: The Implementation of the Principle under Conditions of Supranationality and Citizenship', *Health Policy* 66 (2003), 185–9, and Zwanenberg and Millstone, 'Mad Cow Disease 1980s–2000'. It should be noted that the English language version of the judgment refers to the principle of preventative action, but not explicitly to the precautionary principle. However, other language versions do make reference to the precautionary principle. Compare for example the wording of paragraph 100 of the judgment in the French version which refers to 'principes de précaution et d'action preventive', or the Swedish version which refers to 'försiktighetsprincipen och principen om förebyggande åtgärder'.

[135] For a discussion on the definition and status of the precautionary principle in the EU, see also de Sadeleer, 'The Precautionary Principle in European Community Health and Environmental Law: Sword or Shield for the Nordic Countries?', pp. 12–15.

[136] It was also included in Annex I to the Commission's Communication on the precautionary principle.

[137] See Christoforou, 'Science, Law and Precaution in Dispute Resolutions on Health and Environmental Protection', p. 219.

to emergency measures and that the Court itself emphasised that it was of importance that they be reviewed following a re-examination of the situation.[138] This could imply that, since the measures authorised in the *BSE* case were of interim character, the definition provided might also be restricted to emergency measures with a limited temporal scope. This view would posit the precautionary principle in the EU closer to that in place under the WTO regime[139] than to the wider and more permanent interpretations of the principle employed by some EU Member States.[140]

One important aspect of the Court's statement in the *BSE* case was that it clearly established that the precautionary principle extended beyond the environmental regulation to which Article 191 TFEU (formerly Article 174 EC) – seemingly – confined it. Subsequent case law has put beyond any doubt the fact that the precautionary principle is today applicable to consumer and health protection,[141] and it has been rightly pointed out that it is in this field that it has so far had most significance.[142]

In its subsequent case law, the ECJ has consistently cited the same paragraph introduced in the *BSE* case.[143] In the *Greenpeace* case the Court continued to develop its view on the principle. The background to the case was the referral of two questions to the ECJ by a French court. The questions related to how a Member State is to relate to the authorisation of a GMO by the Commission. In short, is the Member State bound to follow the decision of the Commission, or does it still retain some discretion relating to the authorisation of the GMO?[144] The Court found that some discretion still remained with the Member States and stated:[145]

[138] G. Majone, 'The Precautionary Principle and Regulatory Impact Analysis'. Paper presented at the International Seminar on Regulatory Impact Analysis. Rome 15 June 2001. p. 9, and Scott and Vos, 'The Juridification of Uncertainty'.

[139] See, above, Chapter 3.

[140] Compare with the chapter on the precautionary principle in Sweden below, Chapter 5.

[141] Case C-236/01 *Monsanto Agricoltura Italia and Others* [2003] ECR I-8105, paras. 128 and 133; Case T-13/99 *Pfizer Animal Health SA* v. *Council of the European Union* [2002] ECR II-03305, paras. 139 and 140; Case T-70/99 *Alpharma* v. *Council* [2002] ECR II-3495, paras. 152 and 153; Case T-177/02 *Malagutti-Vezinhet* v. *Commission* [2004] ECR II 827, para. 54; and Case C-41/02 *Commission* v. *Netherlands*, para. 45.

[142] de Sadeleer, 'The Precautionary Principle in European Community Health and Environmental Law', 172.

[143] See, for example, Case C-192/01 *Commission* v. *Denmark*, para. 49, Case 236/01 *Monsanto Agricola*, para. 111 and Case C-41/02 *Commission* v. *Netherlands*, para. 52.

[144] Case 6/99 *Association Greenpeace France and Others* v. *Ministere de l'Agriculture et de la Pêche and Others* [2000] ECR I-1651, paras. 1 and 2.

[145] The Court's reasons for this will be returned to below.

> Observance of the precautionary principle is reflected [...] in the right of any Member State [...] provisionally to restrict or prohibit the use or sale on its territory of a product which has received consent where it has justifiable reasons to consider that it constitutes a risk to human health or the environment.[146]

Once again, the Court emphasised the provisional character of the application of precautionary measures, but the measures at hand in the case in question could hardly be considered to be of an emergency character. The above-mentioned references to the precautionary principle have been echoed by the CFI in a number of cases.[147]

The general definitions of precautionary principle that the Court provided in the *BSE* and *Greenpeace* cases differ in some ways. For example, the *Greenpeace* definition spells out the provisional requirement and holds that 'justifiable reasons' must be provided in order to invoke measures. However, in practice they are used interchangeably to denote the application of the principle.

The precautionary principle in the EU can be used in two significant ways: first, it can be positively used by the EU institutions to issue EU legislation in areas surrounded by scientific uncertainty; second, the precautionary principle can be used by Member States to derogate from free movement provisions or harmonisation measures, provided that they can justify their need to maintain stricter environmental and health regulations than otherwise would be provided for.[148] That it is not mandatory for the Member States to rely on the precautionary principle is relatively clear. However, the issue of whether the EU institutions must rely on the precautionary principle when issuing legislation is more complicated. The CFI held, in the *Artegodan* case, that the precautionary principle is 'a general principle of EU law requiring the competent authorities to take appropriate measures'.[149] In the *Pfizer* case the CFI held that the EU institutions could, in certain cases, be required to act out of precaution.[150] However, it is often far from clear what precautionary action does, in

[146] Case 6/99 *Greenpeace and Others* v. *Ministere de l'Agriculture et de la Pêche and Others*, para. 44.

[147] Case T-199/96 *Laboratoires pharmaceutiques Bergaderm SA and Goupil* v. *Commission* [1998] ECR II-2805, paras. 66 and 67, Case T-70/99 *Alpharma Inc.* v. *Council of the European Union* [2002] ECR II-3495, para. 138 and T-13/99 *Pfizer Animal Health SA* v. *Council of the European Union*, para. 115.

[148] For a discussion on this see Scott and Vos, 'The Juridification of Uncertainty'.

[149] Case T-74/00 *Artegodan*, para. 184.

[150] Case T-13/99 *Pfizer Animal Health SA* v. *Council of the European Union*, para. 444.

fact, involve in a concrete situation. Issuing precautionary regulations in one area may lead to countervailing risks in other areas. When targeting one risk, other, often greater, risks are left unregulated.[151] To require that the EU institutions act in a precautionary manner, it would first have to be defined precisely what the term 'precautionary action' is intended to mean. It is argued here that this should be done rationally, by comparing different regulatory strategies with one another. Only then can it be estimated what a precautionary approach truly is, and only if this were done could the EU institutions be required to act in a truly precautionary manner.[152] At present, as is shown in more detail below, few such requirements are made on the EU institutions. Thus, it appears that the precautionary principle enables the EU institutions to act in situations in which scientific certainty is lacking, but does not require them to do so.[153] However, in one case, *Sweden* v. *Commission*, the CFI revoked a decision to authorise the pesticide paraquat upon the basis that the Commission had not acted in a sufficiently precautionary manner when issuing the authorisation.[154] This would indicate that the EU institutions would, in some cases, be bound to act in a precautionary manner. It remains to be seen if this approach is upheld in the future practice of the courts.

The next section examines how the principle is actually applied, and what requirements the Court has set up for its practical application.

4.3.3 *The role of scientific evidence and the limits of judicial review*

The ECJ has generally proved reluctant to make use of scientific evidence in order to rule on the merits of cases in which issues of scientific uncertainty have been involved. In non-harmonised areas where Member States have taken precautionary measures under Article 36 TFEU, the Court has held that:

[151] These issues were discussed in detail in Sections 2.3.2 and 2.3.3.

[152] For a further discussion on this issue see, above, Sections 2.3.3.

[153] Fisher holds that the EU institutions 'may' rely on the precautionary principle; see Fisher, *Risk Regulation and Administrative Constitutionalism*, p. 212. It has been held by de Sadeleer that the existence of the precautionary principle in EC law requires that the EU institutions 'take a position justifying their decisions'. Thus, they need to reflect on their choices in the light of the precautionary principle, and this is the real benefit of the principle, regardless of its compulsory or non-compulsory status. See de Sadeleer, 'The Precautionary Principle in European Community Health and Environmental Law', 164–5.

[154] Case T-229/04 *Sweden* v. *Commission*, Judgment of 11 July 2007, not yet reported.

> discretion relating to the protection of public health is particularly wide where it is shown that uncertainties continue to exist in the current state of scientific research.[155]

This statement of the Court confirms the consistent previous case law which allows Member States a large amount of leeway when choosing their own levels of protection in cases of scientific uncertainty.[156] As is discussed in more detail below, the Member States do not need to show the existence of a link between the scientific evidence presented and the measure introduced, which is the case under the SPS Agreement. Instead, it is sufficient to demonstrate that uncertainty is present.[157]

Likewise, where broad discretionary powers have been delegated within the EU structure, the ECJ is, in principle, unwilling to investigate the scientific evidence upon which basis a measure has been taken.[158] Since cases in which the precautionary principle is concerned by definition relate to scientific uncertainty, the Court's review is of a limited character.[159] In the EU legal order, this doctrine was developed by the Court in the *Fedesa* case.[160] The fact that judicial review has its boundaries in relation to government action is recognised in most legal systems. The outer limits of judicial review are generally established in constitutional provisions,[161] but pragmatic considerations also restrain courts.[162] Such concerns may include the highly controversial nature of certain policy areas and the lack of expertise on the part of the courts.[163] It should be pointed out that when judicial review is limited by constitutions or treaties, this is a restraint put in place by the legislator. When judicial review is limited due to pragmatic considerations, the restraints are put in place by the courts themselves, and, consequently, can also be modified by the courts themselves. Thus,

155 Case C-41/02 *Commission* v. *Netherlands*, para. 43.

156 See, for example, Case 227/82 *Van Bennekom* and Case 174/82 *Sandoz BV*.

157 Scott and Vos, 'The Juridification of Uncertainty', pp. 264–5.

158 See also, de Sadeleer, 'The Precautionary Principle in European Community Health and Environmental Law: Sword or Shield for the Nordic Countries?', pp. 36–41.

159 That this does not mean that the Court does not in certain cases perform a review of scientific evidence is dealt with below in this section.

160 Case 331/88 *Fedesa*.

161 In the EC legal order the Court is limited by the provisions provided for in the treaties.

162 J. Jowell, 'Of Vires and Vacuums: The Constitutional Context of Judicial Review', in C. Forsyth (ed.), *Judicial Review and the Constitution* (Hart Publishing, 2000), pp. 329–34. Jowell here describes the limits of discretion in common-law systems, but a good case can be made in favour of similar restrictions being in place in the EC legal order. See, for example, E. Fisher, 'Is the Precautionary Principle Justiciable?', *J. Env. L.*, 13:3 (2001), 315.

163 See the general discussion in Jowell, 'Of Vires and Vacuums', pp. 329–34.

there exist two different strands of possible restraints on judicial review. The first one is a formal restraint enacted by the legislator. The second one is a voluntary restraint imposed by the courts using their own discretion. In the case of the EU courts, the limited review in place has been established by the Courts themselves. This is important, since it means that the Courts could also develop its judicial review doctrine, without being bound by statute.[164]

In the *Fedesa* case the ECJ was asked to make a preliminary ruling on the validity of national regulations which aimed to implement Directive 88/146/EC prohibiting the use of a number of hormones in livestock. The UK High Court of Justice asked the ECJ to answer, primarily, whether the ban on hormones was invalid since it was not based upon scientific evidence, and, therefore, infringed a number of legal principles, such as legal certainty, proportionality and equality.[165]

The ECJ began by stating that the aim of the directive in question was to harmonise the divergent national legislations on the matter, and, in doing so, respond to the fears of the consumers of some Member States with regard to the safety of the regulated hormones in question.[166] Fedesa had held that these anxieties lacked any foundation in science, and that, as a consequence, the principle of legal certainty for traders in the hormones was being violated. The Court declined to make a judgment with regard to the scientific evidence underlying the directive, and stated:

> Even if it were to be held, as the applicants in the main proceedings have argued, that the principle of legal certainty requires any measure adopted by the Community institutions to be founded on a rational and objective basis, judicial review must, having regard to the discretionary power conferred on the Council in the implementation of the common agricultural policy, be limited to examining whether the measure in question is vitiated by a manifest error or misuse of powers, or whether the authority in question has manifestly exceeded the limits of its discretion.[167]

In applying this reasoning, the Court effectively confined itself to a narrow, marginal review, concerned with form and not with content, in cases in which broad discretionary powers have been delegated to the EU institutions. The only material review that the Court allows itself is to consider whether a manifest error on the part of the legislator has been committed. If not, every substantive challenge will be disregarded.

[164] It is shown below that for example the US federal courts have developed the margins of their judicial review over time. See, below, Section 7.1.

[165] Case 331/88 *Fedesa and others*, para. 4.

[166] *Ibid.*, para. 6.

[167] *Ibid.*, para. 8.

Having stated this, the Court turned to consider the concept of scientific uncertainty. Here, it appears that the Court was satisfied with the fact that different national authorities had made divergent assessments regarding the risks of scientific uncertainty that was in question/present/at issue:

> In the light of the foregoing, the claim based on the existence of scientific evidence demonstrating the safety of the five hormones in question cannot be upheld. It is not necessary to order any measures of inquiry to verify the accuracy of that allegation; it need merely be stated that, faced with divergent appraisals by the national authorities of the Member States, reflected in the differences between existing national legislation, the Council remained within the limits of its discretionary power in deciding to adopt the solution of prohibiting the hormones in question.[168]

This passage establishes that the Court considers any interpretation of scientific evidence to be within the Council's discretion whenever scientific uncertainty is in question. The limited evaluation of the scientific data performed by the Court merely aims to establish whether the case is surrounded by scientific uncertainty or not. If this is found to be the case, the Court considers itself to be prevented from assessing whether the Council's decision was the appropriate one. Conversely, if the case is found not to be surrounded by scientific uncertainty and the Court finds that the Council had acted contrary to the scientific evidence, a manifest error may have been committed.

It thus becomes important to establish when the Court considers scientific uncertainty to be present. In the *Fedesa* case the fact that different Member States had made different assessments with regard to the risks involved was sufficient.[169] One problem with the Court's restraint when judging under scientific uncertainty is that the issue of the risk is neglected. As was shown in Chapter 2, uncertainty is not necessarily absolute or static. There are different degrees of uncertainty, and research can – partially or completely – help to disperse scientific uncertainty. An uncertain risk can thus be more or less probable, even if an exact probability by definition cannot be assigned to it, depending on the degree of uncertainty it is surrounded by. When the Court stops short of investigating the degree and cause of scientific uncertainty, it also effectively excludes itself from considering whether the risk that measures in question are intended to protect against is probable or not.

[168] *Ibid.*, para. 9. [169] *Ibid.*

The limits imposed here are based upon the Court's own view of its judicial review. Nothing in the relevant Article 263 TFEU limits the Court's powers to this effect. A more activist Court could extend the scope of its review. In the US legal system, for example, the scope of the court's judicial review when judging on scientific evidence has changed over time.[170]

Since the *Fedesa* case, the Court has upheld its view on the limits to its judicial review on a number of occasions. In the *BSE* case[171] the UK held that, due to the inconclusive evidence establishing a link between the outbreak of BSE with the outbreak of Creutzfeld-Jakob disease, the ban on the importation of British beef to the rest of the EU was unjustified. In answering this, the Court began by pointing out that it considered itself to be restricted to investigating if the Commission decision was vitiated by a manifest error or a misuse of powers.[172] This was repeated in the *Safety Hi-Tech* case, which regarded an exemption from the general ban on hydrochlorofluorocarbons (HCFCs) for use in fire-fighting equipment. Safety Hi-Tech held that the Regulation (EC) 2037/2000 on substances that deplete the ozone layer had not taken the available scientific data into account. If it had, Safety Hi-Tech claimed, it would have been clear that other, more dangerous substances, had been left unregulated. Here, the Court held that its task was not to judge, upon the basis of scientific evidence, which risks were to be regulated, and that Safety Hi-Tech had not managed to show manifest error on the part of the Council when issuing the ban.[173]

In the *Pfizer* case the Court of First Instance continued to emphasise its limited review doctrine and stated:

> It follows that, in this case, in which the Community institutions were required to undertake a scientific risk assessment and to evaluate highly complex scientific and technical facts, judicial review of the way in which they did so must be limited. The Community judicature is not entitled to substitute its assessment of the facts for that of the Community institutions, on which the Treaty confers sole responsibility for that duty. Instead, it must confine itself to ascertaining whether the exercise by the institutions of their discretion in that regard is vitiated by a manifest

[170] For an in-depth discussion on this, see Jasanoff, *Science at the Bar*, pp. 69–92.

[171] For a background to the BSE affair, see above and, for example, Antonopolou and van Meurs, 'The Precautionary Principle within European Union Public Health Policy', 185–9, and Zwanenberg and Millstone, 'Mad Cow Disease 1980s–2000'.

[172] Case C-180/96 *United Kingdom* v. *Commission*, para. 60

[173] Case C-341/95 *Gianni Bettati* v. *Safety Hi-Tech* [1998] ECR I-4335, paras. 50–3.

> error or a misuse of powers or whether the institutions clearly exceeded the bound of their discretion.[174]

However, even though the Court in the *Pfizer* case continued to find itself restricted from judging on the merits of the scientific evidence, it indicated that it required the scientific evidence to fulfil certain requirements in order for it to be able to be relied upon. It, thus, held that it would not be sufficient for the EU institutions to take any decision that it pleases, but that decisions had to be backed up by a scientific assessment. How this scientific assessment was to be conducted is subject to scrutiny, but the conclusions drawn from a flawless assessment are outside the reach of the Court. However, the Court failed to make any generally applicable, normative requirements on the scientific assessment, and only stated that it must be 'as thorough as possible'.[175] Some questions regarding the systematic approach of the Court must also be raised. The Communication on the Precautionary Principle holds that scientific evaluation is made up of four constituent parts: hazard identification, hazard characterisation, the appraisal of exposure and risk characterisation.[176] The Court did not test the institutions against these requirements, but was satisfied that the institutions had based their decision upon 'sufficiently reliable and cogent information' and that they had understood 'the ramifications of the scientific question raised'.[177] Nothing is mentioned with regard to how the Court actually reached this conclusion.[178]

Here, it should be pointed out that, when not applying strict requirements, such as those included in the Communication, in the case law, it becomes possible for both the Member States and the EU institutions to circumvent the requirements and the purpose of a risk assessment. This affords the institutions considerable freedom, not only to evaluate scientific advice, but also to decide on how the scientific advice is obtained. In order to avoid situations in which the EU institutions and the Member States devise their risk assessments to obtain particular, preordained results, it would have been helpful if the Court had created strict, formal

[174] Case T-13/99 *Pfizer Animal Health SA* v. *Council of the European Union*, para 169. The Court also referred back to this statement later in the case. See, for example, para. 323.

[175] T-13/99 *Pfizer Animal Health SA* v. *Council of the European Union*, para. 162.

[176] COM/2000/1, p. 13 and Annex III.

[177] T-13/99 *Pfizer Animal Health SA* v. *Council of the European Union*, para 162.

[178] For further discussion on the failure of the Court to test the scientific evaluation against the guidelines laid down in the Communication (including the choice to be guided by the diverging Danish scientific opinion rather than the SCAN), see E. Vos, 'Antibiotics, the Precautionary Principle and the Court of First Instance', *Maastricht Journal* 11:2 (2004), 3–6.

requirements on how the scientific evaluation was to be carried out. This is something that must have been within the Court's reach, without stretching the boundaries of the limited judicial review it has afforded itself. However, as is returned to below, the Court appears to have considered the opinion of the relevant scientific committee, which, at least to some extent, was the basis for the final decision, and in so doing, seems to have respected the Communication requirements, at least in spirit.[179] Clarity, on this point, is an essential starting point for a systematic and science-based application of the precautionary principle.

It also needs to be pointed out that, even though the Court explicitly states that it does not rule on the merits of the scientific evidence or substitute its assessment for that of the other institutions,[180] it still takes the scientific evidence presented into consideration.[181] However, instead of judging on the case, it uses the scientific evidence to establish the existence of scientific uncertainty.[182] How the Court proceeds to perform this is dealt with in detail below.

Furthermore, it should be mentioned that, surprisingly, the CFI diverted from its limited review doctrine in a case regarding the inclusion of the pesticide paraquat in Annex I of Directive 91/414/EC concerning the Placing of Plant Protection Products on the Market.[183] Paraquat had been included in Annex I by a Commission decision following the comitology procedure prescribed in Directive 91/414/EC, and was thus authorised to be placed on the market. Sweden challenged this decision in 2004, and sought to have the decision reversed due to the hazardous effects of paraquat on humans and animals. Without referring to its limited review in cases in which discretion had been given to the EU institutions, the CFI made an expansive reading of the formal requirements on scientific

[179] See, for example, T-13/99 *Pfizer Animal Health SA* v. *Council of the European Union*, paras. 194 and 209.

[180] Case T-13/99 *Pfizer Animal Health SA* v. *Council of the European Union*, para. 393.

[181] See, for example, the statement of the Court in Case 97/83 *Melkunie*, para. 15 and the detailed scientific analysis in the Case T-13/99 *Pfizer Animal Health SA* v. *Council of the European Union* discussed below.

[182] That uncertainty is a precondition for the application of the precautionary principle and therefore has to be established before measures can be introduced has been held consistently by the ECJ. See for example Case 132/03 *Codacons and Federconsumatori* [2005] ECR I-4167, para. 61, and Case C-236/01 *Monsanto Agricoltura Italia and Others*, para. 111. See also, the discussion by van Asselt and Vos in 'The Precautionary Principle and the Uncertainty Paradox', 4, and de Sadeleer, 'The Precautionary Principle in European Community Health and Environmental Law: Sword or Shield for the Nordic Countries?', p. 20.

[183] Case T-229/04 *Sweden* v. *Commission*.

evidence in the relevant legislative act and went on to investigate in detail the scientific reports underlying the Commission's decision.[184] Despite the statements made in previous case law that the European Courts should not substitute the assessment of the EU institutions with that of their own, the CFI in *Sweden* v. *Commission* proceeded to do precisely this. By concluding that the assessments of the scientific evidence by the relevant scientific committee, as well as that of the rapporteur Member State and the Commission, were flawed, the CFI revoked the Decision of the Commission.[185] This judgment is extraordinary since it is apparently inconsistent with the previous settled case law with regard to judicial review of scientific evidence. This creates the highly paradoxical situation in which the ECJ and the CFI, in some cases, limit the possibilities of effective redress for parties affected by precautionary measures with reference to their limited judicial review,[186] and, in other similar cases, are willing to make a detailed assessment of scientific evidence in order to prescribe a more precautionary approach. The possible implications of the judgment for the application of the precautionary principle are returned to below in the section on scientific evidence. Moreover, in *Commission* v. *Parliament* the ECJ also overturned the Commission's decision to exempt the brominated flame retardant decabromodiphenyl ether (DecaBDE) for use in polymers from the general ban on flame retardants of the polybrominated diphenyl ethers (PBDE) family.[187] However, the conclusion of the ECJ was based upon the fact that the Court had concluded that the authorisation meant a de facto total exemption for the substance, whereas the Restriction of Hazardous Substances Directive (RoHS) Directive clearly established that exemptions could only be granted for certain applications of products, not the products per se.[188] Thus, the ECJ avoided examining the issue of the evaluation of the scientific evidence in the case.

Finally, it should be mentioned that the European Court's general approach to judicial review in cases of the precautionary principle differs significantly from how Panels and the AB have approached the subject within the WTO structure.[189] The focus on a threshold-finding of risk is present in the WTO structure, whereas, in the EU structure, it is generally sufficient

[184] *Ibid.*, paras. 108–10 and 174–90. [185] *Ibid.*, paras. 180–1 and 185–6.

[186] See the discussion of, for example, Case 331/88 *Fedesa and others* and Case T-13/99 *Pfizer Animal Health SA* v. *Council of the European Union* above.

[187] Joined Cases C-15/06 and C-295/06 *European Parliament* v. *Commission*, judgment of 1 April 2008, not yet reported.

[188] *Ibid.*, 68–79. [189] See, above, Section 3.3.2.2.

for the party invoking the measures to show that the area is surrounded by uncertainty. The scope of the judicial review is intimately connected to the varying degrees of proof required within the two legal frameworks. Since the WTO Panels look for a link between the scientific evidence and the measures invoked, their review must be deeper in scope. What is required in the EU structure is examined in the following section.

4.3.4 *Risk assessment and risk management*

The decision to enact precautionary measures is a political decision based upon scientific evaluation and advice.[190] The reasons for keeping the decision of whether to issue measures separate from the technical and scientific advice was largely accounted for in Chapter 2. The fact that the Member States have to perform a risk assessment before taking precautionary measures under Articles 34 and 36 TFEU has been consistently stated by the ECJ.[191] In addition, in cases in which Member States wish to derogate from harmonising measures taken under Article 114 TFEU, this must have been preceded by a risk assessment performed by the Member States wishing to derogate.[192]

At EU level, the formal source of scientific advice generally comes from the various scientific committees or, increasingly, the specialised European agencies.[193] This advice is not binding, but when enacting measures which depart from the recommendations of the relevant committee, the EU institutions need to reason why it makes a different assessment of the situation.[194] Thus, the scientific committees essentially perform the scientific assessment, and the Commission or Council act as the risk managers.

[190] For a discussion on the roles of risk assessment and risk management in EC risk regulation see de Sadeleer, 'The Precautionary Principle in European Community Health and Environmental Law: Sword or Shield for the Nordic Countries?', pp. 27–35. See also, Fisher, *Risk Regulation and Administrative Constitutionalism*, pp. 238–41.

[191] Case C-41/02 *Commission* v. *Netherlands*, para. 48; Case C-192/01 *Commission* v. *Denmark*, para. 47; Case C-24/00 *Commission* v. *France* [2004] ECR I-1277, para. 54.

[192] See, generally, Case C-512/99 *Germany* v. *Commission* ECR I-845 and Case C-3/00 *Denmark* v. *Commission* [2003] ECR I-02643.

[193] Vos, 'Antibiotics, the Precautionary Principle and the Court of First Instance', 1.

[194] Case T-13/99 *Pfizer Animal Health SA* v. *Council of the European Union*, paras. 199 and 200. See also Case C-405/92 *Etablissement Armand Mondiet SA* v. *Amement Ijslais SARL* ECR I-6133, paras. 31–2 and Case C-120/97 *Upjohn* [1999] ECR I-223, para. 47.

As was shown earlier, in the *Pfizer* case the Commission was allowed to derogate from the risk assessment presented in the opinion of the Scientific Committee on Animal Nutrition (SCAN) due to the fact that the CFI employed a very restricted review of the Commission's actions in combination with a low level of proof. This indicates that the risk manager (the Commission) has a large margin of appreciation when making use of the scientific assessment (the SCAN Opinion). The CFI also concluded, both in the *Pfizer* and in the *Alpharma* cases, that the purpose of a risk assessment is to inform the decision-maker adequately. In 'exceptional circumstances and where there are adequate guarantees of scientific objectivity in cases of particularly complex scientific facts' the decision-maker is not bound to submit the matter to the relevant scientific committee.[195] However, and remarkably, the CFI did not discuss the matter in any greater detail, and did not suggest how this notion can be reconciled with principles of scientific objectivity, impartiality, proper scientific basis and transparency, which it had emphasised elsewhere in its judgments.[196]

In *Denmark* v. *Commission* the Danish authorities successfully managed to challenge the decision by the Commission upon the basis of it not having taken the relevant scientific committee advice into consideration when issuing its decision.[197] Similar emphasis was put on the opinion of the scientific committee in *Germany* v. *Commission*.[198] Both of these cases concerned requests for derogations under Article 95 EC (now Article 114 TFEU). In deciding on such requests, the Commission has had broad discretionary powers delegated to it. This is in conformity with the Court's case law in similar cases.[199] However, nowhere in either *Denmark* v. *Commission* nor in *Germany* v. *Commission* does the Court signify that it finds its judicial review limited, and it does review the acts of the Commission in far greater detail than it did in the *Pfizer* case, for example. This would seem to indicate that the Court considers the risk

[195] Case T-13/99 *Pfizer Animal Health SA* v. *Council of the European Union*, para. 298 and T-70/99 *Alpharma Inc.* v. *Council of the European Union*, para. 213.

[196] Case T-13/99 *Pfizer Animal Health SA* v. *Council of the European Union*, para. 213. For a more detailed discussion of this and other aspects of the *Pfizer* and *Alpharma* judgments, see Vos, 'Antibiotics, the Precautionary Principle and the Court of First Instance', 7.

[197] Case C-3/00, *Denmark* v. *Commission*, para. 111: 'The contested decision did not take sufficient account of the 1995 opinion. It failed to mention in that connection that the maximum amounts of nitrites set in Directive 95/2 are called into question by the 1995 opinion.'

[198] Case C-512/99 *Germany* v. *Commission* ECR I-00845.

[199] See the discussion on judicial review above in Section 4.2.5.3.

manager (the Commission) less bound by the risk assessment (the opinion of the Scientific Committee) in certain cases than in others.

As mentioned, this inconsistency may be due to pragmatic considerations by the Court. However, the inconsistency with regard to how much force the Court intends to give to risk assessments and how much freedom the risk manager has to act creates serious concerns with regard to legal certainty.

In order to illustrate the awkward results of the failure to maintain a strict separation between risk assessment and risk management, the *Monsanto* case serves as a good example.[200] In the case, Monsanto asked the French authorities to classify bovine somatotrophin (BST) as a substance for which no maximum residue level (MRL) had to be established. This would indicate that the product was considered safe to place on the market from a public health perspective. The matter was passed to the Committee for Veterinary Medicinal Products (CVMP), which expressed an opinion which supported Monsanto's application, in which it stated that it could see no public health risks with BST.[201] However, the Commission, instead of passing the request on to the regulatory committee, which the relevant legislation provided for,[202] took its own decision not to list BST in the Annex of substances for which no MRL would have to be issued. The reason for its action was that a moratorium affecting the placing of BST on the market existed in the EU, which would effectively prevent the marketing of BST regardless of whether it was declared safe or not.[203]

Monsanto brought the Commission before the Court of First Instance to challenge its decision, and the CFI ruled that a marketing ban on a substance did not preclude the inclusion of the same substance in an Annex declaring it safe for public health. The CFI based this primarily upon its previous ruling in an almost identical case, in which it had ruled that the Commission could not base its decision solely upon the existence of a moratorium.[204] The CFI placed special emphasis on the fact that a mora-

[200] Case T-112/97 *Monsanto* v. *Commission* [1999] ECR II-1277.

[201] *Ibid.*, paras. 4–8

[202] Council Regulation (EEC) No 2377/90 of 26 June 1990 laying down a Community procedure for the establishment of maximum residue limits for veterinary medicinal products in foodstuffs of animal origin, OJ L 224, p. 1, 8.8.1990.

[203] Council Decision 90/218/EEC of 25 April 1990 concerning the Administration of Bovine Somatotrophin (BST) and Council Decision 94/936 of 20 December 1994 amending Decision 90/218/EEC concerning the placing on the market and administration of bovine somatotrophin (BST) respectively.

[204] Case T-120/96 *Lilly Industries* v. *Commission* [1998] ECR II-2571.

torium introduced for socio-economic reasons[205] must not be used as an excuse to declare a substance not safe from a public health perspective.[206]

France appealed the judgment to the ECJ. After dealing with procedural issues relating to the case before the CFI, the Court turned to the substantive question. Here, the Court stated that the decision to establish (or the decision not to establish, as it were) an MRL was intrinsically linked to the placing of a product on the market. Thus, the Court argued, that if the placing of a product on the market was effectively made impossible, for example by the existence of a moratorium, the question of establishing an MRL did not need to be answered.[207] In stating this, the Court reversed the ruling of the CFI, which had argued for a separation of the scientific decision regarding the safety of the product from the political decision relating to whether to allow the product on the market or not. In combining the two decisions, the Court blurred the line between what was essentially risk regulation (the issue of whether the substance in question is safe or not) and agricultural politics (whether the substance is to be allowed or not from a socio-economic perspective). The somewhat radical conclusion of the above is that the Court allows a political decision regarding milk quotas to determine the level of health protection to be applied in the EU. In doing so, it becomes possible for the Commission and the Council to enact politically motivated measures which are disguised as measures aimed at regulating risks. If this confusion is maintained in future case law, it risks seriously undermining the systematic approach to risk regulation in general, and the precautionary principle in particular.

This is not to say that the transparency and information flow between the risk assessment and the risk management phase should not be improved. Development in this direction will allow risk assessors to understand the concerns of both the risk managers and the general public, and will also allow the risk managers to understand the reasoning behind, and the limits to, scientific advice.[208] However, it has to be possible to determine on what grounds the decision is motivated, and where scientific assessments end and political decisions begin.[209]

205 More specifically to maintain the milk quotas negotiated between the Member States within the framework of the CAP.

206 Case T-112/97 *Monsanto* v. *Commission*, para. 91.

207 Case C-248/99, *France* v. *Monsanto* [2002] ECR I-1, para. 80.

208 For an in-depth discussion on this issue see above, Section 2.3.1.

209 N. de Sadeleer does not agree with this notion, but seems to argue that by allowing for essentially political considerations in the risk assessment phase, the accountability

4.3.5 Scientific evidence and the introduction of precautionary measures

In the two preceding sections it was discussed that Member States as well as European institutions must perform a risk assessment before invoking precautionary measures, but that the judicial review performed by the ECJ when evaluating the content of such assessments is very limited. It was established that the Court, in general, is content with finding that scientific uncertainty is present and, once this has been established, it leaves the Member States or the institutions considerable leeway in deciding upon what measures to take.[210] The issue of the general problems arising in relation to decision-making under scientific uncertainty were dealt with in detail previously, in Chapter 2.

In the *BSE* case the precautionary ban on British Beef was to stay in place '[p]ending an overall examination of the situation'.[211] In essence, the measure reversed the burden of proof, forcing the UK to show the absence of risk in order to challenge the decision successfully. However, the UK did not strive to prove this; instead, it challenged the Commission's decision upon the basis that it was not sufficiently supported by science. Thus, the reversal of the burden of proof was not dealt with in detail, and only the level of required proof needed to initiate precautionary measures was dealt with. As mentioned above, the Court held that it would apply a limited review of the measure.[212] The Court continued to state:

of decision-makers is increased. This author does not agree with the assertion by de Sadeleer. When scientific reasoning is not distinguished from socio-economic considerations, it becomes difficult to understand the rationale behind decisions. It also increases the possibilities of officials to mask decisions as science-based, when they are really not motivated by concerns relating to risk reduction. Compare de Sadeleer, 'The Precautionary Principle in European Community Health and Environmental Law', 160.

[210] See, for example, Case 132/03 *Codacons and Federconsumatori*, [2005] ECR I-4167, para. 61 and Case C-236/01 *Monsanto Agricoltura Italia and Others*, para. 111. See also, the discussions in Scott and Vos, 'The Juridification of Uncertainty', p. 265; van Asselt and Vos, 'The Precautionary Principle and the Uncertainty Paradox', 4; and de Sadeleer, 'The Precautionary Principle in European Community Health and Environmental Law: Sword or Shield for the Nordic Countries?', p. 20. It was also discussed that the CFI in T-229/04 *Sweden* v. *Commission* appeared to ignore the limited review doctrine in order to overturn a decision of the Commission based on a different reading of the scientific evidence. This issue is returned to below in this section.

[211] Commission Decision 96/239/EC of 27 March 1996 on emergency measures to protect against bovine spongiform encephalopathy, OJ L 078, 28/03/1996, pp. 0047–0048. Article 1.

[212] Case C-180/96 *United Kingdom* v. *Commission*, para. 60.

> In the present case, the publication of new scientific information had established a probable link between a disease affecting cattle in the United Kingdom and a fatal disease affecting humans for which no known cure yet exists.[213]

Thus, first, the Court emphasised that scientific evidence had been produced in support of the measures taken by the Commission. The UK did not submit evidence showing that the risk was not probable, or that a different assessment of the data was more probable. It is unclear how such proof would have been evaluated by the ECJ.

Second, the level of proof here had been set for the Commission to make it 'probable' that the link existed. This is a higher level of proof than a mere suspicion, in that it incorporates an element of likelihood, yet it is not as high as, for example, 'verified' or 'manifest'.[214]

Third, the perceived risk in the *BSE* case was very grave since it was described as life-threatening and incurable.[215] It could be argued that this suggests that the more serious the risk is perceived to be, the more the Court is inclined to allow precautionary measures. Having established this, and once again having emphasised the uncertainty and the gravity of the risk in question, the Court concluded that the Commission had not exceeded its powers when issuing the ban.[216]

In the *Toolex* case, the level of proof imposed by the Court appears to be lower. Where the Court in the *BSE* case relied on the existence of new scientific evidence, the probability of risk and the seriousness of the risk, the Court in *Toolex* mentioned the existence of new scientific evidence with regard to the possibility of trichloroethylene being a carcinogen, but did not require the authorities to show the probability of this risk.[217] Instead, the Court was satisfied with the fact that it was unclear at what exact dose the substance was hazardous for a total ban to be justified. This implies that the requirements for the triggering of precautionary measures in *Toolex* were lower than those used in *BSE*, and that neither of them fully match the vague requirements laid down in the Commission's

[213] *Ibid.*, para. 61.

[214] See definitions in *The New Oxford Dictionary* (Oxford University Press, 1998).

[215] An interesting analogy here is Article 15 of the Rio Declaration in which precautionary measures can be invoked to combat a 'serious or irreversible' risk. It is, however, unclear if the Court in referring to the life-threatening and incurable risk of Creuzfeld-Jakob disease intended to allude to this agreement.

[216] Case C-180/96 *United Kingdom* v. *Commission*, para. 62.

[217] Case C-473/98 *Kemikalieinspektionen* v. *Toolex* [2000] ECR I-5681. AB. paras. 42–5.

Communication.[218] This indicates that the ECJ may be even more lenient to assessments made by the Member States under Article 36 TFEU (formerly Article 30 EC), which was the case in *Toolex*. In more recent case law, the ECJ has downplayed the suggestion in the *BSE* case that a risk would have to be 'probable' in order to trigger the use of the precautionary principle. In *Monsanto Agricoltura* the Court emphasised that what had triggered the precautionary measures in the *BSE* case was 'the uncertainty as to the existence of risks to human health'.[219] However, the measures cannot be based on 'purely hypothetical considerations'.[220]

If the case law so far had failed to provide a uniform interpretation of what is required in order to invoke precautionary measures, things were not much clarified when the CFI fleshed out the use of the precautionary principle in the *Pfizer* case. However, the CFI applied a somewhat more ambitious approach. The case related to the invocation by Denmark of a safeguard procedure included in the Council Directive concerning additives in feeding-stuffs.[221] As support for banning the antibiotic virginiamycin as a growth promoter, Denmark relied on a report from its National Veterinary Laboratory. The Commission submitted the Report to the Scientific Committee on Animal Nutrition (SCAN), which found the link between using virginiamycin as a growth promoter in cattle and resistance towards the use of the antibiotic as a human medicine to be weak. It concluded that the use of virginiamycin did not constitute an immediate risk to health in Denmark. Despite this, the Commission, relying on the precautionary principle, proposed a ban of the product which, following the adequate procedure, was issued by the Council.[222]

Pfizer brought the case before the CFI, where it essentially argued that the decision of the Council was not based on scientific evidence. It argued that a loose interpretation of what needs to be shown when the

[218] The Commission remains vague on the issue of the level of proof required. It merely holds that the precautionary principle cannot be allowed to lead to zero-risk levels being employed and that the triggering of precautionary measures must be preceded by 'identification of the potentially negative effects of a phenomenon' (COM/2000/1, p. 13).

[219] C-236/01, *Monsanto Agricoltura Italia and Others*, para. 111.

[220] C-236/01, *Monsanto Agricoltura Italia and Others*, para. 106; Case C-192/01 *Commission* v. *Denmark*, para. 49; Case C-24/00 *Commission* v. *France*, para. 56; and Case C-41/02 *Commission* v. *Netherlands*, para. 52.

[221] Council Directive 96/51/EC, amending Directive 70/524/EEC concerning additives in feeding-stuffs (1996) OJ L235, p. 39.

[222] Council Regulation 2821/98/EC amending, as regards the withdrawal of the authorisation of certain antibiotics, Directive 70/524/EEC concerning additives in feedingstuffs (1998) OJ L351, p. 4.

precautionary principle is invoked risks leading to a situation of *probatio diabolica*.[223] This would, according to Pfizer, mean a reversal of the burden of proof in all situations involving the precautionary principle, and a party wishing to challenge a measure would always have to show the lack of risk in order to be successful. This is a situation that both the Communication on the Precautionary Principle and the Advocate General have denounced.[224]

The CFI agreed with Pfizer in principle, and held that the EU institutions must show that a risk is not 'purely hypothetical',[225] but 'adequately backed up with scientific data'.[226] These requirements the CFI considered to be fulfilled in the case in question, without mentioning how it had come to this conclusion. However, it must be considered that this initial step, which corresponds with point 5.1.1 Identification of potentially negative effects in the Communication on the Precautionary Principle,[227] apparently only serves to trigger a risk assessment, not precautionary measures. Consequently, the requirements need not be very high; as it is sufficient for a potential risk to be perceived, the likelihood and severity of which is to be examined scientifically.[228] The CFI, then, importantly, pointed out that, in the case in question, a zero-risk approach is particularly inappropriate since it is simply not possible to show scientifically that a future risk does not exist with virginiamycin.[229] Having stated that a potential risk may exist and that a zero-risk approach is not to be applied, the CFI turned to the scientific assessment of the potential risk.

The Court here identified two stages of the risk assessment procedure: ascertaining what level of risk is unacceptable and the scientific assessment of that risk.[230] This is a somewhat confusing and unfortunate

[223] Case T-13/99 *Pfizer Animal Health SA* v. *Council of the European Union*, para. 130.

[224] Opinion of A. G. Mischo in Case C-6/99 *Association Greenpeace France and Others* v. *Ministere de l'Agriculture et de la Pêche and Others*, para. 72.

[225] Case T-13/99 *Pfizer Animal Health SA* v. *Council of the European Union*, para. 143.

[226] *Ibid.*, para. 144. [227] COM/2000/1, p. 13.

[228] Case T-13/99 *Pfizer Animal Health SA* v. *Council of the European Union*, para. 148.

[229] *Ibid.*, para. 144. It should be noted that the Communication actually leaves the door open to the setting of zero-risk levels in stating on p. 13 that '[precautionary measures] must not aim at zero risk [...] However, in certain cases, an incomplete assessment of the risk may considerably limit the number of options available to the risk managers'. This enigmatic formulation indicates that the more limited the risk assessment is and, conversely, the greater the uncertainty, the closer to zero risk the risk levels are allowed to be set. It could be argued that limited scientific data alone should not be a cause for higher risk levels, but that the seriousness of the perceived risk must also be weighed in.

[230] *Ibid.*, para. 149.

distinction. It appears that the CFI here blurs the distinction between what is traditionally seen as risk assessment (scientific evaluation of the data) and risk management (setting the appropriate risk levels).[231] As was stated above,[232] it is not helpful for the clarity of complex decisions in cases of great uncertainty when the terminology is not stringently upheld and is allowed to indicate that political considerations may be a part of what should be a scientific assessment.[233]

As the Court turned to the scientific assessment, which it held must be 'as thorough as possible',[234] the systematic approach of the Court can be questioned. The Communication on the Precautionary Principle holds that scientific evaluation is made up of four parts: hazard identification, hazard characterisation, appraisal of exposure and risk characterisation.[235] As mentioned above, the Court does not test the institutions against these requirements, but is satisfied that the institutions have based their decision on 'sufficiently reliable and cogent information' and that they understood 'the ramifications of the scientific question raised'.[236] It would have been helpful if the Court had created strict formal requirements on how the scientific evaluation is to be carried out. This is something that must have been within the Court's reach without stretching the boundaries of the limited judicial review it has afforded itself. However, it should be borne in mind that the Court appeared to have considered the SCAN opinion, which, at least to some extent, was the basis for the final decision, to live up to the Communication requirements, at least in

[231] In *Ibid.*, para. 151 the Court states that: 'Therefore, determining the level of risk deemed unacceptable involves the Community institutions in defining the political objectives to be pursued under the powers conferred to them by the Treaty.' Thus it would be clear that what the CFI here means is a political consideration which should be found in the risk management stage rather than in the risk assessment stage.

[232] See in relation to the *Monsanto* case above.

[233] As was argued in Chapter 2, scientific assessments are also coloured by the appointments of scientists, their individual convictions, the reigning scientific paradigms etc. However, in order to be as objective as possible, science should strive to rid itself of concerns of a non-scientific nature and therefore it is important that decision-making procedures formally separate what are scientifically based decisions from policy considerations.

[234] T-13/99 *Pfizer Animal Health SA* v. *Council of the European Union*, para. 162.

[235] COM/2000/1, p. 13 and Annex III.

[236] T-13/99 *Pfizer Animal Health SA* v. *Council of the European Union*, para 162. For further discussion on the failure of the Court to test the scientific evaluation against the guidelines laid down in the Communication (including the choice to be guided by the diverging Danish scientific opinion rather than the SCAN), see Vos, 'Antibiotics, the Precautionary Principle and the Court of First Instance', 3–6.

spirit.[237] Clarity on this point is, however, an essential requirement on which to build a systematic and science-based application of the precautionary principle.[238]

With regard to the invoking of the precautionary principle, it appears that this can be done even when scientific evidence shows that the risk is unlikely to occur, or that a different interpretation of the scientific evidence gives a more probable outcome than that proposed by the party invoking the measures. In the *Pfizer* case the Standing Committee on Animal Nutrition concluded that the substance in question, virginiamycin, did not pose an immediate risk to public health in Denmark.[239] In short, the Commission's own risk assessment declared the risk needed for the triggering of precautionary measures to be less than probable. However, it also stated that uncertainty still persists in this field, and this is what the Commission, and later the CFI, considered to be crucial.[240] It was thus not sufficient for the SCAN to have rejected the Danish assessment, on which Denmark had based its original request, and described it as 'misleading',[241] 'made on the basis of a single unsubstantiated statement',[242] 'unsound and without foundation'[243] and 'contains no evidence that existing therapies are likely to be compromised'.[244] Instead, the CFI found the Commission's decision to be justified in the light of the remaining scientific uncertainty.[245] It held that a situation in which the precautionary principle applies by definition coincides with the lack of scientific certainty and that a risk assessment cannot be required to

[237] See, for example, T-13/99 *Pfizer Animal Health SA* v. *Council of the European Union*, paras. 194 and 209.

[238] The failure of the ECJ to identify what is required by risk assessments preceding precautionary action has been criticised by other commentators; see for example de Sadeleer, 'The Precautionary Principle in European Community Health and Environmental Law: Sword or Shield for the Nordic Countries?', p. 26.

[239] Opinion of the SCAN on the immediate and longer-term risk to the value of Streptogramins in Human Medicine posed by the use of virginiamycin as an animal growth promoter (produced at the request of the Commission in response to the action taken by Denmark under a safeguard clause to ban viginiamycin as feed additive) of 10 July 1998.

[240] Case T-13/99 *Pfizer Animal Health SA* v. *Council of the European Union*, paras. 220–32.

[241] Opinion of the SCAN, comment to conclusion 4.

[242] *Ibid.*, comment to conclusion 5.

[243] *Ibid.*, comment to conclusion 6.

[244] *Ibid.*, comment to conclusion 8.

[245] *Ibid.*, final Conclusion II: 'it is of the opinion that a full risk assessment cannot be made until quantitative evidence of the extent of transfer of antimicrobial resistance from livestock sources is obtained and the significance of this within the overall use of antimicrobials for clinical and non-clinical purposes evaluated.'

give conclusive evidence in one way or another.[246] However, the scientific uncertainty surrounding virginiamycin is not absolute. In its report, the SCAN held that it 'firmly' believes that the risk will not materialise in the time it would take to reach full scientific certainty.[247] Thus, the risk does not, in the eyes of SCAN, appear to be acute, but, rather exist in the distant future. In the light of this, it is difficult to see how this risk cannot be viewed as leaning towards hypothetical, something which both the Commission[248] and the CFI[249] have pointed out is not sufficient to enable the precautionary principle.[250] The conclusion of this must be that the proof of scientific uncertainty required for invoking precautionary measures in the *Pfizer* case was even lower than in the case law under Article 36 TFEU (formerly Article 30 EC) described above.

It should be pointed out that the European Free Trade Association (EFTA) Court has made a somewhat more stringent interpretation of what is required for the triggering of precautionary measures. In *EFTA Surveillance Authority* v. *Norway*,[251] the Court stated that a mere lack of perceived nutritional need for a product in a state's population did not suffice to justify trade restrictions.[252] It has to be shown that the measure is motivated by the wish to protect human health. Such a risk to human health must be concrete and cannot be based purely on hypothetical or academic considerations.[253] Like the Commission's Communication, the Court stressed the importance of risk assessment when taking measures to protect human health, and this should include risk identification and an evaluation of the risks.[254] In order to trigger precautionary measures in cases of uncertainty, the EFTA Court positions itself closer to the ECJ in the *BSE* case.[255] The conclusions of the EFTA Court have since been

[246] Case T-13/99 *Pfizer Animal Health SA* v. *Council of the European Union*, para. 142.

[247] Opinion of the SCAN, final Conclusion II: 'SCAN is therefore firmly of the opinion that any risk that might be posed in the future by the use of virginiamycin as a growth promoter will not materialise in the time required to make such an evaluation and most probably not for some years afterwards.'

[248] COM/2000/1, p. 13.

[249] Case 13/99 *Pfizer Animal Health SA* v. *Council of the European Union*, para. 143.

[250] At least, this would be true for the time it would take to finalise the risk assessment. The SCAN Opinion here stopped short of calling the risk 'hypothetical' in that timeframe but said that the risk 'will not materialise in the time required to make such an evaluation and most probably not for some years afterwards'.

[251] Case E-3/00 *EFTA Surveillance Authority* v. *Norway* [2000–2001] EFTA Court Report 73.

[252] *Ibid.*, para. 29. [253] *Ibid.*, para. 29.

[254] *Ibid.*, para. 30.

[255] Or at least established a 'likely considerable harm', which has been argued by some to be close to the probability requirement in the *BSE* case (see Bergkamp, 'Understanding the Precautionary Principle (Part II)', 68).

echoed by the ECJ in a number of more recent judgments under Articles 34 and 36 TFEU (formerly Articles 28 and 30 EC).[256]

The Court of First Instance in the *Pfizer* case separated the triggering of a risk assessment from the triggering of precautionary measures. This corresponds to the systematic approach in the Communication on the Precautionary Principle. In order to trigger a risk assessment the requirements are rightly low. When the CFI turned to the question of what level needs to be shown in the scientific evaluation, it was satisfied with the fact that scientific uncertainty persisted in relevant aspects of the case. The degree and type of this uncertainty was not investigated by the Court. It also declined to turn the formal requirements of the Communication into binding case law. The result of this reluctance on behalf of the Court indicates that the CFI is satisfied with the party wishing to invoke precautionary measures showing the existence of an uncertain risk. However, no thresholds or benchmarks have been established against which this can be measured and each situation is examined on a case-by-case basis.

The EU Courts do not require a link to be established between the scientific evidence presented and a particular risk. Instead, it is sufficient to show that an area is surrounded by scientific uncertainty. However, as discussed above in Chapter 2, complete certainty is unattainable and scientific certainty, generally interpreted as corresponding to 95 per cent certainty, is an extremely high level of proof for a challenging party to meet.[257] The approach assumed by EU Courts is in sharp contrast to how WTO Panels and the AB have approached the issue. Recognising that precautionary measures must be allowed, Panels will investigate scientific evidence to establish a link between the scientific evidence and the risk. This link does not have to be certain, but it needs to be supported by 'sufficient scientific evidence'.[258]

The approach taken by the European Courts, particularly in the *Pfizer* case, has been rightly criticised by Karl-Heinz Ladeur along the same lines as outlined above.[259] The primary problem resulting form the current case law is that the wide discretion for the EU institutions and the Member States in cases of uncertainty leads to situations in which decisions – in practice – become impossible to appeal successfully. As a function of this, the discretion of the EU institutions or the Member States

[256] Case C-192/01 *Commission* v. *Denmark*, para. 54, Case C-24/00 *Commission* v. *France*, paras. 59 and 60 and Case C-41/02 *Commission* v. *Netherlands*, para. 69.

[257] See Sections 2.3.1. and 2.3.2

[258] See, above, Sections 3.3.2.4. and 3.3.2.5.

[259] Ladeur, 'The Introduction of the Precautionary Principle into EU Law', 1455.

becomes virtually boundless, and effective redress for affected parties becomes almost impossible.[260]

In some cases relating to Articles 34 and 36 TFEU (formerly Articles 28 and 30 EC), the Courts may be beginning to recognise the dangers of too loose a requirement on scientific evidence, and too much influence of policy judgements on issues of risk regulation. In his Opinion in *Commission* v. *Netherlands*, Advocate General Poiares Maduro held that the discretion afforded to Member States under Article 34 EFEU (formerly Article 30 EC) 'as regards recourse to the precautionary principle is increasingly restricted the further they depart from scientific analysis and the more they rely on policy judgment'.[261] Unfortunately, this statement was not included in the final judgment of the Court, which, nevertheless, concluded that the Netherlands had not provided 'any scientific studies showing that any intake over the recommended daily allowance of any of the six nutrients in question, regardless of by how much, entails a real risk for public health'.[262] The measures were, therefore, not considered to be justified by Article 34 TFEU (formerly Article 30 EC).

In some other recent cases, however, the ECJ has omitted in its entirety the assessment of the scientific basis for national measures. In two landmark decision relating to the scope of the principle of free movement of goods, as established in Article 34 TFEU (formerly Article 28 EC), the ECJ went to great lengths to answer the highly anticipated question whether restrictions on the use of a product are caught by the free movement provisions. In *Commission* v. *Italy*, the ECJ carefully reasoned why a ban on motorcycle trailers had to be considered a measure having the equivalent effect of a quantitative restriction under Article 34 TFEU (formerly Article 28 EC).[263] However, the ECJ did not discuss the scientific evidence underlying the Italian prohibition before it declared that it could be justified in order to ensure road safety.[264] Similarly, in *Mickelsson and Roos*, a preliminary ruling concerning restrictions on the use of personal watercrafts in Sweden, the ECJ appears to accept that some restrictions, although not a total ban, on the use of personal watercrafts can be considered justified, although it leaves the assessment of the proportionality of

260 Ladeur's proposed 'solution' to the problem includes formal requirements on the decision-maker, cost–benefit analyses and priority-setting. These issues are all dealt with below in this text under the appropriate headings.

261 Opinion of the Advocate General delivered on 14 September 2004 in Case C-41/02 *Commission* v. *Netherlands*, para. 30.

262 Case C-41/02 *Commission* v. *Netherlands*, para. 59.

263 Case C-110/05 *Commission* v. *Italy*, judgment of 10 February 2009 (not yet reported).

264 *Ibid*, paras. 60–8.

the measures to the national court.[265] Nowhere in the judgment does the Court discuss on the basis of what scientific evidence such restrictions could be justified. Similarly, in a recent case concerning the provision of online gambling services, the ECJ declared, without mentioning any scientific evidence, that '[g]ames of chance involve a high risk of crime or fraud, given the scale of the earnings and the potential winnings on offer to gamblers'.[266] In all of these cases it is entirely possible that some types of trailers, waterscooters and gambling activities may have detrimental effects and should be regulated. However, by not discussing the scientific basis for the relevant restrictions at all, the ECJ is moving in a dangerous direction, away from allowing for rational and targeted restrictions and toward a laissez-faire approach, which risks leading to a fragmenting of the internal market and increased legal uncertainty.

Finally, it should be mentioned that in *Commission* v. *Sweden* the CFI appears to have considered that once a certain level of proof has been met, the EU institutions not only have the possibility of issuing precautionary measures, but also have an obligation to do so. In this case, which related to the authorisation of the pesticide paraquat under Directive 91/414/EC, the CFI investigated in detail the scientific evidence on which the relevant Commission Decision was based. In a surprising break with previous case law, the CFI held that the responsible scientific committee and the Commission had misinterpreted some of the scientific evidence. For example, it held that a Guatemalan field study whose results had been extrapolated to address the situation in the EU should have been interpreted to indicate that a more precautionary approach was warranted.[267] In this case, the decision of the CFI appears to be based upon the fact that one person in the Guatemalan test showed test results which exceeded the accepted exposure levels. The fact that the Guatemalan study had been considered by the Scientific Committee, and that the Scientific Committee had found it flawed due to the non-adherence of the test subjects to the relevant safety and hygiene provisions, did not convince the CFI.[268] In a similar vein, the CFI reversed the interpretation of a French study of paraquat, which had also been considered by the Scientific Committee and the Commission.[269] Furthermore, the CFI went on to establish that the Commission had not taken all the studies on the effect of paraquat in

[265] Case C-142/05 *Mickelsson and Roos*, judgment of 4 June 2009 (not yet reported).

[266] Case C-42/07, *Liga Portuguesa de Futeboll Profissional Bwin*, judgment of 8 October 2009 (not yet reported), para. 63.

[267] Case T-229/04 *Sweden* v. *Commission*, paras. 178–81.

[268] *Ibid.*, para. 180.

[269] *Ibid.*, para. 185.

birds and hares into account, and that the substance did pose a significant risk to animals.[270]

The conclusion of *Sweden* v. *Commission* is that the CFI found that if a certain level of proof was met, the Commission must act in a precautionary manner. Interestingly, it appears that the CFI in the case is willing to review the scientific evidence on which a decision to authorise a product is based in order to establish if such a level has been met.

4.3.6 *The burden of proof*

Generally, when Member States wish to maintain or introduce measures to protect the environment or health under Article 36 TFEU, the burden of proving the necessity of such measures rests with them.[271] The fact that this is also the case in situations in which the precautionary principle is concerned has been confirmed by the ECJ in a number of cases.[272] In the rulings of the ECJ, it is emphasised that 'real risks' need to be demonstrated in 'light of the results of international scientific research'.[273] Thus, Member States bear the initial burden to show that precautionary measures can be taken under Article 36 TFEU. However, as shown in the previous section, Member States do not need to show a definite link between the evidence and the risk; instead, it is sufficient to show that the area in question is surrounded by scientific uncertainty. The EU Courts will then evaluate the case brought by the Member State. Since a wide margin of discretion is left to the Member States in non-harmonised areas, it is rare that measures taken in cases of scientific uncertainty will be overturned. In the recent case concerning motorcycle trailers referred to above, the ECJ borrowed from its own case law in the field of taxation[274] in order to emphasise that a Member State's burden of proof is far from absolute when it stated that the 'burden of proof cannot be so extensive as to require the Member State to prove, positively, that no other conceivable measure could enable that objective to be attained under the same conditions'.[275]

270 *Ibid.*, paras. 228–42.

271 See, for example, Case 227/82 *Van Bennekom* [1983] ECR 3883, para. 40 and Case 178/84 *Commission* v. *Germany (Reinhieitsgebot)* [1987] ECR 1227, para. 46, and the discussion in Scott and Vos, 'The Juridification of Uncertainty', 261.

272 C-41/02 *Commission* v. *Netherlands*, para. 47, Case C-192/01 *Commission* v. *Denmark*, para. 46 and Case C-24/00 *Commission* v. *France*, para. 53.

273 *Ibid.*

274 Case C-157/94 *Commission* v. *Netherlands* [1997] ECR I-5699, para. 58.

275 Case C-110/05 *Commission* v. *Italy*, para. 66.

In harmonised areas, Member States can seek to invoke the precautionary principle under safeguard clauses or Article 114 (4) and (5) TFEU.[276] Here, too, it is the responsibility of the Member States to show that the measures that they wish to invoke meet the requirements laid down in the articles or relevant safeguard provisions included in legislative acts.[277] Situations in which Member States wish to maintain measures under Article 114 (4) are similar to situations of derogations under Article 36 TFEU.[278] It is likely that the European Courts will show equal deference in judging on the merits situations concerning Article 114 TFEU as they do under Article 36 TFEU, even though a substantial body of case law is lacking so far. On the other hand, situations in which Member States wish to introduce precautionary measures after harmonisation measures have been established are likely to be scrutinised in more detail. This is due to the fact that the grounds on which Member States can rely are more limited under Article 114 (5),[279] and, particularly, that Member States – according to the requirements in the article – have to support measures with 'new scientific evidence' which shows that the risk the Member State wishes to protect itself from is country-specific. In *Austria* v. *Commission* the CFI made it clear that it agreed with the Commission that the proof provided by the Member States would have to 'cast doubt' on the assessment performed at EU level, and also state why the specific Member State in question would be more vulnerable to the risk in question.[280] The reasoning indicates that, in order to invoke the precautionary principle under Article 114 (5) TFEU, something more than scientific uncertainty would have to be shown by the Member States in order to derogate. This also appears to be the case when Member States wish to invoke safeguard clauses included in harmonising measures. In *Monsanto* the Court emphasised the temporary nature of safeguard measures and held that, in order to trigger them, Member States would have to provide 'detailed grounds' indicating a 'specific risk'.[281]

[276] For a discussion on the distinction between the different requirements in 95 (4) and (5) see Case C-3/00 *Denmark* v. *Commission* and Case C-512/99 *Germany* v. *Commission* [2003] referred to above, as well as the discussion of the cases in Zander, 'The Green Guarantee in Article 95 of the EC Treaty', 65.

[277] Case C-3/00 *Denmark* v. *Commission*, para. 84, and Joined Cases T-366/03 and T-235/04 *Austria* v. *Commission*, para. 63.

[278] With the addition that Member States can rely on the grounds included in Article 30 EC, as well as on grounds relating to protection of the environment or working environment.

[279] Only environment and working environment.

[280] Joined Cases T-366/03 and T-235/04 *Austria* v. *Commission*, paras. 65–8.

[281] C-236/01, *Monsanto Agricoltura Italia and Others*, paras. 108–10.

As was shown above, the level of proof that the EU institutions require when invoking precautionary measures varies on a case-by-case basis. The Court has concluded that broad discretion rests with the institutions not only when choosing measures, but also when deciding on when to invoke them.[282] The ECJ and the CFI have, in several cases, held that, in cases of scientific uncertainty, they are not going to 'second-guess' the actions of the other institutions. The requirement for invoking precautionary measures seems to stop at the prima facie existence of scientific uncertainty. The challenging party is then forced to end the scientific uncertainty and show the non-existence of the risk in order to avoid regulation. This is furthermore complicated by the reluctance of the Court to investigate the scientific evidence upon which the measures are based. That this is the case is supported in detail in the case law of the CFI.

A particular illustration of this interpretation can be found in the *Pfizer* case. Here, the CFI established that a low initial burden of proof lies with the party wishing to invoke precautionary measures. In order to fulfil this, the party needs to demonstrate the existence of a non-hypothetical risk in a field shrouded by uncertainty.[283] Since the Court finds its own judicial review limited, it considers itself prohibited from investigating the matter on the merits of the scientific evidence upon which the measure is based. As discussed above, the Commission's scientific assessment in the *Pfizer* case was carried out by the SCAN. Even though recommending that precautionary measures were not necessary, it also stated that uncertainties with regard to the effects of virginiamycin persisted. The Danish report, on which the Danish authorities relied, went on to emphasise the risks of virginiamycin. These conclusions were refuted by the SCAN.

However, the ECJ considered the Danish report on virginiamycin to show sufficient risk, and the SCAN opinion to show sufficient uncertainty, for precautionary measures to be justified. When challenging these measures, the burden of proof was shifted onto Pfizer to show the non-existence of the risks suggested in the Danish report. The CFI indicated that the level of proof that Pfizer needed to meet was to 'prove conclusively' the non-existence of the risk.[284] It follows logically that the only way to prove something conclusively in the face of uncertainty is to end the scientific uncertainty. Only then does conclusive proof exist in one direction or

[282] See, above, Section 4.2.5.3.

[283] *Ibid.* This is also supported in the Commission's Communication on the Precautionary Principle, p. 16.

[284] Case T-13/99 *Pfizer Animal Health SA* v. *Council of the European Union*, para. 392.

another. Pfizer had, earlier in the case, held that such an interpretation would lead to a situation of a *probatio diabolica*.[285] Thus, it did not endeavour to end the scientific uncertainty, but rather to show that the scientific evidence brought forth by the Danish authorities was 'very unlikely' to be correct and that other 'plausible explanations' existed.[286] This line of argument was based on the findings of the SCAN, and amounted to Pfizer arguing that its interpretation was more probable than that of both the Danish authorities and the Commission. However, this was not sufficient for the CFI.

Given the low levels of proof required by the CFI for the invoking of precautionary measures, and the reluctance of the Court to evaluate scientific data, the only possibility for Pfizer to show manifest error on the side of the Commission was to end the scientific uncertainty. This means that the burden of proof was shifted onto Pfizer, and the level of proof needed was raised to the impossible.[287] Like the SCAN pointed out, a full risk assessment of the effects of virginiamycin on human health was underway but it was time-consuming.[288] Thus it was impossible for Pfizer to rely on a full risk assessment at the time of the case and successfully challenge the decision of the Commission.

The above conclusions of the CFI created a situation of total discretion for the Commission when deciding on precautionary measures. The Court clearly declared that it will neither 'assess the merits of either of the scientific points' nor 'substitute its assessment for that of the EU institutions'.[289] But the CFI maintained that risk and uncertainty must be present for precautionary measures to be introduced. In order to establish this, a limited assessment on the merits of the scientific evidence must surely be performed?[290]

What, then, would be the difference if the CFI had increased the level of proof to requiring the Commission to make a case for the existence of a probable link between the evidence and the measures? First of all, as shown above, the probability requirement was used by the ECJ in the *BSE* case and by the EFTA Court, so it would not mean that the CFI was introducing an alien concept to the precautionary principle

[285] *Ibid.*, para. 130. [286] *Ibid.*, para. 392.

[287] See discussion on the impossibility of conclusive evidence above in Sections 2.3.1 and 2.3.2.

[288] Opinion of the SCAN, final Conclusions II.

[289] Case T-13/99 *Pfizer Animal Health SA* v. *Council of the European Union*, Note 393.

[290] For a discussion on the paradoxical use of scientific evidence in the *Pfizer* case see van Asselt and Vos, 'The Precautionary Principle and the Uncertainty Paradox'.

discussion.[291] If the CFI had, indeed, considered a requirement based on probability, it would have forced the Commission to reason why it had decided to disregard the SCAN Opinion and show that its alternative risk assessment somehow put the SCAN Opinion into question. In this case, the Communities' own risk assessment, as carried out by the scientific committee, severely discredited the Danish report. Nevertheless, the Danish report was interpreted as providing sufficient evidence for the Commission to invoke the measures.

One problem with a strict interpretation of the probability requirement is that, if it were to be made the standard, it might possibly hinder the effective use of the precautionary principle. In order to respond to situations of grave risks and sudden crises, it could be argued that a lower level of proof would be more appropriate. Nevertheless, once the burden of proof had been shifted in the *Pfizer* case, it would have been more appropriate if Pfizer had been faced with the task of making its claim more probable, instead of showing conclusively that the Commission was at fault. This would have forced the CFI to investigate scientific evidence in no greater detail than it already does, but it would force the party wishing to invoke precautionary measures to focus on demonstrating a link between the evidence and the risk, rather than merely showing that the issue is surrounded by uncertainty. It is clear from case law stretching back as far as the *Melkunie* judgment that the ECJ already examines scientific evidence to establish the existence of risk and scientific uncertainty, regardless of its limited judicial review.[292]

As discussed extensively above in Chapter 2, the concept of scientific uncertainty probably covers the majority of situations. In fact, not many situations can be described as meeting the 95 per cent certainty threshold generally accepted as scientific certainty. By allowing for any type of regulation in cases of uncertainty, regardless of the fact that a challenging party may show that different conclusions are more probable

[291] To which the CFI even alludes in para. 143 of Case T-13/99 *Pfizer Animal Health SA* v. *Council of the European Union*. It would thus appear that it considers it at least of interest for the development of the law in this field.

[292] Case 97/83 *Melkunie*, para. 15 where the Court holds that: 'it appears first of all from the documents before the court that the presence of active coliform bacteria in a milk product means that the there is a risk of pathogenic micro-organisms being present and is therefore a direct indication that the product may be a source of real danger to human health.' This conclusion could not have been reached without examining the scientific evidence ('the documents before the court'). For a more recent example, see the reasoning of the CFI in the *Pfizer* case largely accounted for above.

on the basis of the same scientific evidence, the parties are effectively hindered from successfully challenging measures affecting them.

A movement in the direction of decreasing the level of proof when challenging a precautionary measure from 'showing conclusively' a fact, to making it more probable, would be a step towards affording real judicial protection and adequate safeguards for parties affected by precautionary measures. It is also likely to decrease the risk of arbitrariness and inconsistency in the application of the precautionary principle which, despite the Commission's claims to the contrary, continue to tarnish the reputation of the principle.

Finally, it should be pointed out that, in cases where a challenging party does not manage to show that its interpretation of a set of facts is more probable, the measure suggested by the EU institutions should be upheld by the Courts.

4.3.7 Cost-benefit analysis and proportionality

Having discussed what the EU judiciary requires for the triggering and invoking of precautionary measures, it is time to turn to what it requires from the measures per se.[293]

The Commission stresses in its Communication on the Precautionary Principle that precautionary measures should be proportional and based on a cost-benefit analysis. The proportionality requirement has a long history in the EU,[294] and is often divided into two parts.[295] The first part relates to whether a certain level of protection is proportional in relation to the risk it aims to avert, while the second part deals with whether the measure itself is proportional to the aim of ensuring the said level of protection. The Commission's Communication on the Precautionary Principle only deals with the second aspect of the proportionality principle,[296] and the case law in relation to the precautionary principle has also been primarily concerned with considering cost-benefit analysis and the proportionality in relation to the actual measures, and not whether the aims of the measures are proportional

[293] In this regard, see also de Sadeleer, 'The Precautionary Principle in European Community Health and Environmental Law: Sword or Shield for the Nordic Countries?', pp. 37–41.

[294] It is also included in Article 5(3) of the EC Treaty.

[295] Jans, 'Proportionality Revisited', 239, and N. McNelis, 'EU Communication on the Precautionary Principle', *JIEL*, 3:3 (2000), 545.

[296] McNelis, 'EU Communication on the Precautionary Principle', 550.

in themselves.[297] This has been justified by the fact that, as discussed above, the EU Courts apply a limited review of measures enacted by the EU institutions using their discretion. If they were to deal with the first tenet of the principle of proportionality, they would effectively judge on the appropriateness of a particular level of protection, something they have tried to avoid as far as possible.[298]

The reason behind making all trade-restrictive measures subject to a proportionality requirement is the importance afforded to the free movement of goods within the EU. Any limitations of that freedom have to be necessary. In order to determine whether measures are necessary to protect health or environment, they must be balanced against the restraints that they impose on the free movement provisions. Some authors have claimed that a strict application of this reasoning by the ECJ would lead to free movement provisions always prevailing whenever they come into conflict with environmental and health concerns.[299] However, an analysis of the case law of the ECJ does not support this assertion.[300] Instead, Jans claims that it appears that the nature of the interest to be protected is relevant to how the Court determines the necessity of a measure.[301] Since protection of health and the environment are considered to be essential objectives of the EU, the Court is more inclined to look favourably upon restrictions imposed on free movement based on these principles. Furthermore, when it comes to precautionary measures in the face of uncertainty, the test appears to be tilted in favour of the perceived risk.

Jans has identified three cumulative parts of the Court's assessment in relation to the proportionality of a measure: (1) Is the measure suitable? (2) Is the measure necessary? (3) Is the measure proportional *strictu*

[297] For an in-depth discussion on the general development of the principle of proportionality see E. Ellis (ed), *The Principle of Proportionality in the Laws of Europe* (Oxford University Press, 1999), and Jans, 'Proportionality Revisited'.

[298] See for example the reasoning of the ECJ in *Monsanto Agricoltura*, where it states that it finds its performance a proportionality test confined to checking measures against *ultra vires* and manifest error. C-236–01 *Monsanto Agricoltura Italia and Others*, paras. 134–8.

[299] Krämer, *EC Environmental Law*, p. 78.

[300] See, for example, Case C-320/93 *Ortscheit* [1994] ECR I-5243 para. 16, in which the Court held that the health and life of humans rank foremost among the Article 36 (now Article 30) exemptions from the free movement provisions. More recently, in Case T-13/99 *Pfizer Animal Health SA* v. *Council of the European Union*, the CFI stated that the least health concerns always take precedence over economic concerns, which approach is also supported in the Communication on the precautionary principle. This would imply that the scales have tipped in the other direction.

[301] Jans, 'Proportionality Revisited', 246.

sensu?[302] With the third aspect, Jans refers to the balancing of interests between the positive and negative impacts of a disputed measure on intra-EU trade. Such an analysis, the Commission emphasises, is not to take purely economical considerations into account, but also to consider non-economic concerns and the acceptability of a measure to the public.[303] The Commission, however, seems to indicate in its Communication on the Precautionary Principle that the cost–benefit analysis is something that takes place outside of the proportionality test.[304] It should be borne in mind that the proportionality principle is a complicated notion, and that the separation of its different aspects is an abstraction. This division often collapses in the practice of the EU Courts, since they employ the division in some cases and disregard it in others. In the *Pfizer* case the CFI loosely followed the structure advocated by Jans, in that it separated the 'inappropriateness' of the measure',[305] 'the duty to take other, less onerous, measures',[306] the comparison between the objective pursued and the right to property[307] and 'errors in the cost benefit analysis'.[308] However, in dealing with each sub-heading, the Court overlaps and continually refers to the comparison of costs and benefits and the proportionality under every heading.

In addition, when the Court rules on the proportionality of precautionary measures, it also considers itself restricted by the limits it has imposed on its judicial review.[309] Within this confinement, the Court has stated its intent to establish whether the measure introduced is the least onerous one available to the decision-maker.[310] As already mentioned, the Court finds itself restricted to investigating whether the decision by the EU institutions is vitiated by a manifest error.[311]

In the *Hi-Tech* case the ECJ held that the fact that one product is singled out for regulation, while other products with similar risks attached to them are left unregulated, is not necessarily disproportionate. This is allowed if the function of the regulated product can be carried out by 'effective substitutes' and the functions of the similar products cannot.[312]

[302] *Ibid.*, pp. 243–9. [303] COM/2000/1, pp. 18–19.

[304] *Ibid.*, p. 17, where it lists the 'general principle' of 'examination of the benefits and costs of action or lack of action' separately from the 'general principle' of proportionality.

[305] Case T-13/99 *Pfizer Animal Health SA* v. *Council of the European Union*, para. 413.

[306] *Ibid.*, para. 440. [307] *Ibid.*, para. 451.

[308] *Ibid.*, para. 463.

[309] See Case C-331/88 *Fedesa and others*, para. 14 and Case C-180/96 *BSE*, para. 97.

[310] Case C-331/88 *Fedesa and others*, para. 13.

[311] Case C-331/88 *Fedesa and others*. para. 14 and Case C-180/96 *BSE*, para. 97.

[312] Case C-341/95 C-341/95 *Gianni Bettati* v. *Safety Hi-Tech*, para. 55.

In a similar situation in the *Toolex* case, the Swedish authorities had left a system in place for the approval of the banned substance in cases where no other effective substitutes might be available. The ECJ considered this to be sufficient for the requirements of proportionality to be fulfilled.[313] It has been pointed out that it is essential that such a system is open to judicial review by national courts in order for it to be effective.[314] Furthermore, precautionary measures are allowed to be discriminatory towards one Member State, if it can be justified that the risk in question stems only from this particular Member State.[315]

In the *Pfizer* case the CFI specifically turned its attention to the interpretation of the cost–benefit analysis mentioned in the Communication on the Precautionary Principle. The Court stated that it considered cost–benefit analysis to be a 'particular expression of the principle of proportionality in cases involving risk management'.[316] As mentioned above, the Court, in contrast to the Commission, does not view the cost–benefit analysis as something independent from the proportionality requirement. It should be noted that one of the reasons for the Commission to stress the use of cost–benefit analysis was to counter criticism from trading partners that claimed that the current use of the precautionary principle did not take economic concerns sufficiently into account.[317] The inclusion of the cost–benefit requirement in the Communication on the Precautionary Principle may thus aim less at introducing a new legal concept and more at clarifying that the EU also intends to weigh the pros and cons included in risk regulation. The cost–benefit analysis viewed in this fashion could be reminiscent of what Jans calls the proportionality principle *strictu sensu*.[318] However, as shown above, the concept of cost–benefit analysis is usually understood to be a wider concept, which aims to estimate the general efficiency of a measure, rather than provide a mere measure of how certain regulation would have an impact on trade.[319]

If the intention of the Commission was to relieve some fears in relation to measures adopted under the auspices of the precautionary principle, the Court effectively undermined this in the *Pfizer* case. By declaring that health concerns always take precedence over economic concerns, the

313 Case C-473/98 *Kemikalieinspektionen* v. *Toolex*, paras. 47–8.

314 Scott and Vos, 'The Juridification of Uncertainty', p. 262.

315 Case C-180/96 *BSE*, para. 115.

316 Case T-13/99 *Pfizer Animal Health SA* v. *Council of the European Union*, para. 410.

317 For a general discussion on cost–benefit analysis see, above, Section 2.3.2.

318 Jans, 'Proportionality Revisited'.

319 See, above, Section 2.3.2.

Court retorted to rather careless rhetoric.[320] This statement by the CFI can be criticised on several grounds.

First, it is quite clear that, in practice, health concerns do not always take precedence over economic concerns. Society allows for a magnitude of high-risk activities that would be banned if only the health risks attached were considered, and not the benefits associated with them.[321] For example, in his opinion on two cases relating to a ban of the Swedish tobacco product 'snus' Advocate General Geelhoed pointed out that a total ban on smoking would be preferable from a health perspective. However, such a ban 'would lead to unreasonable compliance costs' and the appearance of an 'illegal market'.[322] Clearly, the Advocate General was here alluding to cost–benefit considerations to justify the inaction on smoking with the high costs associated with strict regulation. In the following paragraph, however, he stated explicitly that:

> when it comes to products of small economic importance, [the Commission] can abstain from a difficult cost–benefit analysis and can restrict itself to an assessment of the health effects of the intended measure.[323]

Thus, he seems to imply that a cost–benefit analysis does not need to be performed where the costs associated with regulation are low. Despite this statement, the comparison of costs and benefits is absolutely central to his opinion. In its preliminary ruling, the ECJ did not discuss the issue of the comparison of costs and benefits, it merely considered that, in order to achieve a high level of protection, the EU institutions were acting within their discretion when banning the relevant product.[324]

Second, the question may be raised as to what the use of comparing costs and benefits actually is, if it is clear from the outset that one side will always prevail? The purpose of a cost–benefit analysis is generally understood to be an objective comparison of the pros and cons of the proposed regulation. It has been argued, even by critics of the precautionary

[320] Case T-13/99 *Pfizer Animal Health SA* v. *Council of the European Union*, paras. 456 and 471.

[321] As way of example, activities such as driving, alcohol, cigarettes and sports could be mentioned.

[322] Opinion of the Advocate General delivered in Cases C-434/02 *Arnold André GmbH & Co. KG* v. *Landrat des Kresies Herford* [2004] ECR I-11825 and C-210/03 *Swedish Match AB and Swedish Match AB UK Ltd* v. *Secretary of State for Health* [2004] ECR I-11893, para. 62.

[323] *Ibid.*, para. 63.

[324] Case C-210/03 *Swedish Match AB and Swedish Match AB UK Ltd* v. *Secretary of State for Health*, para. 56.

principle, that cost–benefit analysis is a tool for the decision-maker to use in order to obtain a more complete picture of what is involved in a precautionary decision. The result of the analysis should not necessarily be applied mechanically to the decision on the action to be taken.[325] Instead, it can be used as a tool to allow the regulator to decide on the trade-offs he or she wishes to make, and act accordingly.[326] Thus the cost–benefit analysis is conceived as an analytical tool, rather than as a tool for reaching a certain decision. In order for this analytical tool to be helpful to the decision-maker, it is of essence that the interests to be compared are afforded as real a value as possible, rather than weighted in one direction in order to provide the desired result. In affording health concerns absolute importance, the Court is unfortunately not depicting the complete reality of policy choices made in either the EU or in the Member States. Resources in society are finite and, in order to allocate them appropriately, some form of analysis of costs and benefits must be made.[327] In the same vein, Sunstein has stated: 'No sensible regulatory program […] can be indifferent to cost.'[328]

In order to address these issues, the Commission has, since 2002, begun to place greater emphasis on the performance of impact assessments before issuing regulatory measures.[329] It has since been emphasised that such analyses are a fundamental part of the Commission's better regulation agenda.[330] In 2005 the Commission published Guidelines for

[325] See, for example, Sunstein, *Risk and Reason*, pp. 104–5.

[326] *Ibid.* To illustrate the flaws of a purely mathematical application of weighing costs and benefits, Sunstein uses the example of the whether one would rather lose $20 right away or suffer the 1 per cent possibility of losing $1,980. Statistically the latter choice is somewhat 'less bad', but most people would rather accept the limited certain loss than risking the loss of a greater sum. If a binding cost–benefit analysis were applied, the latter choice would be applied unequivocally.

[327] For a discussion on this see, for example, J. S. Applegate, 'Worst Things First: Risk, Information, and Regulatory Structure in Toxic Substances Control', *Yale J. on Reg.*, 9 (1992), 277.

[328] C. Sunstein, *After the Rights Revolution: Reconceiving the Regulatory State* (Harvard University Press, 1990), p. 90.

[329] Communication from the Commission on Impact Assessment, COM/2002/726, adopted 5 June 2002. For a discussion on the development of an impact analysis in the EU see R. Löfstedt, 'The Swing of the Regulatory Pendulum in Europe: From Precautionary Principle to (Regulatory) Impact Analysis', *Journal of Risk and Uncertainty*, 28:2 (2004), 237. See also, R. Löfstedt, 'The Plateau-ing of the European Better Regulation Agenda: An Analysis of Activities Carried out by the Barroso Commission', *Journal of Risk Research* 10:4 (2007), 423.

[330] Communication from the Commission to the Council and the European Parliament, Better Regulation for Growth and Jobs in the European Union, COM/2005/97 adopted 16 March 2005.

the Performance of Impact Assessments, which held that all items on the Commission's Work Programme, including regulatory proposals, White Papers, expenditure programmes and negotiating guidelines for international agreements, should be preceded by impact assessments.[331] The Guidelines include general considerations to be taken into account when determining the impact of the proposed regulation, but stop short of establishing a concrete system for comparing regulatory alternatives.[332] They do, however, establish a set of basic questions, minimal analytical standards and a common reporting format, and also provide some guidance regarding how to carry out quantitative economic analyses, including discount rates and the value of a statistical life.[333] Cost–benefit analysis and cost-effectiveness analysis are defined and briefly discussed in the annexes to the Guidelines.[334] However, in the text of the Guidelines, multi-criteria analysis, which does not aim to monetise costs and benefits but merely to identify their existence, is highlighted. Cost–benefit and cost-effectiveness analyses are presented as 'alternative approaches'.[335]

Evaluations of EU impact assessments have frequently criticised the Commission for not taking adequate account of the costs and benefits of the regulation in question.[336] A study of 111 EU impact assessments published in December 2007 found that impact assessments had become more informative since 2003, but that important information was still lacking. Primarily, this gap was considered to be due to an inadequate quantification of the benefits and various analytical assumptions.[337]

In conclusion, cost–benefit analysis, as put forth by the Commission in the Communication on the Precautionary Principle and interpreted by the

331 Impact Assessment Guidelines, SEC/2005/791 adopted 15 June 2005, p. 6.

332 For discussions and critique of the Commission's impact assessments see F. Chittenden, T. Ambler and D. Xiao, 'Impact Assessment in the EU', in S. Weatherill, *Better Regulation* (Hart Publishing, 2007), pp. 271–86, and A. C. M. Meuwese, 'Inter-institutionalising EU Impact Assessment', in S. Weatherill, *Better Regulation* (Hart Publishing, 2007), pp. 287–309.

333 Annexes to Impact Assessment Guidelines, pp. 37–42.

334 *Ibid.*, 42.

335 *Ibid.*, 39, and footnote 45.

336 See, for example, Danish Environmental Assessment Institute, *Getting Proportions Right: How Far Should the EU Impact Assessments Go?* (Environmental Assessment Institute, 2006), which found that quantification of data had only been properly carried out in 19 of the 58 impact assessments that had been studied. See also, A. Renda, *Impact Assessment in the EU: State of the Art and the Art of the State* (Center for European Policy Studies, 2006).

337 C. Cecot, R. Hahn, A. Renda and L. Schrefler, 'An Evaluation of the Quality of Impact Assessment in the European Union with Lessons for the US and the EU', AEI-Brookings Joint Center for Regulatory Studies, Working Paper, December 2007, pp. 11–12.

ECJ and the CFI, at least partly seems to be a particular manifestation of the proportionality principle already in effect in European law. As such, it primarily addresses the proportionality of measures contra their effects on intra-EU trade. Other interpretations of the concept of cost–benefit analysis are broader. They stress that the analysis is to be used as a tool to estimate the level of protection that every unit of money buys. In doing so, each measure can be analysed and measured not only against its estimated effect on trade, but also against other measures in the same area and in related areas.[338] Different risk-reducing actions and strategies may then be constructed and ranked according to importance. Such an interpretation goes beyond what the Court seems to find in the proportionality principle. However, since 2002 the Commission has increasingly used regulatory impact assessments (RIA) when elaborating regulatory policy. Most commentators appear to agree that more emphasis needs to be placed on the comparison of costs and benefits when such assessments are carried out.

Furthermore, it appears that a challenge based on the inappropriateness of measures, an erroneous application of the proportionality principle or a flawed cost–benefit analysis are likely to be equally as difficult to make before the ECJ. When it comes to which measures to choose, the Court is also reluctant to criticise assessments made by the European institutions and the Member States using their discretion.

4.3.8 The targeting of risks

In this section the attention is turned to the issue of how much leeway the Member States and the EU institutions enjoy when deciding on which risks to address. Society can never deal with all risks, so regulators inevitably have to make choices as to which risks to target, as well as how to target them. As has been stated repeatedly above, in the EU the Member States and the EU institutions enjoy considerable freedom to take decisions under scientific uncertainty. Generally, what the Member States are required to show in order to take precautionary measures in non-harmonised areas under Articles 34 and 36 TFEU is that the risk that they are addressing is surrounded by a degree of scientific uncertainty. Neither the European Commission nor the European Courts make any requirements on the Member States to target risks comprehensively or efficiently.[339] One

[338] See, above, Sections 2.3.2 and 2.3.3.

[339] This is hardly remarkable when considering the limited role played by cost–benefit analyses and the principle of proportionality in areas surrounded by scientific uncertainty.

example of this is that the ECJ and the EFTA Courts have consistently held that Member States are allowed to prohibit an uncertain risk in one set of products, without having to deal with the existence of the risk in other products, as long as they abide by the principle of proportionality.[340] Thus, it appears that Member States are free to choose to target one source of a risk and leave other sources of the same risk unregulated.

With regard to the situation in harmonised areas, one case in which a Member State wished to maintain protective measures under Article 114 (4) TFEU indicates that the European Commission and the ECJ will make higher demands on consistency in such situations. In *Denmark* v. *Commission*, Denmark wanted to maintain stricter measures in relation to sulphites than those provided for by the relevant EU legislation.[341] The Commission did not allow this derogation, since it considered that Denmark had not justified its wish to allow for sulphites in 16 out of the 61 areas provided for in the harmonising directive, and that the Danish measures would, therefore, be inconsistent.[342] The Commission especially stressed the arbitrary nature of the Danish measures in relation to the fact that Denmark had taken no measures to address sulphites in wine, which appeared to be one of the major sources of sulphite intake in the Danish population. Denmark, instead, wished to address the risk of sulphites by restricting their use in other products.[343] It thus appears that neither the European Commission nor the ECJ considered the 'cumulative effect' doctrine that it had previously elaborated on under Article 36 TFEU (formerly Article 30 EC).[344] The situation in *Denmark* v. *Commission* was similar to that of the *Pfizer* case, in which the Commission chose to combat one source of the risk of resistance to virginiamycin, but left another source unregulated. However, the results in the two cases were diametrically opposed. Moreover, the review of the ECJ is also different in *Denmark* v. *Commission* in comparison to that of the CFI in the

[340] Case E-3/00 *EFTA Surveillance Authority* v. *Norway*, para. 20; Case C-192/01 *Commission* v. *Denmark*, para. 50; Case 41/02 *Commission* v. *Netherlands*, para. 50. In these cases the ECJ and the EFTA Court allowed for the considering of 'cumulative effects' of the use of nutrients and vitamins when issuing precautionary measures under Articles 28 and 30. Thus, a nutrient or vitamin could generally be allowed on the market of a Member State, but import of a specific type of product containing the nutrient or vitamin could be restricted if it is found that the total intake in the population would lead to uncertain risks. In Case 178/84 *Commission* v. *Germany* (*Reinheitsgebot*), the ECJ stated that a total ban on additives in beer could not be considered to fulfil the requirement of proportionality under Article 36 EEC (now Article 36 TFEU) (paras. 44–53).

[341] Case C-3/00 *Denmark* v. *Commission*.

[342] *Ibid.*, para. 86.

[343] *Ibid.*, paras. 89 and 90.

[344] See above, footnote 329.

Pfizer case. Where the Court in *Pfizer* had hidden behind its limited judicial review, the ECJ actively condoned the Commission Decision in the *Denmark* v. *Commission* case:

> the presence of large amounts of sulphites in wine cannot justify, under the procedure laid down in Article 95(4) EC, a general prohibition on the use of sulphites as additives in foodstuffs. To the extent that an applicant Member State requests authorisation to maintain national provision derogating from Directive 95/2 in respect of certain foodstuffs, it has the task of justifying those national provisions in relation to those foodstuffs, and not in relation to other products.[345]

In relation to the actions of the EU institutions, the EU judicature has stressed that it considers its judicial review limited in such a way that it may not 'second-guess' the relevant decision-maker when choosing to regulate risks.[346] That this is the case when deciding upon whether to issue precautionary measures under scientific uncertainty was discussed in detail earlier in this chapter. As long as the regulated risk matches an objective found in the TFEU, the regulator is free to act, on condition that respect is maintained for other provisions provided for in the Treaty, notably the proportionality principle.[347]

In the *Hi-Tech* case it was claimed by Hi-Tech Security that the Commission had chosen to consider only one of many aspects when it banned the product that the company manufactured. If the Commission had taken other ways in which similar substances posed a threat to the environment into account, it would have found that the use of other substances involved greater risks.[348] Since the Commission chose to regulate one substance but left other, more dangerous, substances unregulated, the action did not aim to protect the environment as a whole, but to protect the environment against one particular substance alone.[349] The Court found this to be fully in line with Article 191 TFEU (formerly Article 130r EEC).[350] The conclusion of the *Hi-Tech* case indicates that the ECJ focuses on what the Treaty enables the regulator to do, and not on the effectiveness of the measures introduced. If the aim is to protect the environment, the Court is not going to investigate whether the introduction of the measure does so in a coherent or effective manner. One could argue that such an

345 Case C-3/00 *Denmark* v. *Commission*, para. 89.

346 Case T-13/99 *Pfizer Animal Health SA* v. *Council of the European Union*, paras. 169 and 323.

347 Case C-341/95 *Gianni Bettati* v. *Safety Hi-Tech*, para. 41.

348 *Ibid.*, para 38.

349 *Ibid.*, para. 39.

350 *Ibid.*, para. 41.

interpretation is necessary to ensure that the political control of risk regulation remains with democratically accountable institutions. However, two other examples will show some serious problems with regard to consistency and arbitrariness, not to mention effectiveness, with such a system.

In the *Pfizer* case the Commission revoked the authorisation of virginiamycin as a growth promoter on the basis that the use of the product risked leading to resistance in humans against antibiotics of the streptogramin class.[351] *Pfizer* held that this measure was manifestly inappropriate in that it singled out virginiamycin as a growth promoter, where the real problem with resistance stems from the over-prescription of antibiotics in human medicine.[352] Here, the Council decided to target a small risk, the existence and realisation of which was surrounded by a great amount of uncertainty. At the same time, it left the great and unambiguous risk of the over-prescription of antibiotics in human medicine untargeted. The Court also acknowledged this:

> It appears from the documents before the Court that there is broad consensus among experts that this phenomenon is primarily caused by the excessive and inappropriate use of antibiotics in human medicine.[353]

However, the Court did not find that this prevented the Council from issuing the debated measures. Instead, it held that the issuing of the ban of virginiamycin was one way of addressing the risk, and did not prevent the Council from taking 'further measures', that is combating the over-prescription, at a later date.[354] It is interesting to note that this is a development since the *Hi-Tech* case. Where the Court in *Hi-Tech* held that the institutions are free to choose among several risks which to target,[355] it held in *Pfizer* that, also among several sources of one risk, the institutions are free to decide which ones to address.[356] Effectiveness with regard to

351 Council Regulation 2821/98 amending, as regards withdrawal of the authorisation of certain antibiotics, Directive 70/524/EEC concerning additives in feedingstuffs [1998] OJ L 351/4.

352 Case T-13/99 *Pfizer Animal Health SA* v. *Council of the European Union*, para. 414.

353 *Ibid.*, para. 34. 354 *Ibid.*, para. 419.

355 In the *Hi-Tech* case, the ECJ allowed for the targeting of the substance based on its ozone depletion potential (ODP). The Council did not have to take other risks common in several similar substances into account, i.e., atmospheric lifetime (ALT) or global warming potential (GWP). If this had been done, Hi-Tech claimed, other substances would have posed a greater cumulative risk. See Case C-341/95 *Gianni Bettati* v. *Safety Hi-Tech*, para. 38.

356 The risk stemming from virginiamycin and over-prescription of antibiotics is the same. Both may lead to resistance of streptogramin-type antibiotics.

lowering the overall risk level is not an issue that needs to be taken into account. If it were, measures restricting a low, uncertain risk would not be enacted, while a scientifically certain risk of a much greater magnitude was left unregulated. Later, the CFI further emphasised and clarified that it considered any such efficiency control to lie outside the scope of its judicial review:

> As regards the last aspect, the Court is empowered only to examine whether the opinion contains a statement of reasons from which it is possible to ascertain the considerations on which the opinion is based, and whether it establishes a comprehensible link between the medical and/or scientific findings and its conclusions.[357]

Thus, the Court rests when the link has been established, and no comparison with similar situations will be allowed to influence the interpretation of the appropriateness of the measures.

It must be stressed that this development provides little in the way of consistency and opens the door to arbitrary regulations being passed without any possibility for judicial protection for the affected party. From a regulatory point of view, what the European institutions provide is far from effective risk regulation. If, in the *Pfizer* case, they had seriously intended to combat the very real and immediate risk of resistance in humans towards antibiotics, they should have aimed to regulate the 'main offender', that is over-prescription in human medicine, instead of aiming for the easy target of growth promoters.[358]

Another issue of great significance when using the precautionary principle to target a certain product or set of products is whether the proposed regulation does, in fact, lead to the emergence of other risks. When the precautionary principle is used to enable regulation under uncertainty, but its application does not require consideration of the overall risk level, some controversial results may emerge. As an example of this, it has been shown that the EU regulation of phthalates in children's toys led to the emergence of other substitute products, of which very little was known.[359] Thus, the health benefits brought by the ban on phthalates were offset by the emergence of other health risks.

In summary, it appears that the EU institutions can target risks at will, and that the ECJ will not investigate the consistency or effectiveness of

[357] Case T-74/00 *Artegodan* [2000] II-04945, para. 200.

[358] For a general discussion on the comparison of regulatory alternatives when issuing precautionary measures see above, Section 2.3.3.

[359] Duriodé, 'Plastic Panics', pp. 140–66.

those measures. The same may be true for Member States that wish to invoke the Article 36 TFEU exemptions. However, Member States who wish to derogate from harmonisation measures must show a consistent and effective risk policy and the ECJ will materially review these measures. On the face of it, such disparity seems difficult to reconcile with the requirement of a consistent application of the precautionary principle as advocated in the Commission's Communication on the Precautionary Principle.[360]

The rationale behind the stance of the Courts is, however, not difficult to understand, and has more to do with the politics of European integration than with effective risk regulation: derogations from harmonisation measures threaten the uniform application of the internal market and must be interpreted restrictively.[361] The work of the European institutions, on the other hand, is often subjected to political considerations and sensitive compromises, which the judiciary does not wish to upset. Consequently, it limits its review in these cases. However, these pragmatic concerns aside, consistency and efficiency in risk regulation should be a minimum requirement on the Member States as well as at EU level.

In the Opinion by Advocate General Geelhoed referred to above, the issue was raised of whether the EU was allowed to target certain tobacco products with a ban (in this case, the Swedish product 'snus') while allowing the smoking of cigarettes.[362] It appeared in the case that 'snus' posed a lesser evil to smoking by a ratio of somewhere between 1 to 10 and 1 to 1,000.[363] Despite this, the Advocate General found that the banning of 'snus', while leaving the more hazardous activity of smoking subject to less strict regulation, to be justified. The Advocate General supported this with the fact that the economic costs of banning 'snus' would be much less than banning smoking. He also referred to the long and ambitious European policy towards tobacco products in general and the fact that he believed that 'snus' may work as a stepping-stone towards smoking.[364]

In its preliminary ruling in *Swedish Match AB and Swedish Match AB UK Ltd* v. *Secretary of State for Health*, the ECJ essentially upheld the reasoning of the Advocate General and stated that, due to the fact that

[360] COM/2000/1, p. 18.

[361] See, for example, Opinion of the Advocate General delivered on 30 May 2002 in Case C-3/00 *Denmark* v. *Commission*, para. 76.

[362] Opinion of the Advocate General delivered on 7 September 2004 in Cases C-434/02 *Arnold André GmbH & Co. KG* v. *Landrat des Kresies Herford* and C-210/03 *Swedish Match AB and Swedish Match AB UK Ltd* v. *Secretary of State for Health*.

[363] *Ibid.*, para. 48.

[364] *Ibid.*, paras. 52, 62 and 63.

'snus' has properties detrimental to human health, the ban was justified.[365] Thus, the Court agreed with the Advocate General that the EU institutions can choose to target one source of a risk, and are not obliged to assume a consistent policy towards the same risk stemming from different sources. Once again, the conclusion must be that, when targeting risks, the institutions are not bound to target them in an effective or comprehensive manner.

4.4 Conclusions

The purpose of this chapter has been twofold. First, it has described the development of precautionary thinking and the precautionary principle in EU law. In the main, this is a story of the growing attention that environmental and health concerns received during the last decades of the twentieth century, and especially since the mid-1970s until the present time. But it also makes it clear that precautionary considerations date back to a time long before the precautionary principle came into vogue. It has been shown that precaution was the basis for regulation as far back as the 1960s, even though it was not mentioned explicitly. During the 1970s newfound environmental concerns meant that precaution began to be mentioned explicitly in Environmental Action Programmes and other instruments of soft law character. In the second half of the 1980s, sustainable development and the precautionary principle began to become concepts to be considered in legislative form. In 1992, the same year in which the Rio Declaration was agreed upon, the precautionary principle became a part of the EC Treaty. During the second half of the 1990s, the precautionary principle became a much used and hotly debated instrument in the wake of the *BSE* crisis and other highly publicised food and health crises. Now, in the first decade of the twenty-first century, precaution seems to have matured into the precautionary principle and begs to be applied as a legal instrument.

This leads to the second, and more important, part of this chapter: What does the precautionary principle entail and how is it to be applied and challenged before the European Courts? To investigate this issue, the attention was turned to the European Commission's Communication on the Precautionary Principle and to the case law before the ECJ and the CFI. Having found the Communication to be vague and without either a

[365] Case C-210/03 *Swedish Match AB and Swedish Match AB UK Ltd* v. *Secretary of State for Health*, paras. 52–8.

definition or unequivocal guidelines for the practical use of the precautionary principle, the main attention was focused on the case law.

The Courts, recognising the controversial nature of risk regulation at EU level, have largely applied a hands-off approach to cases relating to the precautionary principle. A lot of leeway has been granted to the Member States and the other European institutions when applying the principle. The definitions provided for in the *BSE* and the *Greenpeace* cases have been echoed by the Courts in subsequent case law. However, the two definitions differ on important issues. Whereas the *BSE* definition merely establishes that Member States may take protective measures before the reality of risks are apparent, the *Greenpeace* case holds that such matters must be provisional and backed up with 'justifiable reasons'. In limiting its judicial review to dealing with the form of a decision, rather than the substance of the same, the Court effectively prevents itself from providing more detailed definitions.

The ECJ and the CFI have consistently held in their case law that the precautionary principle is to be applied under uncertainty, provided that a 'non-hypothetical' and 'science-based' risk is present. This implies that a threshold of scientific evidence needs to be met in order to invoke precautionary measures. However, in practice the limited review carried out by the Courts means that the party, Member State or EU institution invoking the precautionary principle does not need to show a link between the scientific evidence presented and the proposed risk. It is sufficient for the specific area to be surrounded by varying degrees of uncertainty. It appears that the Courts are at their most deferential in cases in which Member States invoke the precautionary principle under Article 36 TFEU and when the EU institutions issue precautionary regulations using the powers which they have had delegated to themselves. The Courts will make higher requirements on Member States which wish to derogate from existing harmonisation measures under safeguard procedures and Article 114 (5) TFEU.

Once the Court has established the existence of scientific uncertainty, the burden of proof is shifted onto the regulated party. Since the Court finds its judicial review limited to ruling on errors in form and manifest error, the level of proof needed to challenge the original decision is very high. In fact, the only way to show manifest error on the part of the regulator under conditions of uncertainty is to prove the regulator wrong by ending the uncertainty. In most cases involving deep uncertainty, this is not possible, and thus a successful challenge against precautionary measures often appears futile.

Since unequivocal rules for what level of scientific evidence is required for precautionary measures to be invoked are lacking, and the possibilities for an affected party to seek redress are severely limited, the EU institutions enjoy a near absolute discretion when issuing precautionary measures. The same is true for Member States acting under Article 36 TFEU. The Courts have allowed for this by restricting their judicial review to an absolute minimum, instead of taking a more active role in evaluating the scientific evidence brought before it. Here, the European courts differ greatly from the deep review performed within the framework of WTO dispute settlement. In direct contrast to the above, the judgment of the CFI in *Sweden* v. *Commission* entailed a detailed review of the scientific evidence underlying the decision to authorise the pesticide paraquat. Based on this assessment, the CFI overturned the decision of the Commission, which had authorised the pesticide. It remains to be seen what the implications of the judgment are for the future of the judicial review of precautionary measures in the EU.

The European Courts have also proved deferential in relation to the quality of the measures introduced. In the Commission Communication on the Precautionary Principle it is established that cost–benefit analysis is to be an integral part of precautionary decision-making. However, this requirement is neither binding nor elaborated upon. In the case law this requirement appears, for the most part, to be substituted by a traditional proportionality test. However, whereas the proportionality test aims to test the effects of regulation on intra-EU trade, a cost–benefit analysis is generally perceived as having a wider scope. In assigning values to all aspects of regulation, the total costs and benefits can be compared. This includes effects on trade, but the primary focus is on how many lives every monetary unit saves. If cost–benefit analysis were elaborated on and integrated into EU risk regulation, it would provide an invaluable tool for the decision-maker when deciding upon precautionary measures.

The same is true for risk trade-off analysis, which so far has not been elaborated upon at EU level. Every risk that is mitigated leads to the emergence of other risks. Taking this into account when deciding on risk policies is essential and can help to avoid situations such as that which arose from the regulation of PVC plastics.[366]

Science, despite its shortcomings, remains the basis upon which precautionary legislation must be founded. Nothing except science can provide us with something which resembles objective information with regard

[366] Duriodé, 'Plastic Panics'.

to the nature of uncertain risks. In the European legal order, the independence of science has been safeguarded by the separation of risk assessment and risk management. In order to maintain a systematic approach to risk regulation, such an approach must be maintained. Only then does it become clear what constitutes scientific considerations and what constitutes political considerations, and only then can regulatory regimes be understood and explained. In the EU risk assessment is primarily carried out by scientific committees consisting of independent experts. The exact effects of their advice remain unclear. In the case law of the ECJ and the CFI, it appears that the EU institutions are not bound by their advice in the least, as long as they provide a very rudimentary reasoning in order to explain why they do not wish to follow the advice. Member States, on the other hand, can rely on the opinions of the scientific committees in order to avoid being bound by harmonisation measures.

In summary, the application of the precautionary principle as an independent legal instrument in the EU is still in its infancy. However, its development has so far been somewhat worrying: it is unclear what is required for the invoking of the principle, which, in turn, leads to a limited redress for regulated parties; the burden and standard of proof are not clearly defined; and the consequences of regulation are not appropriately considered. In order for the precautionary principle to be successfully used and justified in the international arena, it needs to find its place in a more systematic framework. An improved and stricter model for regulatory impact assessment needs to be established. Such a model should be built on risk assessment, cost–benefit analysis and risk trade-off analysis. This will show the state of the risk and the consequences of regulation. This RIA can then be presented to the decision-maker. In the decision-making phase, the general public should be involved and have full access to the results and methods behind the RIA.[367] Only when such a systematic approach is in place can risk regulation in general, and the precautionary principle in particular, be successful and reach its aim of protecting the environment and the public from uncertain risks.

[367] See, for example, Löfstedt, 'Risk Communication and Management in the Twenty-First Century', 335–46, in particular, at 340–1.

5

The precautionary principle in Sweden

5.1 The development and status of the precautionary principle in Sweden

5.1.1 *Introduction*

Precautionary considerations have a long tradition in Sweden. Some commentators have traced a 'precautionary principle' as far back as the 1940s in some areas of Swedish law.[1] Since the late 1960s something resembling a modern precautionary principle guiding all environmental and health regulation has been in effect. Despite this, the precautionary principle, and risk regulation in general, has attracted little attention from scholars in the area. One reason for this is that nature and health protection is historically an uncontroversial issue in Sweden. The culture of spending free time in the forests and on the coast is very strong. The Nordic countries are the only ones in the world which sustain the concept of 'allemansrätt', which means that everyone enjoys a total freedom of access to nature. The fact that Swedish decision-making is traditionally consensus-based may also have been an influence. This is further emphasised by the unique political stability that Sweden has enjoyed, which has enabled the government to pursue a steady and continuous dialogue with industry.

Even though the precautionary principle rarely functions as an individual legal principle in Swedish law, but, instead, functions together with other legal norms, an attempt is made here to isolate the influence and the meaning of the principle. The same structure as was followed when discussing the precautionary principle in EU and WTO law is followed

[1] G. Michanek, 'Chapter 7: Sweden', in N. de Sadeleer (ed.), *Implementing the Precautionary Principle: Approaches form the Nordic Countries, EU and USA* (Earthscan, 2007), p. 121. For an outline of earlier Swedish regulation of hazardous substances, see Karlsson, 'The Precautionary Principle, Swedish Chemicals Policy and Sustainable Development', 344–7.

here. Thus, the constituent elements of the principle, such as its definition, the role of scientific evidence and judicial review, and the burden of proof are examined and discussed separately. Preparatory work to legislation is an important source of Swedish law and relevant parts are dealt with in detail. The scarce case law that exists has been identified and analysed. Special emphasis has been put on cases from the new system of environment courts. Finally, the views of legal commentators are accounted for and discussed.

5.1.2 Background and development

Rudimentary precautionary reasoning in Swedish legislation has been traced back as far as the Swedish Water Code of 1941.[2] However, explicit reference to a 'precautionary principle' as such was never made in Swedish law. Despite this, an essentially precautionary approach has guided all Swedish environmental and health regulation since the Swedish Environmental Protection Act was introduced in 1969.[3] Its Article 5 held:

> Anyone who takes or intends to take an environmentally harmful action shall take the safety measures, accept the limits put on the enterprise, and observe the general precautionary measures that can reasonably be demanded to prevent or repair any detriment. The responsibility to remedy detriment remains also after the action has occurred.
> The extent of the requirements, according to the first paragraph, is determined by the basis in what is technically feasible for an enterprise of the kind in question and with regard to public, as well as private, interests.[4]

The wording of the article does not make reference to scientific uncertainty. However, Article 5 was, to a certain degree, elaborated on in the

[2] G. Michanek, 'Att väga säkert och vikten av att säkra', in E. M. Basse *et al.* (eds.), *Fågelperspektiv på rättsordningen: Vänbok till Staffan Westerlund* (Iustus Förlag, 2002), p. 74 and note 23.

[3] Miljöskyddslag (SFS 1969:387).

[4] 'Den som utövar eller ämnar utöva miljöfarlig verskamhet skall vidtaga de skyddsåtgärder, tåla den begränsning av verksamheten och iakttaga de försiktgihetsmått I övrigt som skäligen kan fordras för att förebygga eller avhjälpa olägenhet. Skyldigheten att avhjälpa olägnehter kvarstår även efter det att olägneheten upphört.'Omfattningen av åligganden enligt första stycket bedöms med utgångspunkt i vad som är tekniskt möjligt vid versamhet av det slag som är i fråga och med beaktande av såväl allmänna som enskilda intressen. (Translations of Swedish texts in this chapter are by the author unless otherwise stated.)

preparatory work to the Environmental Protection Code.[5] It should be mentioned that preparatory work is an important source for the interpretation of Swedish law, second only to the actual text of the law. Both the courts and the administration make constant reference to the often very detailed reasoning found in Bills (Proposition (hereinafter Prop.)) presented to the Swedish Parliament and expert reports preceding Bills (Statliga Offentliga Utredningar (hereinafter SOU)).[6]

The Minister of the Environment, who presented the Bill which proposed an Environmental Protection Code to the Swedish Parliament, included the following rationale without explicitly mentioning precaution:

> I consider it clear that uncertainty regarding the hazardousness of a substance shall not be to the detriment of the public at large, but should instead be to the detriment of the party emitting the substance into the air or into the water. This fundamental thesis is also of the utmost practical importance. It implies that one does not have to wait for damage to occur before action can be taken. It also implies that the party wishing to emit an insufficiently known substance which one has a reasonable cause to presume to be harmful must be able to show that inconvenience does not need to be feared.[7]

In relation to how science should be taken into account when regulating risks, the Minister of the Environment only pointed out that all action should be based upon current scientific findings and the best technology available.[8]

On the face of it, Article 5, read in the light of the preparatory work, appears to be a manifestation of the strictest form of the precautionary principle. It lays the responsibility for all uncertainty relating to a substance squarely on the side of the party wishing to carry out an activity. It is sufficient that a 'reasonable cause to presume' the realisation of a risk

[5] Proposition 1969:28. Förslag till Miljoskyddslag, p. 210.

[6] For an introduction to the sources of Swedish law, see B. Lehrberg, *Praktisk juridisk metod* (Iustus Förlag, 2001). For a brief introduction to the jurisprudential foundation of the Swedish legal system, see G. MacCormack, 'Scandinavian Realism', *The Juridical Review*, 1970, 33–55.

[7] Proposition 1969:28, p. 210. 'Jag anser det natruligt att den osäkerhet som kan råda rörande ett ämnes farlighet inte skall gå ut over allmänheten utan bör drabba den som släpper ut ämnet I luften eller vattnet. Denna grundsats är också av största praktiska betydelse. Den innebär att main inte behöver vänta med aett ingripande till dess skador har inträffat. Den innebär också att den som vill släppa ut ett otillräckligt känt ämne som man har grundad anledning anta är skadligt måste kunna visa att olägenhet inte behöver befaras.'

[8] Proposition 1969:28, pp. 218 and 219.

exists for this to be the case. Furthermore, the realisation of the risk does not have to bring 'harm' but rather 'inconvenience'.[9] This implies that not only more or less quantifiable degrees of damage are considered, but also varying degrees of nuisance.

However, Michanek has pointed out that the Environmental Protection Act was created on the basic presumption that industry generally brings economic welfare. Thus, the reigning opinion was that industrial activity should, in principle, be allowed and only major environmental or health disturbances should trigger prohibition. In practice, prohibition was very rare and the precautionary approach was, instead, used to minimise the effects of hazardous activities on nature and health.[10] This is also emphasised by the Minister of the Environment in the preparatory work, where he stated that 'all means' must be exhausted with the aim of avoiding inconvenience to industry while maintaining high protection. Examples of compromise solutions include cleansing options for emissions and the recycling of waste. If prohibition is the only remaining option, it has to be made effective after a time period which allows the industry in question to adjust its production methods.[11] It was also emphasised that all production methods, as well as the demands made by the government on industry, had to be based on the best available technology. In considering what this entailed, regard had to be given to what was economically feasible for an individual enterprise.[12] This appears to have been something akin to an objective proportionality principle. The main focus does not seem to have been on the detriment that a certain activity may have caused as compared to the benefits it might have brought. Instead, it centred on the economic conditions of the individual party that wished to carry out an activity. If an enterprise was found to have economic difficulties when fulfilling an environmental norm, exemption from the norm could be granted. The apparent aim and result of this approach was that economically more successful undertakings were judged according to a stricter environmental and health standard than less successful undertakings.[13]

[9] The Swedish word 'olägenhet' must be interpreted to be wider in scope than the word 'skada' (harm). Compare with the wording of Chapter 9, Article 3 of the Swedish Environmental Code, which holds that an 'olägenhet' for human health is anything that may pose a negative effect to human health that is not negligible or temporary.

[10] Michanek, 'Att väga säkert och vikten av att säkra', pp. 74 and 75 and Michanek, 'Chapter 7: Sweden', pp. 121–3.

[11] Proposition 1969:28, p. 216. [12] *Ibid.*, p. 218.

[13] Michanek, 'Att väga säkert och vikten av att säkra', pp. 75 and 76 and Michanek, 'Chapter 7: Sweden', pp. 121–3.

In addition to the proportionality requirement, a cost–benefit analysis was applied in certain individual cases. Where the costs associated with an enterprise abiding by binding environmental norms were excessively high and the benefits small or negligible, exemptions were sometimes granted. The discrepancy between costs and benefits in such cases had to be so great that it could not be considered 'environmentally justifiable'[14] to demand that the higher standard be abided by.[15]

The early version of the precautionary principle or approach in Swedish legislation certainly looks strict on paper. It is important to note that precaution is not singled out and treated as a concept to be applied solely in cases of scientific uncertainty, which the 'modern' version of the precautionary principle mandates.[16] Instead, it appears to have developed as a wider concept which encompassed other aspects of risk regulation such as prevention.

However, the strict statement of the Minister of the Environment in the preparatory works was not upheld in practice, but appears to have served more as a policy than a legally binding norm. It appears that the Swedish government wanted to retain the possibility of applying a strict precautionary approach but also of granting exemptions where it saw fit. The reluctance of the government to bind itself to a coherent and general application of the environmental norms in the Environmental Protection Code was emphasised by the Minister of the Environment in the preparatory work. For example, he explicitly rejected the laying-down of legally binding emission standards. Instead, he favoured a system of recommendations and individual permits, thus maintaining both flexibility in the system and the power in the hands of the Swedish government to decide in each individual case.[17] The power of the government was further emphasised by the fact that the government itself was, in many cases, the highest authority to which administrative decisions could be appealed.[18] This allowed for a limited influence of the courts in the interpretation of

[14] 'Miljömässigt motiverat.'

[15] Michanek, 'Att väga säkert och vikten av att säkra', p. 76. For a more detailed analysis of this very limited practice see S. Westerlund, *Miljöskyddslagen: En analytisk lagkommentar* (Åmyra Förlag, 1990), Section 7.21 and S. O. Svärd and T. Rahm, *Prövning enligt miljöskyddslagen* (Gebers Förlag, 1975), p. 166.

[16] For a discussion on the development of the precautionary principle in international law, see Sections 3.2. and 3.3. For a discussion on the definition of the precautionary principle in EU law, see Section 4.2.5.2.

[17] Proposition 1969:28, p. 219.

[18] This issue is discussed in more detail below in relation to judicial review; see Section 5.1.4.

Swedish environmental law, and allowed the government an almost unlimited flexibility in applying the law.

5.1.3 *The current context and definition of the precautionary principle*

In January 1999 the Swedish Environmental Code came into force.[19] The Code[20] is the first attempt to collect all Swedish environmental legislation into one law. Previously, environmental legislation was not systematically treated as one independent area of the law but, instead, environmental regulations were to be found in separate laws. Fundamental to all environmental laws was the Environmental Protection Act discussed above, but more specific regulations were scattered across the legal system. Topical issues here, such as pesticide regulation and radiation protection, were found in unrelated laws. The result of this was that the environmental legislation became increasingly difficult to oversee. The introduction of the Environmental Code sought to respond to this issue by co-ordinating the legislation, but it also aimed to update, clarify and strengthen relevant parts of the legislation.[21]

The scope of the Environmental Code includes the protection of the environment as well as the protection of human health.[22] Section 2 of the Code establishes the general rules and principles relating to the respect and protection of both the environment and human health. It is interesting to note that already Article 1 of Section 2 establishes that it is the party wishing to carry out an activity that has to show that the rules and principles are abided by. The onus of protecting the environment is thus put squarely on anyone performing an activity.[23]

In Article 2:2 it is specified that anyone intending to carry out an activity must acquire the knowledge needed considering the nature of the activity. The precautionary principle finds its reflection in the successive Article. The first paragraph of Article 2:3 is virtually identical to Article 5 of the Environmental Protection Act referred to above. However, an effort to modernise the language ever so slightly has been made, and the fact that 'best technology available' is to be used when professional activity is

[19] Lag (SFS 1998:808) om införande av miljöbalken, Article 1.

[20] Miljöbalken (SFS 1998:808). [21] See Proposition 1997/98:45 Miljöbalk, p. 1.

[22] Miljöbalken (SFS 1998:808), Article 1:1.

[23] That this is an intentional construction aimed at reversing the burden of proof is emphasised in the preparatory works. See Proposition 1997/98:45, pp. 209 and 650.

concerned has been mentioned explicitly in the text of the law. The first paragraph of Article 2:3 states in its present form:

> Persons who pursue an activity or take a measure, or intend to do so, shall implement protective measures, comply with restrictions and take any other precautions that are necessary in order to prevent, hinder or combat damage or detriment to human health or the environment as a result of the activity or measure. For the same reason, the best possible technology shall be used in connection with professional activities.[24]

In the second paragraph, one sentence has been added to emphasise that it is not just preventive measures that fall under the scope of the Article:

> Such precautions shall be taken as soon as there is cause to assume that an activity or measure may cause damage or detriment to human health or the environment.[25]

In the preparatory work it is pointed out that the aim of this inclusion is to reflect the precautionary principle in the Code explicitly.[26] However, when it comes to fleshing out the content and use of the principle, neither the Code nor the preparatory work offer much assistance. As mentioned briefly above, bills presented by the government to the Swedish Parliament provide not only the text of the law but also explanations and discussions on the purposes behind the legislation. In the case of Proposition 1997/98: 45, the discussion is divided into two parts. The first part is a general discussion on the purposes and principles behind the Environmental Code. The second part is article-specific, and introduces the text of the law followed by an explanation of each article and how it is to relate to the previous legislation. In relation to Article 2:3, the government holds that the aim is to replace Article 5 of the Environmental Protection Act.[27] In substance, few changes are intended, except the fact that the scope of the Environmental Code is wider than the scope of the Environmental

[24] 'Alla som bedriver eller avser att bedriva en verksamhet eller vidta en åtgärd skall utföra de skyddsåtgärder, iaktta de begränsningar och vidta de försiktighetsmått i övrigt som behövs för att förebygga, hindra eller motverka att verksamheten eller åtgärden medför skada eller olägenhet för människors hälsa eller miljön. I samma syfte skall vid yrkesmässig verksamhet användas bästa möjliga teknik.' Official English translation, available at: www.regeringen.se/content/1/c4/13/48/385ef12a.pdf (accessed 1 March 2010).

[25] 'Dessa försiktighetsmått skall vidtas så snart det finns skäl att anta att en verksamhet eller åtgärd kan medföra skada eller olägenhet för människors hälsa eller miljön.' (Official English translation, available at www.regeringen.se/content/1/c4/13/48/385ef12a.pdf (accessed 1 March 2010)).

[26] Proposition 1997/98:45, p. 656. [27] *Ibid.*, p. 653.

Protection Act.[28] Thus, the elaboration on precaution by the Minister of the Environment in the preparatory work to the Environmental Protection Code is still relevant. However, degrees of precautionary action have been introduced to correspond to the varying levels of harm that the different activities bring. In Article 2:9 it is stated that, if it is feared that an activity or measure may lead to 'an injury or nuisance of significant importance', only the government can issue an authorisation, and only where there are 'particular reasons' for the action to take place. If an activity or a measure risks leading to a situation in which 'a large number of people' may see their 'living conditions significantly altered,' it may not be carried out at all.[29] This is what is known as the 'stop-rule'. In the preparatory work, it is indicated that Article 2:9 can only come into play when the other rules in Chapter 2 of the Environmental Code are not sufficient.[30] As examples of when the rule might be employed, the preparatory work offers the building of new roads or the establishment of harbours as examples.[31] This may indicate that the rule is primarily intended for large-scale projects of a planning nature. However, the preparatory work also holds that, when applying the principle, all other Chapter 2 rules must have been exhausted. Here, the substitution principle (or product choice principle) is mentioned, which may indicate that hazardous substances may also be concerned. It should be borne in mind that a large number of people must be affected by the risk for it to be of any relevance. What is clear is that the greater the risks are perceived to be, the stricter the precautionary principle becomes in its application. It could be argued that precautionary action under Article 2:3 of the Environmental Code is aimed at mitigating risk, rather than prohibiting activities associated with it. This is supported by the indicative list of measures which may be considered by the regulator under Article 2:3.[32] If this were the case, and no case law exists to draw the line between the two articles, it would mean that Article 2:3 lays down the basic 'precaution requirement' in Swedish law and functions as a legal basis for setting precautionary standards for an otherwise allowed activity.[33] In order to ban a product or an activity, the requirements of

[28] This is of no particular relevance for the discussion here. The Environmental Protection Act concerned only hazardous activity stemming from the use of the ground, buildings or constructions. In the Environmental Code the source of the disturbance is of no relevance. Thus, for example transport falls under the provision. See *ibid.*, pp. 653–4.

[29] Miljöbalken, Article 2:9, para. 2.

[30] Proposition 1997/1998:45. p. 583.

[31] *Ibid.*, p. 665.

[32] *Ibid.*, p. 656.

[33] It is indicated in the preparatory works that a keyword for what would be allowed under this article is 'best available technology'. If it can be shown that the best available

Article 2:9 would have to be fulfilled, that is, an injury or nuisance of significant importance would have to be present. However, that the line is not so clear-cut is clarified in a case from the Environmental Appeals Court, in which it was held that the 'precautionary principle' allows the Environmental Court of Appeal to ban a practice that it finds to be in violation of the principle.[34] Since Article 2:3, according to the preparatory work, embodies the precautionary principle in Swedish legislation, this implies that Article 2:3 is referred to. Instead, it has been held that Article 2:9 should be interpreted to encompass a definite burden for the proponent of an action to show that the action is without risk when the risk associated with the action is likely to cause significant damage or inconvenience to human health or the environment. If this burden is not met, and even if other protective and mitigating measures have been taken, the action may not be carried out.[35]

The precautionary principle is defined as a broad, individual responsibility to act with care. This includes the fact that it can enable prohibitive action as in EU law or under the SPS Agreement. However, it is not restricted to being used by the regulator in such instances, even though this is what will be focused on here. The notion of precaution is closely connected to other rules in Chapter 2 of the Environmental Code, such as the knowledge requirement, the substitution principle, and the principle of economising the use of natural resources.

5.1.4 *Judicial review*

Since 1999 many decisions relating to human health and the environment have been appealed in a specialised court system. Instead of being brought before the general administrative courts, appeals are raised before a new category of environmental courts (Miljödomstolar).[36] The environmental courts consist of at least one judge and one environmental expert. Complex cases may require an additional judge, an additional expert and up to two advisors with a background in regional issues, industry or the environment.[37] The procedure in the environmental courts follows the general procedure in the administrative courts.[38] In some cases civil suits

technology is being used, the activity should be allowed, unless the requirements for a ban under Article 2:9 are fulfilled. *Ibid.*

[34] Judgment of the Environmental Appeals Court of 12 May 2005, case no. M 3225–04, p. 9.

[35] Michanek, 'Chapter 7: Sweden', pp. 128–9.

[36] Miljöbalken, Article 20:3.

[37] *Ibid.*, Article 20:4 and 20:5.

[38] *Ibid.*, Article 20:3.

may be brought before the environmental courts. In such cases rules relating to civil procedure are applied. The composition of the court remains as described above.

Judgments from the environmental courts can be appealed to the Environmental Court of Appeal (Miljööverdomstolen). Svea Hovrätt, one of the regional appeals courts, hosts this Court.[39] The composition of the Environmental Court of Appeal is made up of at least three judges and one environmental expert.[40] The role of the environmental experts and advisors to the courts is to make a substantial review of complicated technical issues possible.[41] That the environmental courts have an obligation to investigate the facts is stipulated in the Environmental Code.[42] It has been pointed out in the preparatory works that this is to be interpreted as a 'complete obligation'[43] to investigate the facts of a case.[44] This follows the general Swedish rule in administrative court procedures.[45] Based on this, it could confidently be argued that the scientific evidence on which a case is based should be analysed in detail by the Swedish environmental courts. Furthermore, this is underscored by the technical expertise included in the courts' composition. Judgments of the Environmental Court of Appeal relating to administrative decisions by regional governments or government authorities cannot be appealed further.[46]

Despite the 'state-of-the-art' Swedish environmental court system, some important decisions have been exempt from this procedure. Importantly for this research, decisions relating to the approval of pesticides, as well as planning decisions, have, until recently, been appealed to the government in the last instance, using the ordinary administrative courts, rather than in the new courts.[47] With regard to pesticides, the appeal procedure was changed in 2006, and decisions by the Swedish Chemicals Agency (Kemikalieinspektionen (hereinafter KemI)) are now

[39] For an overview of the Swedish court system, see the publication Regeringskansliet. The Swedish Judicial System – A Brief Presentation, published by the Swedish Ministry of Justice. Available at: www.regeringen.se/content/1/c6/01/62/79/023155e4.pdf (accessed 1 March 2010).

[40] Miljöbalken, Article 20:11.

[41] Prop. 1997/98:45, p. 460

[42] For the Environmental court see Miljöbalken Article 22:11 para. 2 and for the Environmental Appeals Court see Article 23:4.

[43] 'Fullständig utredningsskyldighet.'

[44] Proposition 1997/98:45, p. 459.

[45] Förvaltningsprocesslag (SFS 1971:291) Article 8.

[46] Miljöbalken, Article 23:8.

[47] With regard to appeals of decisions by the Swedish Chemicals Agency relating to requests for approval of pesticides see Article 29, Förordning (SFS 1998:947) om

subjected to appeal in the environmental courts.[48] Planning permission, which is highly relevant for the regulation of base stations and is discussed in more detail in the following chapters, is still subject to appeal within the traditional administrative court system.[49] Planning and zoning decisions are subject to more deferential review in accordance with Chapter 10 of the Local Government Act.[50] The often highly political nature of planning and zoning rules means that local government enjoys wide discretion when taking such decisions, and appeals can only be made on the basis of procedural mistakes, ultra vires action on the part of the local government or the illegality of the decision.[51]

The reason for maintaining, until recently, the Swedish government as the highest level of appeal for decisions by public authorities in relation to, for example, pesticides, was that such decisions often entail sensitive political considerations and/or the exercise of powers within a certain margin of discretion afforded to a government authority. In such cases, it was not considered appropriate for courts to 'second-guess' the decisions taken by essentially political bodies.[52] This resembles the reasoning that the ECJ has brought forth in its case law.[53] In the late 1980s, however, the possibility for a limited review of government decisions by the Supreme Administrative Court was created. This review was limited to instances in which citizens' rights or obligations were involved.[54] In the preparatory works to the Act on the Judicial Review of Certain Administrative Decisions, it appears that Swedish courts may go further than the ECJ has allowed itself.[55] Where the ECJ refrains from trying the facts on which a decision is based, the Swedish Supreme Administrative Court (Regeringsrätten) is explicitly authorised to review 'the appraisal of the facts which is the foundation for application

bekämpningsmedel. Planning permission relating to base stations for mobile telephone systems are appealed through the traditional system of administrative courts; see Chapter 13 of Plan- och Bygglag (SFS 1987:10).

48 Article 53, Förordning (SFS 2006:1010) om växtskyddsmedel, which refers to the environmental court procedure in Chapter 19 of Miljöbalken.

49 Plan- och Bygglag, Chapter 13, Article 2.

50 Plan- och Bygglag, Chapter 13, Article 1, which refers to Kommunallagen (SFS 1991:900), Chapter 10.

51 I.e., what in Sweden is known as 'rättsprövning'.

52 Proposition 1987/88:69, pp. 23–5.

53 See, above, Section 4.2.5.3.

54 See Lag (SFS 1988:205) om rättsprövning av vissa förvaltningsbeslut.

55 Even if the preparatory works relate directly to appeals to the Supreme Administrative Court, there is no reason to assume that what is stated there is not analogously applicable to other appeal courts. Nothing indicates that the judicial review by the Regeringsrätten should be construed as wider or narrower than the review of other courts.

of regulation'.[56] Excluded from the judicial review are issues that are of a clearly political nature. But also in such cases the Swedish Constitution, Regeringsformen, stipulates in Article 7:2 that decisions taken by the government must be prepared adequately, and can be overturned if they are shown not to have been. However, it appears that the courts are content with a formal, rather than substantial, review of whether such adequate preparation can be said to have been carried out.[57]

In certain cases, the Supreme Administrative Court can refer an appealed decision back to the responsible authority for review.[58] In relation to an appealed decision stemming from KemI, the Supreme Administrative Court made it clear that the Agency was bound to try the case again, and that it would have to review all the facts that the challenging party had brought forth. The review carried out by KemI was not allowed to be restricted to considering whether the original decision was vitiated by a manifest error,[59] but a full review on the merits needed to be carried out.[60] In a case relating to the expansion of a power station in the north of Sweden, the Supreme Administrative Court also decided to overrule the decision that the government had taken, based on a different interpretation of the facts of the case.[61] However, in other cases the Supreme Administrative Court has held that a delegation of discretionary powers from the government to another authority restricts its substantive judicial review.[62]

The Supreme Administrative Court has showed considerable restraint in reviewing measures and has failed to elaborate on the scope and means of its review in cases in which wider appreciation of the facts has been performed. However, similar restraint has not necessarily hindered other courts. In one of several controversial cases relating to permitting the resumption of the construction of a tunnel in southern Sweden (Hallandsåsen), the relevant environmental court substantially reviewed and amended the initial decision of the relevant regional authority.[63] Upon appeal, the Environmental Court of Appeal

[56] Proposition 1987/88:69, p. 23.

[57] See, for example, the judgment of the Supreme Administrative Court in RÅ 1999 Ref. 76 regarding the highly political decision of when and where to begin the termination of Swedish nuclear plants.

[58] Förvaltningslag (SFS 1986:223) Article 27.

[59] 'Uppenbart oriktigt'.

[60] Judgment of the Supreme Administrative Court in case RÅ 1996 Ref. 51, p. 208.

[61] Judgment of the Supreme Administrative Court in case RÅ 2004 Ref. 14.

[62] See, for example, the judgments of the Supreme Administrative Court in cases RÅ 1999 Ref. 76 p. 430; RÅ 2001 note 6 and RÅ 2003 Ref. 43.

[63] Judgment of 14 February 2002 of the Environmental Court at Växjö Tingsrätt, Case no. M 13/99.

modified certain purely technical and construction methodological requirements that the environmental court had imposed.[64] Thus, both courts showed a clear inclination to review the substance of a decision made by a government authority using its discretion. However, it does not seem to be an unequivocal rule that the environmental courts make full use of their duty of investigation. In one case relating to the uncertain health risks posed by a wood burner in a populated area, the environmental court reversed the ban from the relevant government authority on the grounds that the scientific evidence had not been sufficiently substantiated. This was reversed by the Environmental Court of Appeal, which did not make any effort to resolve the scientific uncertainty. Instead, it held that any ambiguities in the scientific evidence should be to the detriment of the party wishing to carry out an action.[65] In summary, it could be argued that Swedish courts have a relatively far-reaching possibility of performing a deep judicial review. However, in practice it appears that the courts often let pragmatic considerations restrict them, and it appears that, in cases of uncertainty, they are inclined to place the responsibility for resolving uncertainties on the applicant, rather than seeking to resolve them.[66]

5.1.5 Risk assessment and risk management

How scientific advice should be assessed, and the role that risk assessment should play in risk regulation is not generally established in Swedish law. However, a government report stemming from 1993 holds that decisions relating to risks to human health or the environment are to be based on risk assessments.[67] A strict separation between risk assessment and risk management is, nevertheless, not upheld. As a matter of fact, debate on the issue is non-existent in preparatory work and legislation as well as in the case law and the literature.[68]

[64] MÖD 2003:100. [65] Mål no. MÖD 2005:9.

[66] In this regard, it should be emphasised that very few judgments exist in Sweden regarding the content and application of the precautionary principle. Thus, any conclusions on the basis of case law necessarily reflect this. See also, Michanek, 'Chapter 7: Sweden', pp. 133–4.

[67] Miljöbalk – förslag till åtgärder för ett framtida miljöskyddsarbete, SOU 1993:27, p. 201.

[68] See, for example, Michanek, 'Chapter 7: Sweden', p. 133, who holds that risk assessment and risk management should function as separate phases, but does not elaborate on how this should be maintained in practice.

A system of strictly scientific bodies, as found at the European level, is lacking in Sweden. Assessment as well as management tasks are generally entrusted to government authorities or agencies. The authorities relevant for this study are the National Radiation Protection Authority (Statens Strålskyddsinstitut (hereinafter SSI)) and KemI.[69] In the internal workings of the institutes, the border between what are risk assessment choices and what are risk management choices appears to be blurred. In interviews at KemI, it transpired that most authorisation decisions were taken after ad hoc meetings between various civil servants in the agency. Since the burden of information is put squarely on the applicant, the independent investigation of KemI is limited. 'Common sense' plays a big role in the decision-making.[70] In an interview in the Ministry of the Environment, it transpired that it is the Minister of the Environment who decides when scientific research needs to be commissioned or evaluated by a committee. A lot of the day-to-day decision-making is based upon ad hoc contacts with industry and environmental organisations.[71] Somewhat paradoxically, it was pointed out by representatives of KemI that upholding a clear separation between political decisions and scientific evaluation was essential.[72]

In cases of activities for which permits are needed, the requirement of risk assessment appears strict and based on the knowledge requirement in Article 2:2 of the Environmental Code. Thus, it is for the applicant to carry out the risk assessment. What needs to be shown is returned to below.

5.1.6 *Scientific evidence and the introduction of precautionary measures*

It has been argued above that the precautionary principle in Swedish legislation is not exclusively an instrument that the regulator can make use of when a certain set of requirements has been met. Instead, it is a broad, overarching obligation that anyone carrying out activities which may have an impact on the environment or on human health must abide

[69] Their tasks, organisation and composition are laid down in Governant ordinances. See Förordning (SFS 1988:525) med instruktion för Kemikalieinspektionen and Förordning (SFS 2006:524) med instruktion för Statens strålskyddsinstitut. This will be returned to in some more detail in relation to the case studies below.

[70] Interview with officials at KemI, 9 September 2004.

[71] Interview with official at the Swedish Ministry of Environment, 8 September 2004.

[72] Interview with officials at KemI, 9 September 2004.

by. However, in determining what constitutes a breach of the principle, the issues of at what level an obligation to act in a precautionary manner ensues and what needs to be shown to counter a demand for precautionary action needs to be discussed.

The issue of the level of proof required for precautionary action is not established in the Environmental Code or elaborated on in any detail in the preparatory work. In the Government Report, which is part of the preparatory work to the Environmental Code, scientific uncertainty was recognised, but only defined as a situation in which:

> science as of yet has not reached the point where the connection between an action or activity and nuisance or damage to human health and environment can be proven, even if a causal relationship can be assumed to exist.[73]

As noted above, the Minister of the Environment held, in the preparatory work to an earlier version of the environmental legislation, that uncertainty should be to the detriment of a proponent of an activity.[74] In the preparatory works to the Hazardous Substances Act 1973, it was acknowledged that some properties of chemicals are inherently uncertain and it can thus not be deemed appropriate that a manufacturer or importer of a chemical is forced to show 'the complete absence of harm'[75] in a product.[76]

Judgments that directly deal with the issue of level of proof in cases of scientific evidence are lacking. However, in one judgement, the Swedish Supreme Court has stated that, in order to have standing before the court when challenging a hazardous activity, the risk associated with the activity cannot be allowed to be theoretical or entirely insignificant.[77] Whether this only relates to the issue of legal standing in civil suits or whether it could be applied by analogy to other issues of risk is not clear.

Nothing in more recent legislation has been designed to counter the statement by the Minister of the Environment in 1973, but the interpretation of the Environmental Code indicates that a stricter view may be taken on the matter. Article 2:2 of the Environmental Code entails a knowledge requirement. It states that anyone carrying out an activity needs to acquire

[73] 'där vetenskapen ännu inte har kommit så långt att sambandet mellan en åtgärd eller verdsamhet och olägenheter eller skador på människors hälsa och miljön kan bevisas, även om ett orsassamband kan antas föreligga.' SOU 1996:103 Del 1. Miljöbalken: En skärpt och samordnad miljölagstiftning för en hållbar utveckling', p. 239.

[74] Proposition 1969:28.

[75] 'fullständig frånvaro av skadlighet.'

[76] Proposition 1973:17, p. 92.

[77] NJA 2004 s. 590.

the relevant knowledge associated with the activity. The Article applies to anyone carrying out an activity that may have effects on the environment or on human health. However, it has been indicated in the preparatory work that industry might have a more extensive burden than individuals.[78] That this Article functions together with the precautionary principle in Article 2:3 to create a general duty of care provision has been elaborated on by the Environmental Court of Appeal in a few cases.[79]

In a case relating to the use of chemicals in the forest industry, the Environmental Court of Appeal held that the Article, when read in the light of the precautionary principle, led to the conclusion that all chemicals that may pose any detriment to the environment or health should be kept in a register by the user. If uncertainties pertain to the effects of the chemicals used, the party using them is obliged, if necessary, to conduct its own laboratory experiments to establish the effects.[80] This construction creates a situation in which it appears that the company has something resembling strict liability for all chemicals that are not ancillary to its production.[81] The level of proof needed to be met by the company is very high, considering that the chemicals at issue were legally imported or manufactured on the Swedish market. The recognition of scientific uncertainties emphasised in the preparatory works of 1973 are thus played down.[82]

In a more recent case from the Supreme Court, a limit of sorts to the knowledge requirement was elaborated. In the case, the Environmental Court of Appeal had prescribed that the Swedish pharmaceuticals company Akzo Nobel could not use chemicals for which there is a lack of 'documented knowledge concerning detriment for health and the environment based on poor biodegradability, chronic toxicity and potential for bioaccumulation' in its production. On appeal to the Supreme Court, Akzo Nobel argued that such conditions were impossible to fulfil, and that it would not be prudent to regulate matters of product choice in such

[78] Proposition 1997/98:45, p. 210.

[79] That the precautionary principle and the knowledge requirement are intrinsically connected are indicated in the SOU 1996:103, pp. 239–41. The two requirements are here dealt with under the same heading.

[80] See, for example, Judgment of 30 March 2005 of the Environmental Appeals Court in case no. M 9408–03, pp. 10–11; and Judgment of 12 May 2005 of the Environmental appeals court in case no. M 3225–04, pp. 9–10.

[81] In case no. M 9408–03, the Environmental Appeal Court drew a line between chemicals which were to be used in the general production and chemicals of more ancillary or marginal use.

[82] The database was probably required by the court in order to facilitate and enforce the use of the substitution principle, which requires an individual or undertaking to use a less hazardous substance if possible.

a fashion. Furthermore, it claimed that the ruling of the Environmental Court of Appeal would be in breach of Article 34 TFEU (formerly Article 28 EC). In its judgment, the Supreme Court found that the requirements in Chapter 2 of the Environmental Code have, in general, been considered too vague to be coupled with penal sanctions.[83] The conditions that the Environmental Court of Appeal had established, that is, 'documented knowledge', the Supreme Court considered to be 'lacking any precision' and stated that it would be impossible for the complainant to identify what and how the conditions should be met. It concluded that the conditions stipulated by the Environmental Court of Appeal could not be upheld in their present shape, but it did not exclude that they could be changed to be sufficiently precise.[84] In a follow-up case to the case outlined above, the Environmental Court of Appeal found that a specific list of issues that Akzo Nobel needed to be able to verify with relevant documentation was acceptable.[85]

In the literature it has been pointed out that the guidance in Swedish legislation and case law relating to levels of proof in cases of scientific uncertainty is so scarce that it makes any conclusions hypothetical at best.[86] Discussions in the literature have, instead, built on analogous reasoning built on civil or criminal procedure,[87] or tort law.[88]

Nilsson quotes the preparatory works to the Hazardous Substance Act as the best definition of the precautionary principle in Swedish law. From this definition, she concludes that the level of proof required to trigger precautionary measures is that a 'reasonably well-founded suspicion' of a risk should be sufficient to trigger precautionary measures.[89] Lindblom holds that the general rule in the Environmental Code is not that the regulator needs to show suspicion of risk, but rather that a party wishing to carry out an activity needs to show that the risk will not materialise. This indicates that the level of proof needed to be met by the regulator is low and the burden of the regulated party subsequently is high.[90] This is returned to in the following chapter. Westerlund has held that it should never be the case that the regulator would have to show that a risk stemming from a chemical product is more than 50

[83] Proposition 1997/98:45, 524. [84] NJA 2006 s. 310.

[85] MÖD 2009:5.

[86] A. Nilsson, *Att byta ut kemikalier* (Nerenius och Santerus, 1997), p. 258.

[87] P. H. Lindblom, *Miljöprocess Del II* (Iustus, 2002).

[88] B. Bengtsson, *Miljöbalkens återverkningar* (Norstedts, 2001).

[89] 'rimligt välgrundad misstanke.' Nilsson, *Att byta ut kemikalier*, p. 271.

[90] Lindblom, *Miljöprocess Del II*, pp. 493–9.

per cent probable.[91] In conclusion, there seems to be a general consensus among Swedish commentators that the level of proof needed to be met in order to trigger precautionary measures is generally low. This is also supported by the wording of Article 2:3 of the Environmental Code, where it is held that precautionary measures shall be taken on any occasion where there is 'reason to presume' the existence of risk.[92]

5.1.7 The burden of proof

As shown in the previous chapters, the application of the precautionary principle in international, WTO and EC law owe a lot to the definition in the Rio Declaration. The triple negative definition found there states that the lack of scientific evidence is not to mean that precautionary measures cannot be taken where there is a threat of a serious or irreversible nature. This implies that the 'normal' state is that an activity is allowed. Only when a suspicion of risk is raised may precautionary measures be taken.[93] The burden of proof is then shifted, and the proponent of the activity has the onus to show that the risk will not materialise. It appears that, under this definition, precautionary measures are an option open to a regulator, rather than an obligation.[94]

The rules included in the Swedish Environmental Code appear, at first glance, to put the onus entirely onto the proponent of an activity.[95] The applicant is obliged to consider precautionary measures before any activity is initiated. However, as was shown in the previous section, this obligation does not arise until it is 'presumable' that the proposed activity entails a risk.[96] The effect of this is that, when the regulator exercises its power to judge on the admissibility or prohibit an activity or product, it must first show that the risk is presumable. Thus, the burden of proof is not initially reversed, but, instead, rests with the regulator. Once the 'presumable level' has been reached, the burden is shifted onto the proponent of the activity to show that concerns relating to the risk are unfounded.

91 S. Westerlund, *Lag om kemiska produkter: Kommentar* (C. E. Fritzes Förlag, 1985), p. 179.

92 'skäl att anta'. The Swedish word 'anta' [to presume] has the same basis as the word 'antagligt' used as an adjective by Ekelöf.

93 For a discussion on the burden of proof under the WTO regime see above, Section 3.3.2.2 and, in EC law, above, in Section 4.3.3.

94 Even if it appears that the European Court of Justice has sidestepped this notion in the recent Case T-229/04 *Sweden* v. *Commission*, Judgment of 11 July 2007, not yet reported. The issue is returned to below.

95 Miljöbalken Article 2:3, para. 1.

96 Miljöbalken Article 2:3, para. 2.

What looks like a strict obligation to take precautionary measures is, in practice, made contingent upon the meeting of a low level of proof on the part of the regulator. Thus, the 'real', or initial, burden of proof lies with the regulator rather than with the proponent of the activity or product.[97]

The issue that remains to be discussed is thus what actually needs to be shown by the proponent of an activity or a product, once the challenging party has fulfilled its burden of proof. The text of the law does not give any guidance with regard to the level of proof required after the initial burden of proof has been met. The preparatory work to the Environmental Code cites the preparatory works to the 1969 Environmental Protection Act.[98] There it was stated that:

> the party wishing to release an insufficiently known substance of which one has founded reason to presume is hazardous must be able to show that detriment need not be feared.[99]

This wording re-establishes that the proponent of an activity has the burden of proof shifted onto it, once a 'founded reason' has been established, and that it is not sufficient for it to show that the non-existence of the risk is as likely as the existence of the risk. The fact that it must be shown that the realisation of the risk 'need not be feared' suggests that a higher level of proof must be met by the proponent of an activity or a product, once the burden of proof has been shifted. This is further emphasised by the preparatory work to the Chemicals Act 1973. In relation to risks stemming from chemicals, it is held here that it must be shown:

> as far as is possible based on the current state of scientific research that the suspicion is unfounded.[100]

[97] This situation of a low level of proof on one side which triggers a higher level on the other side may lead to the conclusion that the burden of proof rests with the regulated party form the outset. That this is not the case has been elaborated on by Ekelöf (P. O. Ekelöf, *Rättegång* IV (Norstedts, 2009). He has named situations such as this one as subjected to a 'false burden of proof' on the side of the regulated party, whereas the real burden of proof actually needs to be fulfilled by the regulator. That the burden then is shifted onto the regulated party follows normal procedural rules. Once one party has fulfilled its burden of proof, it is for the other party to prove this wrong. What makes the situation precarious is the marked difference in levels of proof required.

[98] Proposition 1997/98:45, p. 209.

[99] Proposition 1969:28, p. 210. 'den som vill släppa ut otillräckligt känt ämne som man har grundad anledning anta är skadligt måste kunna visa att olägenhet inte behöver befaras.'

[100] Proposition 1973:17, p. 97. 'så långt det på vetenskapens nuvarande ståndpunkt är möjligt att misstanken är ogrundad.'

Nilsson sees two ways in which this requirement can be fulfilled: Either the original scientific method used can be discredited and another, more accurate, proposed, or it can be shown that the results proposed in the original research do not adequately depict reality. Thus, what needs to be shown here, according to Nilsson, is that the line of evidence presented by the regulator is flawed, not that a different set of scientific facts solves the issue.[101] It cannot be sufficient to show that other equally possible solutions exist.[102] The fact that this notion is further emphasised if the decision is taken by using the discretion bestowed upon a government authority has been supported in some of the case law.[103] The level of proof needed to be met by the party affected by precautionary regulation is then significant. However, it appears lower than in EC law, where a challenging party needs to show 'manifest error' on the part of the regulator.[104] Lindblom has pointed out that, once the burden of proof has been reversed, the required level of proof is likely to be more stringent in the new Environmental Code than in the previous legislation.[105] Support for this is to be found in the preparatory work.[106] He suggests that the level to be met now would be to prove or verify that a risk will not materialise.[107]

5.1.8 *Cost–benefit analysis*

According to Article 2:7 of the Environmental Code, Articles 2:2 to 2:6 are subject to a reasonableness requirement, with special emphasis on costs

[101] Nilsson, *Att byta ut kemikalier*, p. 272. She here uses the term 'eliminationsbevis', which means literally 'elimination proof'. What needs to be eliminated is the evidence brought forth by the other party.

[102] *Ibid.*

[103] See, for example, the judgment of the Supreme Administrative Court in RÅ 1998 Ref. 56 relating to whether a decision regarding the authorisation of herbicides by KemI was in conformity with EC law; and the judgment of the Supreme Administrative Court in RÅ 2003 Ref. 43, where the issue of the spread of allergens from stables for horses to a neighbouring housing development was under scrutiny. Here, recommendations from the Swedish National Board of Health and Welfare were relied upon by the court, despite the fact that they lacked support in specific scientific evidence. Instead, the recommendations were based on analogous recommendations for smell, noise, flies and other allergies. Older, as well as more recent, albeit inconclusive, scientific evidence found in reports by other government agencies and reputable researchers in the field advocating a less onerous policy were rejected by the court.

[104] See discussion above in Section 4.3.3.

[105] Lindblom, *Miljöprocess Del II*, note 87, p. 494.

[106] Proposition 1997/98:45 Del 2, p. 454.

[107] Lindblom, *Miljöprocess Del II*, note 87, p. 494.

and benefits.[108] These Articles include inter alia the knowledge requirement (Article 2:2), the precautionary principle (Article 2:3), the 'best location' requirement (Article 2:4), the principle of economising (Article 2:5) and the substitution principle (Article 2:6). The stop-rule in Article 2:9 of the Environmental Code is thus explicitly excluded from the scope of any cost–benefit considerations.

The aim of the article is to provide a safety valve for measures for which the environmental benefits do not correspond to the costs associated with them.[109] The article does not establish a general cost-efficiency test for environmental regulation, nor does it function as a traditional proportionality test. Instead, it establishes that the costs incurred may not be 'unreasonable' in relation to the environmental benefits. The article was intended to be a codification of already existing practice.[110] It has been pointed out by Nilsson (in relation to chemicals regulation) that guidance as to how the weighing of interests is to be carried out is, in practice, too scarce to allow for almost any conclusions at all.[111] This rings true also outside of chemicals regulation. Two interesting points have, however, been made by Bengtsson. First, what needs to be considered is whether a measure is 'unreasonable'.[112] This implies a higher level of proof than, for example, 'unproportional'. Second, the burden of showing the unreasonable nature of a measure rests with the party making the claim.[113] The interests included in the balancing act not only include monetary values contra environmental benefits, but also general 'societal interests'.[114] In relation to the demand for best available technology in Article 2:3, the reigning opinion is that it is 'unreasonable' to demand better technology from an undertaking than is economically feasible.[115]

5.1.9 The targeting of risks

As shown above, the Swedish courts have often proved reluctant to 'second-guess' government authorities when acting within a margin of discretion. This makes the issue of challenging a decision upon the basis that it does not follow a rational prioritisation of risks difficult,

[108] For a brief discussion on the cost–benefit balancing in the Environmental Code see Michanek, 'Chapter 7: Sweden', pp. 129–30.
[109] Proposition 1997/98:45, p. 231. [110] *Ibid.*, p. 662.
[111] Nilsson, *Att byta ut kemikalier*, p. 137. [112] 'Orimligt'.
[113] Bengtsson, *Miljöbalkens återverkningar*, p. 70.
[114] 'Samhällsintressen'. Proposition 1997/98 Del 2, pp. 232 and 459.
[115] Bengtsson, *Miljöbalkens återverkningar*, p. 71.

since the decision of which risks to target is eminently a political one.[116]

However, in 1999 the Swedish Parliament ratified a Bill which laid down a set of environmental quality objectives designed to ensure sustainable development.[117] A Council for co-ordinating, following up and informing the public about the environmental quality objectives was established. It maintains a website on which the various ministries and authorities responsible for the different objectives continuously report their progress and the problems they experience in fulfilling the objectives.[118] In April of 2001 the Swedish Parliament ratified a Bill laying down seventy-one concrete partial goals, or benchmarks, and time-frames for the realisation of the objectives.[119] The environmental quality objectives are of a policy character, but provide an interesting insight into the transparent targeting of risks employed by the Swedish government. However, it should be borne in mind that the environmental quality objectives were not chosen upon the basis of a regulatory impact assessment, or upon the basis of any other cost-efficiency modelling, but are political in nature. They also treat risks in absolute, rather than in relative terms. Thus, all the identified risks to human health or the environment should be treated simultaneously, and are not ranked in order of priority.

Of prime interest for this study are objectives Number 4 'A Non-Toxic Environment' and Number 6 'A Safe Radiation Environment'. As stated above, the environmental quality objectives are not ranked from one to fifteen, but are, instead, to be fulfilled simultaneously within one generation.[120] The benchmarks established to enable each objective to be fulfilled are supposed to guide Swedish legislation, assist in the interpretation of the existing legislation and function as guidelines for Swedish negotiations in the international arena.[121] In summary, the environmental quality objectives show a clear, and, in some cases, very detailed, agenda for the targeting of risks in Sweden. The specific goals of the reduction of chemicals and radiation protection will be discussed in detail below.

[116] This is also the case at EC level. See, above, Section 4.3.8.

[117] Proposition 1997/98:145.

[118] Miljömål: Environmental Objectives Portal. Summary of environmental policy goals. Available at: www.miljomal.nu (accessed 1 March 2010).

[119] Proposition 2000/01:130. [120] Proposition 1997/98:145 Part 3, p. 4.

[121] Proposition 2000/01:130, p. 16.

5.1.10 *Conclusion*

The precautionary principle has developed in Sweden in a relatively unchallenged manner. Instead of being singled out as an independent legal principle, it functions together with other legal norms, such as the principle of preventive action, the requirement to economise on the use of natural resources and the 'knowledge-requirement' to create, in many cases, a broad duty of care. However, it is possible to identify instances in which Article 2:3 of the Environmental Code has been instrumental in bringing about decisions based on strictly precautionary considerations.

The precautionary principle in Sweden is a broad and individual requirement directed at all citizens to act in a precautionary manner. It is not presented as just a tool to be used by the regulator in cases of uncertainty. The principle works together with, and inspires, the other norms in Chapter 2 of the Environmental Code. The result of the broad scope of the principle is that, in many cases, it can be viewed more as a general duty of care than the precautionary principle that has been previously discussed at international and European levels.

Even though the judicial review of government decisions in Sweden appears broader than in EC law when considering the reasoning in the preparatory work, courts have often proven reluctant to review substantially decisions taken by agencies and the government using their discretions. However, the new environmental courts appear to be more inclined to perform more detailed reviews of scientific evidence. In several cases, they have based their judgments on a very detailed scientific analysis of evidence put before them.

Sweden does not conduct a clear separation of risk assessment and risk management. In cases in which agencies have had decision-making powers delegated to them, the scientific analyses and the management decisions are often taken by the same people. 'Common sense' plays a major role in the decision-making. The applicant must, in most cases, carry out the scientific assessment. When the government needs general scientific advice, reports are commissioned on an ad hoc basis.

The presumption in Swedish law appears, at first, to be that it is the party carrying out an activity or manufacturing a product that has the burden to show the absence of risk. However, this is a 'false' burden of proof. When issuing a precautionary measure, the initial burden of proof actually rests with the regulator. The level of proof required is low. Some support exists for the notion that it must be based on scientific evidence and not be theoretical. The text of the law, in the light of preparatory work, case law and commentators, seems to put the level needed at 'presumable'. Once this low level of proof has been met by the regulator, it shifts onto the party facing regulation, and is subsequently high. Commentators have held that what is required is to disprove the chain of evidence provided by the regulator, rather than proving the activity or product to be safe. This seems to lead to a situation in which the level of proof required is slightly lower than that required by the ECJ. It probably rests at 'proven, verified'. However, guidance on this issue from the preparatory work, as well as the case law and the literature, is scarce at best. Future research in this area should centre upon investigating the basis of government decisions and judgments from lower courts.

According to Article 2:7 of the Environmental Code, measures issued must be 'reasonable'. The article seems to encompass what courts previously referred to as the requirement that measures must be 'environmentally justifiable', as well as some form of proportionality requirement. The term 'environmentally justifiable' was a form of cost–benefit requirement aimed at hindering measures to improve the environment when the benefits were so small that they could not justify the great costs associated with them. It is important to note that this was not developed as a traditional cost-efficiency requirement, but was, instead, used to prevent cases of extreme differences between costs and benefits. The reasonableness requirement is also not based on cost-efficiency reasoning. Instead, it appears to be aimed at focusing on the regulated party and its economic ability to sustain the regulation in question. Thus, economically successful undertakings have been deemed to be able to accept stricter regulation than weaker ones. Once again, it has been pointed out by commentators that information regarding how this balancing is carried out by courts is, in practice, generally lacking.

Swedish policy on the environment and human health regulation is laid down in fifteen broad environmental quality objectives which are to be reached within one generation. These are, in turn, divided into partial goals, which have been given deadlines for completion. This system provides for a broad, but relatively detailed, system of communicating which risks will be targeted. This prioritisation is not based on rational concepts, such as risk assessment and cost–benefit analysis, but appears strictly political in nature.

5.2 Swedish pesticide regulation

5.2.1 Introduction

The Swedish version of the precautionary principle is wide in scope and often strict in application. In the previous section it was discussed how the principle has come to be defined in the law, in the preparatory work and in the case law. It was concluded that the concept of precautionary action is firmly integrated in the Swedish legal system, but that it is difficult to establish with any degree of authority when and how a strict version of the principle will be applied. This is largely due to the lack of discussion on decisions taken under conditions of uncertainty by the legislator, as well as in the case law and the literature.

The aim of this section is to narrow the focus and centre on how the principle influences the regulation of pesticides. Sweden has a long history of low tolerance towards chemicals and, in recent years, this has culminated in the policy goal: 'A Non-Toxic environment'. The explicit aim of this policy is to eradicate all man-made toxic substances from the environment by the year 2020. The inherent uncertainty associated with chemicals has been stated as the basis for this radical policy. Thus, it appears that the Swedish regulatory aim in relation to chemicals is precautionary in the extreme, since it does not distinguish between substances with which real risks are associated, but, instead, aims at banning a whole category of products, regardless of the actual risks associated with them under normal use. However, the precautionary principle in the area of chemicals is far from limited to influencing policy. It also functions as a legal concept in its own right, and is then closely connected to the principle of substitution. In this section the different parts of the precautionary principle are analysed and discussed in relation to how they have been applied in the area of pesticide regulation. The issues touched upon include the structure and independence of the relevant government agency, how risks are assessed, the level and burden of proof, and cost–benefit analysis.

Since pesticide regulation today is largely carried out at EU level, the radical Swedish approach is compared with the relevant EU legislation. The discrepancies between the two levels are highlighted, and the problems that Sweden will encounter if it wishes to maintain its highly precautionary approach to pesticides are discussed. As an example, the regulation of the pesticide paraquat in Sweden and the EU is mentioned. This issue is particularly interesting since it shines light on some of the conflicts that different approaches to the precautionary principle can lead to at European and international levels.

5.2.2 *The legal framework*

In 1962 the first comprehensive regulation of chemicals was introduced into Sweden with the Chemicals Act. The same year a new decree on pesticides was established.[122] Interestingly, the Pesticides Decree only affected chemical products and not biological ones.[123] For chemical pesticides, a general approval scheme was established and KemI was appointed the relevant authority to which requests for authorisation were to be made.[124] Biological pesticides, on the other hand, were not originally subjected to a general authorisation scheme. Instead, they were allowed on the market unless they were in breach of any of the general provisions in the Environmental Protection Code.[125] It was not until 1992, and the Biological Pesticide Authorisation Act, that the scope was widened to include also the pre-authorisation of biological pesticides.[126] It is interesting to note that the stricter measures here were originally taken towards what were considered man-made substances, rather than against 'natural' substances, despite the fact that there is nothing to suggest that a chemical substance is intrinsically more hazardous than a naturally occurring substance.[127]

The division between biological and chemical pesticides was maintained when the Environmental Code (Miljöbalken)[128] was issued in 1998.[129] However, since the entry of Sweden into the EU, pesticides have been subject to the authorisation regime laid down in Directive 91/414/

[122] Förordning (SFS 1985:836) om bekämpningsmedel (SFS 1985:836). For a summary of earlier Swedish chemicals policy see Karlsson, 'The Precautionary Principle, Swedish Chemicals Policy and Sustainable Development'.

[123] *Ibid.*, Article 1. [124] *Ibid.*, Article 3.

[125] Miljlöskyddslagen (1969:387).

[126] Lag (SFS 1991:639) om förhandsgranskning av biologiska bekämpningsmedel.

[127] Sunstein refers to the common mistake of believing man-made substances to be more hazardous than naturally occurring ones as a symptom of loss aversion. See Sunstein, *The Laws of Fear*, p. 44. See also P. Rozin, 'Technological Stigma: Some Perspectives from the Study of Contagion', in J. Flynn, P. Slovic and H. Kunverther (eds.), *Risk, Media, and Stigma: Understanding Public Challenges to Modern Science and Technology* (Earthscan, 2001), p. 38. Rozin holds that, on the whole, more lives are lost to naturally occurring disasters than to man-made ones.

[128] Miljöbalken (1998:808).

[129] The term 'chemical pesticide' is defined in Article 5 of the Environmental Code and the term 'biological pesticide' in Article 6.

EC, which does not distinguish between the two types.[130] The fact that all types of pesticides, chemical as well as biological, need to be authorised by either another EU country or KemI is also made clear in Article 14:13 of the Environmental Code.

The procedure for the authorisation of pesticides is established in Chapter 14:14 of the Environmental Code, and fleshed out in Directive 91/414/EC and the updated Decree on Plant Protection Products.[131] Directive 91/414/EC subjects the authorisation of active substances in pesticides to a comitology procedure in accordance with Council Decision 1999/468.[132] The application for the authorisation of a pesticide on the Swedish market is to be made to KemI as mentioned above.[133] Provided that the provisions in Article 4 of Directive 91/414/EC are fulfilled, the request can be granted. In short, this means that the product's active substances must be listed in Annex I to the Directive, that the product is sufficiently effective and that it does not cause unnecessary suffering to animals or has unacceptable harmful effects on human or animal health or on the environment. Furthermore, the physical and chemical properties of the product have to have been established and considered appropriate for the intended use. The product cannot be authorised until maximum residue levels have been provisionally established by KemI and notified to the Commission. Once this has been done, EU-wide maximum residue limits are established using the regulatory committee procedure provided for in Article 19 of the Directive.

In situations in which a pesticide is introduced onto the Swedish market and has already been authorised in another Member State, the principle of mutual recognition comes into effect.[134] Sweden, or any other country, cannot, as a general rule, require renewed testing of a product already authorised. However, if it can show that, for example, particular dietary requirements in its country's population demand a stricter policy,

[130] For a discussion on the relationship between Swedish and EU legislation in the area of chemicals, see A. Nilsson, 'The Precautionary Principle in Swedish Chemicals Law and Policy', in N. de Sadeleer, *Implementing the Precautionary Principle: Approaches from the Nordic Countries, EU and USA* (Earthscan, 2007), pp. 295–313.

[131] Förordning (SFS 2006:1010) om växtskyddsmedel.

[132] Council Decision 1999/468 of 28 June 1999 laying down the procedures for the exercise of implemeting powers conferred on the Commission. OJ L 184, p. 23, 17/07/1999. As referred to in *ibid.*, Article 19.

[133] Chapter 14: 13 of the Environmental Code and Article 4 of the Förordning (SFS 2006:1010) om växtskyddsmedel.

[134] Article 10 of Directive 91/414/EC and Articles 17–19 of the Förordning (SFS 2006:1010) om växtskyddsmedel.

it can restrict the authorisation of a product.[135] If a Member State refuses to recognise tests performed in another Member State or an authorisation in its entirety, the Commission should be notified and the matter will be brought before the regulatory committee provided for in Article 19.[136]

A safeguard clause has been included in Article 11 of the Directive. This clause enables Member States provisionally to restrict or prohibit the use of a product it has authorised or is under obligation to authorise. In order to do so, the Member State must have 'valid reason' to consider the product 'a risk to human or animal health or the environment'.

5.2.3 *The institutional framework*

Sweden employs an intricate system of independent governmental authorities (statliga myndigheter), which are entrusted with various advisory, controlling and regulatory functions. At present, around 478 agencies exist.[137] They are established by law or decree, and their budgets and tasks are controlled by the government through the so-called budget control documents.[138] Agencies are protected from direct interference by the government in individual cases by the Swedish constitution.[139]

KemI was set up by decree in 1985.[140] The original decree was replaced by another decree in 1988, which is still in force.[141] KemI's mandate is to function as the main agency for issues relating to the risks to health and the environment posed by chemicals. More specifically, KemI's tasks include exercising control over, and issuing authorisations for, chemical and biotechnological substances, registering and keeping the records of chemical and biotechnological substances, and investigating, compiling and documenting activities and risks in the area.[142]

The authority is led by a Director General, who leads the work in the agency and chairs KemI's board, which consists of eight members.[143] Both the Director General and the board members are appointed by the government.[144] The Director General is not formally subordinated to the

135 *Ibid.*, Article 1. 136 *Ibid.*, Article 3.

137 According the Swedish Agency for Public Management. See www.statskontoret.se/upload/Publikationer/2007/2007103.pdf (accessed 1 March 2010).

138 'Regleringsbrev'. 139 Regeringsformen, Article 11:7.

140 Förordning (SFS 1985:643) med instruktion för kemikalieinspektionen, Article 2.

141 Förordning (SFS 1988:525) med instruktion för kemikalieinspektionen.

142 *Ibid.*, Article 2.

143 For a list of current members of the board see http://kemi.se/templates/Page____3154.aspx (accessed 1 March 2010).

144 Verksförordning (SFS 1995:1322), Articles 32 and 33.

board but can decide which issues the board should decide and which issues the Director General can decide on his/her own.

In general, Swedish government agencies decide on their own internal organisation.[145] However, with regard to KemI, it has been established that a 'Toxicological Council'[146] is to function as an expert organ for consultation and advice with regard to toxicological questions.[147] The role of the council is strictly advisory and nothing indicates that KemI is bound to consult it in individual cases. On the contrary, it is emphasised that the council functions as an 'informal' consultation body.[148] A large part of the work of the council is made up of organising seminars and training activities, as well as functioning as a bridge between academic researchers and KemI. The Toxicological Council is chaired by the Director General, and its members consist of representatives of other agencies, as well as independent researchers. The government appoints the members of the toxicological council for a specified period of time.[149]

KemI is formally an independent agency, and the government is prevented from interfering in individual cases brought before it. Despite this, the government retains influence over the agency through the possibility of appointing the Director General and the members of the board, assigning its budget and giving detailed instructions in the budget control documents. Furthermore, the government has reserved the power to approve the scientific experts in the Toxicological Council. This is further emphasised by the fact that the government-appointed Director General chairs the expert council. In short, it could be argued that the lack of a formal separation between risk assessment and risk management, in combination with the strong controlling powers retained by the government, raises some concern as to the true independence of the agency, as well as the objectivity of its decisions.

5.2.4 *Risk assessment and risk management*

The authorisation of pesticides is dealt with in Chapter 14 of the Environmental Code, which is the general chapter dealing with chemicals. Chapter 14 of the Environmental Code does not repeat or specify the

[145] *Ibid.*, Article 17. [146] 'Toxikologiska rådet'.

[147] Förordning (SFS 1988:525) med instruktion för kemikalieinspektionen, Article 7.

[148] See the website of KemI, www.kemi.se/templates/Page____2832.aspx (accessed 1 March 2010).

[149] Förordning (SFS 1988:525) med instruktion för kemikalieinspektionen, Article 8.

general precautionary principle or the substitution principle, which were both established in Chapter 2 of the Environmental Code.[150] However, Article 14:7 develops the knowledge requirement established in Chapter 2 of the Environmental Code into a requirement for a risk assessment when handling chemicals. It thus appears that that anyone handling any chemical product in Sweden is responsible for the fact that 'satisfactory investigation for the assessment of what health or environmental damage the product or organism may cause'[151] exists.

The investigation to be carried out is more specific in Article 14:7 of the Environmental Code than in the general requirement in Article 2:2 of the Environmental Code. While Article 2:2 of the Environmental Code makes a general demand on anyone carrying out an activity which may cause environmental damage to have 'knowledge' about this activity, Article 14:7 of the Environmental Code holds first that the investigation must be done 'in accordance with science and verified experience.'[152] Second, it makes four requirements with regard to what the investigation needs to fulfil. First, the substances or organisms which may give the product hazardous properties must be identified. Second, the kind and degree of the hazardous properties must be distinguished. Third, actions to protect health and environment should be suggested. Finally, actions required to deal with waste from the product or organism must be spelled out. The risk assessment is not limited to situations in which a hazardous substance, or a substance with new or unknown properties, is introduced onto the Swedish market, but is a general requirement for all chemical products.

The risk assessment is the foundation on which further measures can be based. With regard to precautionary measures, the substitution principle plays an important role. In short, the substitution principle means that substances which pose a risk to health or to the environment should be replaced by less risky substances whenever such an alternative exists. It is explicitly stated in the preparatory work to Article 2:7 of the Environmental Code that not only should substances that have been confirmed as hazardous be affected by the rule, but also substances surrounded by uncertainty.[153]

KemI does not employ a formal separation of risk assessment and risk management. The officials working for the agency are responsible for

150 For discussion on this see the previous section.

151 'Tillfredsställande utredning för bedömning av vilka hälso- eller miljöskador produkten eller organismen kan orsaka'.

152 'I enlighet med vetenskap och beprövad erfarenhet'.

153 Proposition 1997/98:45, 659–660.

assessing scientific information as well as for deciding on the appropriate measures to be taken upon the basis of the scientific information.[154] The burden of investigation and information about chemical products is put squarely on the applicant.[155] Little independent research is carried out by KemI itself, and the Toxicological Council has a strictly advisory function, which KemI is not required to use.[156] Often decisions on, for example, authorisations are taken in a highly informal manner by civil servants and KemI, in which 'common sense' plays a large role.[157]

The procedures that KemI adheres to when approaching pesticides are summarised in a Report from 1992 (hereinafter the KemI Report).[158] The KemI Report, which was drawn up by the agency itself, establishes a substantive framework for regulating the risks from pesticides. As mentioned above, a strict separation of risk assessment and risk management is not in effect in Swedish pesticide regulation. The Report is essentially of risk management character, in that it lays down a framework for how to deal with risks provided that certain thresholds are met. However, as mentioned earlier, the report is drafted by KemI itself, and KemI is also the body which assesses and classifies individual risks on a case-by-case basis. Thus, it appears that KemI functions both as risk manager and risk assessor. The tasks are not separated within the agency, either. Instead, the risks are evaluated on an ad hoc basis, often in meetings of an informal nature.[159]

The KemI Report separates the health effects of pesticides from effects on the environment. Health effects are considered with regard to acute toxicity, eye and skin irritation, sensitisation, subacute, subchronic and chronic toxicity, carcinogenicity, mutagenicity and reproductive toxicity. Environmental effects are looked at from the point of view of degradability, bioaccumulation, mobility and effects on non-target organisms. In each category a substance's 'danger class' is established. The four classes

[154] The lack of formal requirements on the collection and evaluation of scientific evidence can also be discerned from Nilsson, 'The Precautionary Principle in Swedish Chemicals Law and Policy', p. 302. She holds that the 'Swedish approach […] allows for listening to the facts presented by the scientists, evaluating them as facts and then acting in accordance to what is appropriate from a legal or political point of view.'

[155] Article 5 of Förordning (SFS 2006:1010) om växtskyddsmedel.

[156] For an account of the composition and tasks of the Toxicological Council see www.kemi.se/templates/Page____2988.aspx (accessed 1 March 2010).

[157] Interview with officials at KemI, 9 September 2004.

[158] Principles for Identifying Unacceptable Pesticides, KemI *Report*, No. 4/ 92 (National Swedish Chemical Inspectorate, 1992).

[159] Interview with officials at KemI, 9 September 2004.

are: (1) Extremely dangerous products (Very toxic), (2) Very dangerous products (Toxic, highly corrosive), (3) Dangerous Products (Corrosive, harmful, irritant), and (4) Moderately dangerous products (Moderately harmful).[160] The danger class to which a pesticide is assigned prescribes how the pesticide will be treated. Below, the focus will be primarily on health effects.

The most controversial aspect of the KemI Report is the establishment of so-called 'cut-off criteria'. For many categories of risk, certain levels of acceptable hazard are established. If a pesticide exceeds this level, it is automatically not allowed on the market. For example, with regard to acute toxicity, the cut-off criterion is when a pesticide is considered 'very toxic', that is, when Danger Class 1 is met.[161] For corrosive and carcinogenic substances, Class 2 is sufficient.[162]

In a case regarding a pesticide with the active ingredient mancozeb, a so-called ethylenebisdithiocarbamate (EBCD)-substance, which reached the Swedish Supreme Administrative Court (Regeringsrätten), Rohm and Haas, the manufacturer, appealed against a decision by KemI to decline a request for prolonged authorisation for the substance.[163] A large part of the case dealt with the principles of precaution and substitution, and this will be discussed in more detail below. However, the Court also touched generally on how the cut-off criteria are to be interpreted in the light of EU legislation, primarily Directive 414/91/EC. Rohm and Haas argued that Directive 91/414/EC and its annexes give clear guidance as to how the risk assessment and the approval of substances is to be carried out. KemI's use of cut-off criteria was in violation of this, since it allowed KemI to refuse the authorisation of pesticides without trying the substance under 'normal conditions'.[164] This is the case since the conclusion of the system of cut-off criteria, according to the plaintiffs, is that it is sufficient for a substance to be present in the pesticide for action to be taken by KemI in order to prevent it from reaching the market. This would be the case even if the actual circumstances in which the pesticide was being used would prevent handlers or the environment from being adversely affected by the substance. Furthermore, decisions on the basis of the cut-off criteria in cases in which a substance has been approved in another Member State

[160] Principles for Identifying Unacceptable Pesticides, KemI *Report*, No. 4/ 92, 9.

[161] *Ibid.*, 11. [162] *Ibid.*, 12 and 15.

[163] The Judgment of the Supreme Administrative Court in Case RÅ 1998 Ref. 56.

[164] Compare with Article 4 of Directive 91/414/EC: 'and having regard to *all normal conditions* under which it may be used.' (Emphasis added).

is, according to the plaintiff, in violation of the mutual recognition clause included in Article 10 of the Directive.[165]

At the time when the case was decided, Annex 1 of the Directive had not yet come into force. This circumstance was doubtless of relevance to the Court's decision to uphold KemI's original assessment. In the absence of Annex 1, the Court found that it was possible to maintain stricter rules than the ones provided for in the Directive. Mutual recognition under Article 10 is dependent upon the inclusion of the substances in Annex 1. Interestingly, the Court did not clearly state that this was the case only before Annex 1 came into force. Instead, it held that this must be the conclusion, 'at least'[166] under the present circumstances.[167] Thus, the Court did not limit the possibilities of maintaining stricter rules for issues stemming from the period before Annex 1 was in force. This does not mean that the Court explicitly undermined the mutual recognition clause, but it indicates that it did not necessarily view it as absolute. One related issue is that it appears that the Court did not view the area of pesticides as being harmonised before the Annexes to the Directive were in place. This is clear since it continually refers to the possibilities of refraining from respecting requests for mutual recognition upon the basis of Articles 30 and 36 EEC (currently Articles 34 and 36 TFEU) instead of Article 100a EEC (currently Article 114 TFEU), which would have been the appropriate legal basis had the area been harmonised.

In a more recent case, the Supreme Administrative Court reversed decisions by KemI and, on appeal, reversed decisions by the government, which subjected a pesticide containing the active ingredient glyfosate to a more limited authorisation than was the case at EU level.[168] The Court here put great emphasis on the fact that the substance had been included in Annex 1 to Directive 91/414.[169] Thus, the Court reasoned, a risk assessment at EU level had declared the substance safe for use. KemI and the Swedish government, on the other hand, claimed that releasing the substance into the ground water in particularly sensitive areas would entail risks. The Supreme Administrative Court found little evidence to support this assertion. Instead, it stated that the decision seemed to be based upon 'a general opinion on the part of the government authorities that pesticides containing glyfosate bring a risk of pollution of water' rather than

165 The Judgment of the Supreme Administrative Court in Case RÅ 1998 Ref. 56, p. 1–2.
166 'i vart fall'.
167 The Judgment of the Supreme Administrative Court in Case RÅ 1998 Ref. 56, p. 10.
168 The Judgment of the Supreme Administrative Court in Case RÅ 2005 Ref. 2.
169 *Ibid.*, p. 6.

concrete risk assessments.[170] Owing to this fact, it declared the stricter Swedish rules to be disproportionate.[171] By holding that the measures were 'disproportionate', rather than in direct violation of the Directive, the Court carefully avoided discussing in any detail the issue of whether stricter, proportional measures can be maintained even though a substance has been included in Annex I.

5.2.5 *Scientific evidence and the introduction of precautionary measures*

The general rules of care in Article 2 of the Swedish Environment Code apply in all areas in which environment or health protection is concerned. Thus, the precautionary principle is the foundation upon which Swedish pesticide regulation is built. As discussed above, the Swedish interpretation of the precautionary principle is generally wide, and, in many cases, resembles a general duty of care more than the precautionary principle discussed at European and international levels.[172] The principle of substitution, or product choice as it is also commonly known, can be viewed as a particular manifestation of the Swedish version of a strict precautionary principle. Nowhere is this more tangible than in the area of chemicals regulation. An example of how far-reaching precautionary considerations are in the area of chemicals regulation is the ambitious environmental quality objective of 'A Toxic Free Environment' established by the Swedish Parliament in 1999.[173] The explicit goal stated there is to eliminate all man-made substances with toxic properties from the environment, and to restore naturally occurring substances to their original levels by the year 2020. This goal was further specified in another bill before parliament in 2001 titled 'A Chemicals Strategy for a Non-Toxic Environment'.[174] The explicit reason for this radical policy was that the lack of knowledge and the persistent uncertainty associated with many chemicals calls for a precautionary approach.[175] The environmental quality objectives are of a policy character and are thus not legally binding.[176] However, the impact

170 *Ibid.*, p. 7. 171 *Ibid.*

172 Karlsson, 'The Precautionary Principle, Swedish Chemicals Policy and Sustainable Development', 350.

173 Proposition 1997/98:145.

174 Proposition 2000/01:65. 'Kemikaliestrategi för en giftfri miljö', pp. 9–14.

175 *Ibid.*, p. 7.

176 See also, Karlsson, 'The Precautionary Principle, Swedish Chemicals Policy and Sustainable Development', 352–3.

that the objective 'A Toxic Free Environment' has had for the work of KemI seems difficult to exaggerate.

As mentioned above, the main requirement for authorisation of a pesticide at European level is the inclusion of its operative components in Annex I of Directive 91/414/EC. The information needed to be supplied together with an application is detailed and laid down in Annex II to the Directive. For the authorisation for pesticides, the required documentation needed is established in Annex III. The general principles which need to be fulfilled for authorisation are laid down in Annex VI. In summary, the EU rules establish absolute rules for how pesticides are to be authorised. If a pesticide is considered safe for use under the relevant rules, it can be allowed on the market regardless of other products existing there already. The Swedish substitution principle, on the other hand, establishes that it is not sufficient for a pesticide to be safe for use. If there are other products of equal efficacy on the market which are associated with less risk, the pesticide should not be allowed.[177] Thus, the Swedish rules are not absolute, but, instead, establish a moving target. Essentially, this is a very strict application of the precautionary principle with the explicit aim that, in the end, only risk-free and near risk-free products will exist on the market. This appears to be a concrete legal application of the environmental quality objective 'Toxic Free Environment'. If such an application is upheld in practice, it is likely to be in conflict with the relevant harmonised EU legislation. The question then arises as to what level of proof must be provided by KemI before taking precautionary action, including invoking the substitution principle.

In previous chapters it was stated that the guidance as to what needs to be shown by the regulator when intending to take precautionary measures is generally weak in Swedish legislation, as well as in the preparatory work and the case law.[178] With regard to pesticides, the clearest embodiment of precautionary considerations is the establishment of cut-off criteria. The use of cut-off criteria focuses the risk assessment on the individual substances which are found in a given product. If a substance is sufficiently risky, the product is automatically banned. Often, this appears to be the case without real consideration of the actual risks that the product poses when used in practice.[179]

[177] *Ibid.*, p. 661. [178] See, above, Section 5.1.

[179] For an argument in this direction see R. Nilsson, 'Science and Politics in the Regulation of Chemicals in Sweden', in M. Gough (ed.), *Politicizing Science: The Alchemy of Policymaking* (Hoover Institutions Press, 2003), pp. 227–59.

In case 4704–1996, KemI based its decision not to prolong the authorisation for a pesticide with the active substance mancozeb upon the grounds that it was considered carcinogenic. The basis for this conclusion appears to have been primarily two scientific reports of Swedish and Finnish origin, and unspecified 'studies on rats'.[180] Other mainstream reports, indicating that mancozeb does not possess significant carcinogenic properties even at high doses, and no carcinogenic effects at doses which may be present in the actual use of the pesticide in question, were disregarded. In fact, at the time of the decision, it appears that no animal studies suggested that mancozeb was carcinogenic at the low dosages which normal use of the pesticide would entail. Despite this, KemI referred to 'studies in rats' to justify its decision.[181] Which studies this referred to was not further specified. Most other European countries had, at the time, authorised mancozeb, and the clear majority of scientists considered the product in question safe for use.[182] KemI recognised the uncertainty surrounding the product, but stated that the ban was warranted by the precautionary principle. This was based on the fact that uncertainties pertained to the substance in question, even though the current state of scientific research held that the risks of the pesticide under normal use were negligible.[183] In the light of the low risk associated with the product and the weak scientific evidence supporting the assertion of any risk at all when used under normal circumstances, the KemI measure must be considered to be an application of a remarkably strict form of the precautionary principle.

In the more recent case regarding pesticides containing the active substance glyfosate, KemI gave a limited authorisation, instead of the full authorisation sought by Monsanto. Glyfosate is included in Annex I to Directive 91/414/EC. If a Member State wishes to maintain or introduce a conditional authorisation of a substance which has been included in Annex I, it has to conduct an individual assessment of the product, which indicates a particular risk for its population. In the case in question, KemI had not supported its decision with any detailed assessment of the product. Instead, the Swedish Supreme Administrative Court found

180 KemI Memo (P. Bergkvist), PM inför beslut om fortsatt godkännande för EBDC-medel. 8 November 1994 (not published), p. 2.

181 In the material obtained here, the reference was not more specific than 'studies in rats'.

182 See the following subsection.

183 KemI Memo (P. Bergkvist), Yttrande over kompletterand material från Rohm and Haas och Elf Atochem Agri angående överklagandena av Kemikalieinspektionens beslut den 9 november 1994 i fråga om bekämpningsmedel innehållande EBDC, 23 August 1995, p. 1–2.

that the decision seemed to be based on a 'general opinion on the part of the authority that pesticides containing glyfosate entail risk'.[184] Thus, it appears that the Court here required more scientific evidence for precautionary measures to be justified than in the earlier case. Mere policy concerns were not considered acceptable here. It remains unclear exactly where the courts will draw the line. However, what does appear to be clear is that conflict will arise between KemI's precautionary stance, which focuses on a chemical's intrinsic properties, and the EU approach under Directive 91/414/EC, where the focus is on the risks posed by a pesticide under normal use.

A good example of this conflict is the controversy surrounding the pesticide paraquat, which was dealt with recently by the Court of First Instance in Case T-229/04.[185] Paraquat was added to Annex I of Directive 91/414/EC in December 2003, by using the regulatory committee procedure provided for in the Directive. However, paraquat has been banned in Sweden since 1983 due to its acute toxicity when ingested. That the substance is toxic is an uncontested fact among both its proponents and its critics.[186] Paraquat has been known, for example, to be used as a suicide agent, primarily in developing countries.[187] The risk of paraquat poisoning is scientifically certain and can, therefore, best be described as a 'certain risk' as previously discussed in Chapter 2. The high toxicity of paraquat led to it being included in the 'Dirty Dozen' list of hazardous pesticides established by the Pesticide Action Network (PAN) in 1982. The PAN has actively lobbied for global bans of the 'Dirty Dozen' pesticides since the list was drawn up. Today, the list includes eighteen pesticides, and of these, all but three have been listed under the Rotterdam Convention for Prior Informed Consent (PIC).[188] Paraquat is not among them.

[184] Judgment of the Supreme Administrative Court in Case RÅ 2005 Ref. 2.

[185] Case T-229/04 *Sweden* v. *Commission*, Judgment of 11 July 2007, not yet reported.

[186] See, for example, Syngenta's webpage: www2.syngenta.com/en/products_brands/downloads/paraquat_jh_nov_03.pdf (accessed 1 March 2010; International Program on Chemical Safety (INCHEM) Poison Information Monograph (PIM) No. 399. Available at: www.inchem.org/documents/pims/chemical/pim399.htm (accessed 1 March 2010); J. Madeley, 'Paraquat – Syngenta's Controversial Herbicide', SSNC Report (2002), pp. 10–12. Available atwww.chemsec.org/images/stories/news_publications/rap-inter-syngenta-paraquat.pdf (accessed 1 March 2010); B. Dinham, 'Why Paraquat Should Be Banned', *Outlook on Pest Management* (December 2004), p. 269.

[187] Madeley, 'Paraquat – Syngenta's Controversial Herbicide', p. 12.

[188] The Convention on the Prior Informed Consent Procedure for Certain Hazardous Chemicals and Pesticides in International Trade, Rotterdam, 10 September 1998. Available at: www.pic.int/ (accessed 1 March 2010).

Syngenta, which manufactures the paraquat-based pesticide Gramoxone, and other proponents of the use of paraquat, hold that, despite its acute toxicity when ingested orally, the substance's general toxicity is not worse than other household chemicals or pesticides. In order to prevent accidental oral ingestion, Syngenta has added a blue dye, a strong odour and an agent to induce vomiting to Gramoxone. This is in line with measures taken to prevent poisoning in relation to other products.[189] There is general agreement that the addition of the above agents have substantially mitigated the risk of poisoning through oral ingestion of Gramoxone. It has been pointed out that the risk still pertains with regard to independent manufacturers of paraquat who do not add the agents.[190] Some other reports have suggested that paraquat exposure may lead to Parkinson's disease.[191] Scientific uncertainty surrounds this issue. A relationship appears to exist between the pesticide and disease, but a causal link remains to be found.

It is not entirely clear on what scientific evidence the Opinion of the relevant regulatory committee, in this case the Standing Committee on the Food Chain and Animal Health (hereinafter the Standing Committee), was based. Requesting an opinion from a scientific committee is not mandatory under Directive 91/414. However, the Commission had put four questions relating to the safety of paraquat before the Scientific Committee on Plants in 2001, and it is likely that the reply to these questions influenced the Commission.[192]

The Opinion of the Scientific Committee focused entirely on the toxicity of paraquat. No mention was made, for example, of reduced lung capacity or Parkinson's disease. With regard to the toxicity, it admitted that paraquat is very toxic when orally ingested or when absorbed by the lungs. However, skin and eyes were not deemed particularly sensitive

189 See, for example, www2.syngenta.com/en/products_brands/gramoxone_page.html (accessed 1 March 2010).

190 Dinham, 'Why Paraquat Should Be Banned'.

191 M. Thiruchelvam *et al.*, 'The Nigrostriatal Dopaminergic System as a Preferential Target of Repeated Exposures to Combined Paraquat and Maneb: Implications for Parkinson's Disease', *The Journal of Neuroscience*, 20:24 (2000), 9207; J. Peng *et al.*, 'The Herbicide Paraquat Induces Dopaminergic Nigral Paoptosis through Sustained Activation of the JNK Pathway', *J. Bio. Chem.*, 279:31 (2004), 32626; and M. Thiruchelvam *et al.*, 'Overexpression of Superoxide Dismute or Glutathione Peroxidase Protects against the Paraquat + Maneb-induced Parkinson Disease Phenotype', *J. Biol. Chem.*, 280:23 (2005), 22530.

192 Opinion of the Scientific Committee on Plants on Specific Questions from the Commission Regarding the Evaluation of Paraquat in the Context of Council Directive 91/414/EEC, SCP/PARAQ/002-Final, 16 January 2002.

to exposure.[193] The results were based on animal studies. The Scientific Committee concluded that the risks with paraquat can be adequately dealt with through the use of protective equipment for operators. Some predictive exposure models that the Scientific Committee reviewed indicated that operator exposure might exceed the acceptable operator exposure level (AOEL).[194] However, the Scientific Committee considered such models to 'markedly over-estimate the actual exposure to paraquat in real working situations'.[195] Instead, it decided to rely on field exposure studies, which indicated exposure to paraquat within the AOEL. It is highly interesting to note that the Scientific Committee relied on the risks of paraquat during normal use, and not the risks associated with the substance in isolation. Clearly, a degree of scientific uncertainty surrounded the situation of how the product would affect handlers, but this was not deemed to be sufficient for the Scientific Committee to suggest a more precautionary use of the pesticide.

Sweden, on the other hand, focuses on the fact that the active substance paraquat, in isolation, has highly toxic properties. In February 2004 Sweden took action against the European Commission under Article 263 TFEU (formerly Article 230 EC), claiming that the decision to authorise paraquat was in violation of the precautionary principle as formulated in Article 191 TFEU (formerly Article 174 EC).[196]

The primary focus of the Swedish challenge was the scientific evidence reviewed by the Regulatory Committee and the Commission when deciding upon the authorisation. Sweden emphasised that the scientific reports based on predictive exposure modelling had unequivocally shown that exposure in excess of the AOEL would always be a fact.[197] Furthermore, Sweden held that the field studies that the decision was ultimately based upon were flawed and did not provide sufficient evidence to conclude that paraquat could be used safely. Sweden held that, in two of the three field studies existing on paraquat, one person had suffered exposure which exceeded the AOEL.[198] The Swedish Government also held that, by ignoring the data existing in relation to the potential link between paraquat and Parkinson's disease, the

193 *Ibid.*, pp. 6–12. 194 *Ibid.*, p. 12.

195 *Ibid.*

196 Swedish Ministry for Foreign Affairs (EU Legal Secretariat), Letter to the Registrar at the Court of Justice of the European Communities, 26 February 2004, 2 and 19 (available from the author).

197 *Ibid.*, p. 20. 198 *Ibid.*, pp. 20–1.

decision was arbitrary and once again in violation of the precautionary principle.[199] Finally, it held that the risk reduction measures suggested by the Commission in its decision with regard to animals were unlikely to be upheld in practice. For these reasons, Sweden considered the rules laid down in Directive 91/414/EC with regard to the protection of animal health to be breached.[200]

In summary, it can be held that, in its challenge, Sweden firstly considered paraquat to be so toxic that it should not be allowed under any circumstances. Second, Sweden held that the uncertainties pertaining to the risk posed by the normal use of paraquat warrants a more precautionary approach than that advocated at EU level. The case of paraquat is significant in that it highlights the different approaches adopted by Sweden and the EU when considering scientific evidence and the precautionary principle. Swedish civil servants have described the case as 'the environmental case of the decade'.[201]

In its judgment of 11 July 2007, the Court of First Instance found in favour of Sweden. In short, it concluded that the Commission had neglected formal requirements under Directive 91/414/EC when adding paraquat to Annex I, as well as having underestimated the risks of the substance to human health as well as to the environment.[202]

The case is interesting since it shows that serious conflicts between national legal systems and the EU institutions exist in the field of pesticide regulation. Furthermore, it is interesting to note that, in this case, Sweden managed successfully to bring an action against the Commission for not being precautionary enough. Thus, it would appear that the Commission is under an obligation under EU pesticide legislation to act in a precautionary manner. However, it is important to note that, in this case, the Court of First Instance did not vindicate the general Swedish policy towards pesticides, but merely ruled that, in the case before it, the Commission had opted for a level of protection that could not be considered sufficiently precautionary in the light of the existing scientific evidence.

[199] *Ibid.*, pp. 24–5. [200] *Ibid.*, p. 27.

[201] Statement by Swedish Foreign Office official Anders Kruse, quoted in Swedish daily newspaper *Svenska Dagbladet*, 24 August 2004. Available at: www.svd.se/dynamiskt/inrikes/did_8027374.asp (accessed 1 March 2010).

[202] For a further discussion on the substance of the case see above, Chapter 4.

5.2.6 *The burden of proof*

As discussed above, it is for the applicant to support a request for the authorisation of a pesticide with the relevant scientific information.[203] This information must show that the pesticide in question corresponds to the level of protection prescribed by the relevant legislation accounted for above, and specified in the KemI Report. However, in each individual case KemI has the burden of showing that the application is not in conformity with the relevant rules.[204] What the applicant is required to show, in order to challenge KemI's decision, has been dealt with, to some extent, in the scarce and rather inconsistent case law on pesticides.

In the case regarding the mancozeb-based pesticide discussed above, the plaintiff, Rohm and Haas, questioned the scientific evidence on which KemI had based its decision. Among other things, it held that KemI had not paid sufficient attention to certain scientific evidence. Furthermore, it held that the government, which had tried the case before it was appealed to the Supreme Administrative Court, had not been able to conduct a complete investigation. This was due to the fact that the decision by KemI, which clearly was a party to the suit, was the only expertise that the government relied on for its decision.[205] The Court held that it found that mancozeb can be 'suspected on scientific grounds'[206] to possess properties that can be hazardous for both human health and the environment.[207] Once this has been established, the burden of proof falls on the plaintiff to disprove this assertion. The Court found support for this in the preparatory work to the Chemical Substances Act.[208]

The scientific evidence brought forth by Rohm and Haas was not considered sufficient by the Court to reverse the decision by the agency. KemI based its decision primarily upon two Swedish scientific reports which suggested that mancozeb contained carcinogenic properties.[209] Rohm

[203] Article 5 of Förordning (SFS 2006:1010) om växtskyddsmedel.

[204] See, for example, the conclusions of the Supreme Administrative Court in its judgment RÅ 2005 Ref. 2, which was discussed above.

[205] The Judgment of the Supreme Administrative Court in RÅ 1998 Ref. 56.

[206] 'På vetenskapliga grunder misstänkas'.

[207] The Judgment of the Supreme Administrative Court in RÅ 1998 Ref. 56.

[208] *Ibid.* Since the case was decided before the Environmental Code entered into force, the Court here refers to the Chemical Substances Act and its preparatory works. However, as was mentioned above, no material change was intended in this regard when the Environmental Code was adopted. If anything, the present regulation may be stricter.

[209] KemI Memo, 8 November 1994, pp. 2–4.

and Haas disputed this evidence in a memo attached to a letter to KemI in 1995.[210] Since the Swedish Ministry of the Environment decided to make this report confidential, it is not possible to make a valid judgment on the scientific evidence it contains. However, it can be assumed that it would be along the lines of the Danish Ministry of the Environment Report which functioned as the risk assessment for Denmark regarding mancozeb in 1993 and was also relied on by Rohm and Haas in the case.[211] The conclusion of this report was that mancozeb did not possess any considerable carcinogenic properties and was safe for use. The Danish conclusions were based upon a plethora of reports from various European, as well as non-European, countries. As discussed above, KemI held that studies in rats had shown that mancozeb was carcinogenic.[212]

However, the Danish report held that carcinogenicity had only been indicated in one out of three animal studies. And, even then, it only became measurable at much higher doses than the normal use of the product in question would involve.[213] The Court did not consider it sufficient that Rohm and Haas had presented the fact that no carcinogenic effects had been shown to stem from the substance at the dosages involved in the normal application of the pesticide. KemI, the government and the Court were all unmoved by the exhaustive (at the time) evidence presented by Rohm and Haas. It is difficult to imagine what more Rohm and Haas could have done to disprove the assertion of the government. The focus on the risks associated with the active substance of a pesticide, rather than the product under normal use, makes it very difficult for a party to challenge a precautionary decision. Since all substances pose risks at a sufficiently high level of exposure, the approach adopted by KemI gives it nearly unlimited powers to regulate substances, and, subsequently, it is nearly impossible for a challenging party to appeal against a decision successfully.

However, since the coming into force of Annex 1 to Directive 91/414, the leeway of KemI is being restricted. In the more recent case relating to Glyfoasate referred to above, the Court seems to put more emphasis on the need for scientific evidence in cases relating to pesticide regulation. However, in that case, the reversed burden of proof was not touched

[210] N. Gocha, Rohm and Haas, Comments on DeZäta Granulat Registration withdrawal by KemI, 18 January 1995 (not published).

[211] Miljøstyrelsen, Vedrørende Dithane DG, 9 June 1993 (available from the author).

[212] Bergkvist, 'PM inför beslut om fortsatt godkännande för EBDC-medel', p. 2.

[213] Miljøstyrelsen, Vedrørende Dithane DG.

upon since the Court found that KemI had not reached its initial level of proof.[214]

5.2.7 *Cost–benefit analysis*

In relation to pesticides, the general 'unreasonable test' in Article 2:7 of the Environmental Code must be considered to be applicable. As discussed above, this is neither a traditional cost–benefit analysis nor a proportionality requirement. Instead, it aims to codify the earlier Swedish practice which stated that actions to protect the environment should only not be taken where this is not 'environmentally motivated'. The presumption is thus that measures should be taken. Only where great costs would result in only marginal environmental improvement should they not be taken.

In the preparatory work to the Environmental Code, the weighing of costs and benefits when considering authorisations for pesticides is emphasised, but not elaborated on in any detail.[215] However, interestingly enough, the substitution principle is connected to this requirement. The preparatory work held that, if other similar or more efficient but less risky substances exist, this should influence the estimation of the benefit that the pesticide would bring.[216] This has further been elaborated in the case law. In Case 4704–1996, Rohm and Haas held that, by making use of the substitution principle, KemI had utilised comparative evaluations which were not in conformity with Directive 91/414/EC. However, the Court found that the substitution principle could be found to be in conformity with the general cost–benefit considerations which should be performed under the Directive:

> It does not appear as incompatible with the Directive to, as a part of the benefit evaluation, pay some notice to the fact that other products with the same area of use but differing health and environmental effects exist on the market.[217]

Furthermore, in this case Rohm and Haas argued that a strict application of the cut-off criteria discussed above is in violation of Directive 91/414/EC, since the Directive calls for a more nuanced evaluation of pesticides.

[214] The Judgment of the Supreme Administrative Court in RÅ 2005 Ref. 2.

[215] Proposition 1997/98:45, 812. [216] *Ibid.*

[217] *Ibid.*, p. 11. The decree guiding the work in government agencies also states that agencies must consider the costs of planned regulation and make such evaluations available (Verksförording, Article 27).

Annex VI to the Directive calls for a weighing of the costs and benefits of the proposed regulation, as well as an evaluation of the pesticide under the conditions in which it will actually be used. Thus, it is not exclusively concerned with the intrinsic hazardousness of a particular substance.[218] The application of cut-off criteria, on the other hand, allows for a bypassing of this. Once the criteria are met, the product should not be allowed on the market, regardless of the circumstances under which the pesticide is to be used and potential benefits which it may have. That the evaluation is based solely on the basis of the intrinsic properties of the substance in question is made clear in the KemI Report:

> Pesticides which fulfil the cut-off criteria are regarded as unacceptable within the meaning of the Ordinance solely on the basis of their intrinsic properties.[219]

However, in the case at hand the Court found that KemI had still performed a more extensive risk assessment, and that the cut-off criteria had not exclusively determined the result. Thus, the Court found the decision to be in conformity with the Directive. The Decision was upheld in this part, too. In particular, before the entry into force of Annex I, the Court found it possible to maintain stricter rules than the Directive provided for. As mentioned above, later case law may indicate that the Directive is now given more weight than the Supreme Administrative Court gave it here.[220] This could lead to a situation in which the costs and benefits of a pesticide will be weighed by KemI, rather than just the risk being considered. However, in interviews at KemI, it transpired that cost–benefit analysis is not a part of the decision-making process in relation to pesticides. On the direct question of how costs and benefits were evaluated, officials replied unequivocally that economic considerations did not influence the decision of whether to authorise pesticides or not.[221]

5.2.8 Conclusion

A strict version of the precautionary principle lies at the heart of Swedish pesticide regulation. This is clearly illustrated by the policy goal 'A Non-Toxic Environment', which states that all man-made substances with

[218] See Section 1.8 of Annex VI to Directive 91/414/EC.

[219] KemI Report, p. 7.

[220] See the judgment of the Supreme Administrative Court in Case RÅ 2005 Ref. 2.

[221] Interview with officials at KemI, 9 September 2004.

toxic properties are to have been removed from the environment by the year 2020, regardless of whether they are toxic at normal use or not.

KemI is the agency entrusted with issuing authorisations and detailed rules regarding pesticides. The agency is formally independent, but the government retains important powers of control and command through the issuing of instructions and appointments. No formal separation of risk assessment and risk management is in effect. Instead, decisions are often taken after informal deliberations between civil servants in the agency.

One concrete effect of the precautionary nature of Swedish pesticide regulation is the focus of KemI on the risks posed by individual substances included in a product, rather than on the risk posed by the whole of a product under normal use. This approach leads to a situation in which the level of accepted risk is extremely low. As a product of this, the level of proof needed to be met by KemI when taking precautionary measures is low. Earlier case law seems to have indicated that mere policy considerations could, in some cases, be sufficient to warrant restrictive measures. However, in one case which stems from after the coming into full effect of Directive 91/414/EC, the Supreme Administrative Court held that the mere 'opinion' of the agency was not enough to justify precautionary measures. Some form of scientific basis is required.

Once the low burden of proof on the part of the regulator (KemI) is met, the burden is shifted onto the regulated party to show that the agency was mistaken. The level of proof which the regulated party needs to meet is high. In one case in which KemI had referred to 'studies on rats' in order to justify a precautionary ban, it was not sufficient for the contesting party to show that no studies on rats that indicated any risk from the product in question under conditions of normal use existed.[222] The Supreme Administrative Court agreed here with KemI that it did not matter that the product as such could be used at a comparably low risk. In order to be successful, the challenging party must show that each constitutive substance which makes up the product is without risk. In the case mentioned here, the plaintiff argued that it would be unreasonable to have to show that the substance in question (mancozeb) was safe for use at high doses, since normal use of the product would only concern significantly lower doses. KemI and the Court were unmoved by these pleas. The conclusion with regard to the burden of proof, once it has been shifted onto the regulated party, is thus that it is not sufficient to show that a pesticide is

[222] The Judgment of the Supreme Administrative Court in RÅ 1998, Ref. 56.

without risk when used under normal circumstances. Instead, each substance which comprises the pesticide must be shown to be risk-free. Such reasoning ultimately leads to a situation in which only zero-risk products are tolerated.

The substitution principle is connected to the issues discussed above of what level of risk is tolerated. This principle, in practice, constitutes a particular manifestation of the precautionary principle, and means that if safer pesticides of equal effect than the one in question exist on the market, it should not be allowed. Thus, it establishes comparative assessments, instead of absolute requirements with regard to what level of risk should be tolerated. Such moving targets are difficult to comply with, and may lead to situations of considerable legal uncertainty. The Supreme Administrative Court has held that the substitution principle is a part of the balancing which should be carried out before regulation is issued.

As discussed in the previous chapters, the proportionality principle in Swedish environmental law is not quite comparable to the one at European level. The Swedish version aims to address situations in which it is not 'environmentally motivated' to take action. The threshold for this to be the case appears to be high. Thus, it does not establish an unequivocal cost–benefit requirement, or even a proportionality test in the EU sense, but is limited in scope to situations in which the environmental benefits are very low and the costs associated with regulation very high. In contrast, it appears that when KemI evaluates substances, it only considers the risk and not the benefits associated with the product. This enables it to apply a very restrictive policy towards products. If the benefits of a substance are not considered, the risks associated with it are impossible to justify. This is returned to within the framework of US pesticide legislation.

The issue of the regulation of paraquat at EU level provides a clear illustration of how different views on the precautionary principle can lead to conflicts not only within the EU, but also in regulation at international level. By focusing its assessment of pesticides on individual substances, rather than on the risks posed by the actual product under normal use, Sweden's approach is strict and is in conflict with the position assumed by most other EU Member States and the Commission. Furthermore, by making different conclusions to those of the Commission and emphasising scientific reports which suggest risks not taken into account when drawing up the decision to authorise paraquat, Sweden made it clear that, in its view, the precautionary principle is to be interpreted strictly at EU level, too. Whether this is also the view of the ECJ remains to be seen.

5.3 Swedish regulation of base stations

5.3.1 Introduction

Where the regulation of chemicals has an impressive history ranging back to the middle of the eighteenth century, the regulation of radiation in general and mobile telephony in particular, is, for natural reasons, a much more recent area of regulation. Only since 1976 has the emission of radio waves been regulated at national level in Sweden. This is due to the fact that radio technology itself, as well as the awareness of the potential risks stemming from it, is relatively new. With regard to radiation in general, some types of risks are well documented. Most famously, this is the case with regard to ionising radiation, which is lethal even in very small doses. But non-ionising radiation can also be severely hazardous.

In Sweden, as in the other legal systems investigated here, the legal framework in relation to base stations is complicated due to the fact that radiation protection rules have to be used in combination with general environmental rules as well as planning provisions. The distinction between the different laws is not clear-cut and different provisions often overlap. Furthermore, the terminology is not unambiguous. The meaning and application of concepts central to the precautionary principle seem, in some cases, to be interpreted differently, for example, in relation to base stations than they were in relation to pesticides. To complicate things further, the government has not established legally binding safety standards, but the relevant agency has, instead, opted for the issuing of non-binding recommendations. So far, little case law exists to clarify the situation, especially from the higher courts. Below, an attempt is made to interpret the main provisions and considerations which influence the Swedish policy towards radiation from base stations. An attempt is made to clarify the roles of the relevant government agency and local government when deciding on safety standards and the placement of base stations. The definition of the precautionary principle in relation to radiation is examined in detail, as are the concepts of relevance to its implementation, such as risk assessment and risk management, cost–benefit analysis and the role of the public.

5.3.2 The legal framework

A number of laws regulate the area of base stations for mobile telephones in Sweden at present. With regard to the type and the amount of radiation that is allowed to be emitted, the Radiation Protection Act (Strålskyddslagen) (hereinafter SSL) is relevant.[223] The placement of base stations is regulated in the Planning and Building Act (Plan- och bygglagen) (hereinafter PBL)[224] and, to some extent, the Environmental Code. The amount of exposure to which workers who routinely come into contact with radiation are allowed to be exposed is established in the Work Environment Act (Miljöskyddslagen).[225] However, the special rules relating to workers will not be investigated in detail below.

First, it should be pointed out that releasing ionising, as well as non-ionising, radiation should be considered an 'environmentally hazardous' activity according to the Environmental Code, if it 'may cause detriment to the surroundings'.[226] 'Detriment to human health' is defined as meaning 'any disturbance that is liable to have adverse effects on health in medical or hygienic terms which are not minor or temporary'.[227] Considering the wording of the relevant provisions, it would appear that it is not only activities whose effects are scientifically certain that should be considered as environmentally hazardous. This is in conformity with the general precautionary principle in Article 2:3 of the Environmental Code. The main effect of declaring an activity environmentally hazardous is that it can be subjected to an authorisation procedure and subjected to various degrees of governmental control.[228] The establishment of base stations is not subjected to an authorisation procedure, except for the procedures at local level provided for in the planning regulations. However, Swedish courts have declared that the release of radiation from third-generation mobile telephony base stations should be subjected to control, primarily at local level.[229] It is interesting to note that the Environmental Court of Appeal considers the radiation emanating from base stations to be environmentally hazardous, especially in the light of the fact that, as is discussed in more detail below, courts generally have not been inclined to pay any particular attention to uncertain risks in relation to the issue. However, the practical effects of declaring an activity to be environmentally hazardous are unclear. In drafting the Environmental Code, it was pointed out by some stakeholders that the term was unnecessary since the Code makes all activities, regardless of whether they are declared environmentally hazardous or not, subject to the general duty of care (including the

[223] Strålskyddslag (SFS 1988:220). [224] Plan- och bygglagen (1987:10).
[225] Arbetsmiljölag (SFS 1977:1160). [226] Miljöbalken Article 9:1.
[227] Miljöbalken Article 9:3. [228] Miljöbalken Article 9:6.
[229] Judgment of the Environmental Court of Appeal in Case MÖD 2005:55.

precautionary principle) included in Article 2 of the Environmental Code.[230]

The general framework provision for regulating the radiation emanating from base stations is the SSL. The purpose of this act is to protect humans, animals and the environment from the detrimental effects of radiation.[231] The term 'radiation' in the Radiation Protection Act encompasses ionising as well as non-ionising radiation.[232] Following the general approach in Swedish environmental and health regulation,[233] the onus is on the performer of an activity involving radiation to avoid or minimise detrimental effects.[234] This also includes the responsibility to make sure that installations emitting radiation have been equipped with the appropriate safety arrangements.[235] It is explicitly stated that precautionary measures are included in this obligation also under the Radiation Protection Act.[236]

The government, or any authority appointed by it, may establish further detailed provisions regarding protection and control of radiation.[237] The Swedish Radiation Protection Agency (SSI) was originally given this task, but since June 2008 the agency has been renamed Strålsäkerhetsmyndigheten (hereinafter to as SSM).[238]

As mentioned above, the establishment of base stations for mobile telephony is not subjected to a centralised authorisation scheme like the one which is in place for pesticides. Instead, the establishment falls under general planning rules. Such rules are laid down in the Planning and Building Act, according to which, it is up to local authorities to plan the use of land and water.[239] The relationship between the Environmental Code, which lays down the general environmental rules, and the Planning and Building Act, which is *lex specialis* for planning and building issues, is not entirely clear.[240] The rules are, to a large extent,

230 See the opinion of Koncessionsnämnden in Proposition 1997/98:45, 325–6.

231 Article 1 of the Radiation Protection Act.

232 Article 2 of the Radiation Protection Act.

233 Compare with the general duties of care in Chapter 2 of the Environmental Code accounted for above in Section 5.1.

234 Strålskyddslagen Article 6.

235 Strålskyddslagen Article 6.

236 Strålskyddslagen Article 6, point 1.

237 Strålskyddslagen Article 12.

238 Förordning med instruktion för Statens strålskyddsinsititut (SFS 1988:295) and Förordning (2008:452) med instruktion för Strålsäkerhetsmyndigheten. See also www.stralsakerhetsmyndigheten.se/Allmanhet (accessed 1 March 2010).

239 Plan- och Bygglagen Article 2.

240 See, for example, B. Bengtsson, *Speciell fastighetsrätt* (Iustus, 1999), p. 20 and M. Band Eriksson, 'Placera masten rätt: Miljö och naturskyddskrav på 3G-utbyggnaden', *Svenska Naturskyddsföreningen Rapport* (Svenska Natuskyddsföreningen, 2003), p. 10.

of similar character and can be said to complement each other. However, some serious discrepancies are apparent. For example, the Planning and Building Act does not include anything resembling the precautionary principle included in the Environmental Code. According to the preparatory works to the Planning and Building Act, the rules relating to the care for common interests are of a deliberate general character, in order to allow for interpretation in the light of societal changes.[241] This would indicate that the precautionary principle, as well as the other general considerations in Chapter 2 of the Environmental Code, could be of relevance for the Planning and Building Act. This is further supported in the preparatory works, where it is held that the Environmental Code is to influence the interpretation of the Planning and Building Act and that that the increasing importance of environmental concerns and the considerations in Chapter 2 of the Environmental Act should play a role in the application of the Planning and Building Act.[242] This issue has been discussed in some highly interesting cases in which the plaintiffs have relied on the precautionary principle to challenge planning permission relating to base stations before the Swedish courts. This issue is returned to below.

In order to establish a base station, planning permission from the relevant local authorities needs to be sought.[243] Every local government should have a building committee which receives applications and grants planning permissions.[244] In general, planning permission regarding base stations is granted provided that the provisions in Articles 3:1–3:3 and 3:10–3:13 of the Planning and Building Act are fulfilled.[245] These articles include primarily aesthetic requirements, both with regard to other buildings in the area and nature. Only in Article 3:2 of the Planning and Building Act is it pointed out that installations are not to interfere negatively with traffic or otherwise pose 'significant detriment' for the surroundings.[246]

241 Proposition 1994/95:230, p. 30.

242 Proposition 1997/98:90, Följdlagstiftning till Miljöbalken mm., p. 158.

243 Plan- och Bygglagen 8:2, p. 5 establishes that radio or telephone masts or towers are subjected to a general requirement of authorisation.

244 Plan- och Bygglagen Article 8:19.

245 Plan- och Bygglagen, Article 8:11, p. 3 in conjunction with Plan- och Bygglagen Article 3:14. Slight differences in the procedure are in effect depending on whether the area where the installation is to be established has been included in a detailed area plan or not (compare Plan- och Bygglagen Articles 8:11 and 8:12). However, these differences are of no relevance for the discussion here.

246 'Betydande olägenheter'.

5.3.3 *The institutional framework*

Base stations for mobile telephony are regulated on two main levels in Sweden. First, the Radiation Protection Act establishes that the government, or an authority set up by the government, may issue binding regulations for radiation safety.[247] This task entails the setting of safety standards for different types of radiation, including mobile telephony. This has been delegated by decree to SSM.[248] Second, local governments have been authorised, through the Planning and Building Act, to decide on the location of base stations. This separation of competences indicates that SSM is the appropriate body to set the level of risk allowed in relation to base stations. Local authorities should thus be exempt from taking a more or less precautionary stance than that set by SSM. Their task, provided that the request for a building permit meets the SSM standard, is limited to ordinary planning concerns. The fact that this issue is far from uncontroversial is returned to shortly.

Like the Swedish agency responsible for chemicals, KemI, which was discussed above, SSM is an independent agency and the government is prevented from interfering with its work directly. However, and also as in the case of KemI, the government retains important possibilities to exert control over SSM and direct its work through both appointments and budget control documents.

SSM has a number of tasks relating to research and advice. It is the co-ordinator of radiation research in Sweden, and is responsible for providing information to the public in this field.[249] It may inform and issue recommendations regarding radiation protection,[250] and it has been given the power to issue prohibitions or other conditions where this is necessary for the fulfilment of the objectives in the Radiation Protection Act.[251] As with other Swedish agencies, the work of SSM is led by a Director General who is appointed by the government.[252] The Director General is seconded by, but not subordinated to, a board whose members are also appointed by the government.[253] A Transparency Council assists the Director General and the Board in the management of SSM.[254] A number of advisory boards

[247] Strålskyddslagen Article 12.
[248] Förordning (2008:452) med instruktion för Strålsäkerhetsmyndigheten.
[249] *Ibid.*, Articles 2 and 7. [250] *Ibid.*, Article 2, point 6.
[251] Strålskyddslagen Article 32.
[252] Förordning med instruktion för Statens strålkskyddsinstitut, Article 25.
[253] Verksförordning (SFS 1995:1322), Articles 32 and 33.
[254] Förordning (2008:452) med instruktion för Strålsäkerhetsmyndigheten, Article 19.

assist SSM on different matters under its jurisdiction.[255] Otherwise, SSM, like other agencies, is free to decide on its internal organisation.[256] Furthermore, as in the case of KemI, no apparent separation of risk assessment and risk management is in effect.

Furthermore, SSM has the power to issue binding regulations within the framework outlined above,[257] in order to ensure that the aims of the Radiation Protection Act are fulfilled. Such regulations are frequently issued by the agency. With regard to base stations, the SSM has stopped short of issuing binding requirements. Instead its predecessor, SSI, opted for issuing a non-binding General Advice.[258] It is stated in the General Advice that it aims to provide guidance for the interpretation of Article 6 of the Radiation Protection Act.[259] As mentioned above, Article 6 of the Radiation Protection Act establishes that radiation should be allowed only when it is at a level that is safe for humans, and that it should include a precautionary approach. This issue is returned to below. Even though not formally obliged to, it appears that courts have, so far, proven inclined to rely extensively on the SSI General Advice when deciding on issues relating to the risks stemming from base stations.[260] For the most part, this may be due to the fact the General Advice is the only comprehensive advice issued by a government body and that it effectively codifies the international standards in the area.

Since the General Advice is not binding, local authorities are not formally bound by it when dealing with planning matters which come before them. This could potentially create some leeway for authorities to make their own risk assessments. However, the Government, and, on appeal, the Supreme Administrative Court, have in three judgments stated that

[255] www.stralsakerhetsmyndigheten.se/Om-myndigheten/Organisation/Rad-namnder/ (accessed 1 March 2010).

[256] Verksförordning, Article 17. [257] 'Föreskrifter'.

[258] Statens strålskyddsinstituts föreskrift SSI FS 2002:3, Allmänna råd om begränsning av allmänhetens exponering för elektromagnetiska fält (General Advice on Limiting the Public's Exposure to Electomagnetic Fields (hereinafter the General Advice).

[259] The General Advice, Article 1.

[260] See the following judgments from the Swedish Supreme Administrative Court: Judgment of the Supreme Administrative Court in Case no. 1163–2001 delivered on 12 July 2001 (not reported); Judgment of the Supreme Administrative Court in Case no. 3635–1999 delivered on 14 October 1999 (not reported) and Judgment of the Supreme Administrative Court in Case no. 3650–1999 delivered on 14 October 1999. The latter was also reported in a note in the yearbook of the Swedish Supreme Administrative Court (RÅ 1999 note 217). See also, judgments from lower Environmental Courts in, for example, M 390–05, Växjö Tingsrätt, Judgement of 3 October 2005 and M 3123–04, Östersunds Tingsrätt, Judgment of 12 January 2005.

planning permissions regarding base stations cannot be challenged on health grounds when the radiation emitted from them is in line with the General Advice.[261] In Case 1163–2001 the County Administration[262] in Dalarna went as far as to state:

> The County Administration must, when deciding upon the appeal, consider what can be considered scientifically proven at the time of the decision.[263]

The decision of the County Administration was upheld in both the Government Decision and the Decision of the Supreme Administrative Court. It thus appears relatively clear from the case law that the courts consider the SSI General Advice to be the relevant provision against which to judge the risks stemming from base stations. In the light of this, local authorities should be exempt from taking a more precautionary stance than the one advocated there. That this is also the view of the Government was emphasised in a recommendation from the National Board of Health and Welfare,[264] which was addressed to local governments, in which it was stated that Swedish courts have found that the precautionary principle cannot be used by local governments to limit the planning of base stations.[265] The National Board of Health and Welfare has a stake in the regulation of base stations since it is the agency which is generally responsible for health issues in Sweden.

Finally, it should be mentioned that since radiation from base stations is considered environmentally hazardous according to Chapter 9 of the Environmental Code, local governments have the authority to oversee the fact that the activities are safe. However, when doing so, it appears that they should use the SSI General Advice as a benchmark.[266] This issue is returned to in the following section.

[261] *Ibid.*

[262] 'Länsstyrelsen' is the first instance of appeal of decisions taken by local government.

[263] 'Länsstyrelsen måste vid sin prövning av överklgandet ugtå från vad som kan anses vara vetenskapligt belagt vid tidpunkten för prövingen.' (Regeringsrättens dom i mål no. 1163–2001).

[264] Socialstyrelsen.

[265] Meddelandeblad från Socialstyrelsen till kommunala miljö- och hälsoskyddsnämnder, länsstyrelser och andra berörda myndigheter, June 2008. Available at: www.fas.se/upload/dokument/särskilda_områden/eloverkanslighet/2008-1-11_2008112.pdf (accessed 1 March 2010).

[266] See the reasoning by the Environmental Court of Appeal in Case MÖD 2005:55.

5.3.4 *Risk assessment and risk management*

Due to the importance given to the SSI General Advice by courts, as well as the administration, it can safely be argued that it codifies the Swedish approach to radiation from base stations. A closer look at the General Advice and the risk assessment on which it is based is, therefore, warranted.

As has been mentioned throughout this chapter, a separation of risk assessment and risk management is generally not in place in Sweden.[267] With regard to radiation, the SSM has the task of preparing research as well as taking decisions based upon it. With regard to base stations, the General Advice is not based upon an original risk assessment by SSI. Instead, the aim is to incorporate the EU Council Recommendations dating from 1999 in Sweden.[268] Thus, SSI substituted the major part of its risk assessment with that which has been accepted by the Council. It, in turn, merely incorporates the recommendations issued by the International Commission for Non-Ionising Radiation Protection (ICNIRP), which is based upon a literature study performed by ICNIRP.

The actual risk assessment performed by the SSI is limited to establishing whether the radiation emitted in Sweden falls within the scope of the ICNIRP recommendations. This assessment is performed in an SSI Report which precedes the issuing of the General Advice.[269] In this report, SSI presents findings from empirical studies of emitted radiation relating to mobile telephones in different situations. The study centres solely on the amount of radiation emitted, not on the actual or perceived effects it has or may have. The values measured are then compared with the ICNIRP values and declared to be in conformity with them.[270]

5.3.5 *Scientific evidence and the introduction of precautionary measures*

The precautionary principle is intended to guide all areas of environment and health protection in Sweden. This is stated in Article 2:3 of the Environmental Code, which has been discussed extensively above. That the precautionary principle is also meant to guide issues relating

[267] See, above, Section 5.1.4. [268] Article 1.1, SSI FS 2002:3.

[269] U. Bergqvist *et al.*, 'Exponering för radiofrekventa fält och mobiltelefoni', SSE rapport: 9 (April 2001).

[270] *Ibid.*, pp. 32–3.

to radiation is established in Article 6 of the Radiation Protection Act, in which it is held that 'precautionary measures' should be considered in order to mitigate the risks from radiation. In the preparatory works, it is emphasised that areas of scientific uncertainty should be covered by the provision.[271] The Planning and Building Act does not include any explicit reference to the precautionary principle. The courts and some government agencies have made it clear that local governments should not be guided by precautionary considerations when issuing planning permission for base stations. Thus, it appears that the precautionary principle is not a tool that can be applied at local level, but can be applied only by the central government or its agencies.

In 1996 a number of government agencies, including SSI, issued a joint statement (hereinafter the Joint Statement) with the aim of clarifying the application of the precautionary principle for low-frequency electromagnetic radiation.[272] The prime target for the Joint Statement was the 'ordinary sources' of electromagnetic radiation, such as household electricity, but it was indicated that it should also be used by analogy in other situations in which radiation may be a concern. In the commentary to the General Advice, SSI has explicitly referred to the Joint Statement in relation to base stations for mobile phones.[273] Thus, it can be assumed that the Joint Statement indicates how the precautionary principle is to be applied also in the field of mobile telephones. In the Joint Statement, the precautionary principle is defined as follows:

> If measures, which generally reduce exposure, can be taken at reasonable costs and other consequences, one should strive to reduce fields that deviate significantly from what can be considered normal in the local environment. As regards new electrical installations and buildings, one should strive at the planning stage to shape and place these so as to reduce radiation.[274]

[271] Proposition 1987/88:88, p. 65.

[272] Myndigheternas försiktighetsprincip om lågfrekventa electriska och magnetiska fälten vägledning för beslutsfattare. Available at www.av.se/dokument/publikationer/adi/adi_477.pdf (accessed 1 March 2010).

[273] Kommentarer till Statens strålskyddsinstituts allmänna råd (SSI FS 2003:3) om begränsning av allmänhetens exponering för elektromagnetiska fält, p. 2. Available at: www.ssi.se/forfattning/pdf/2002_3Komment.pdf (accessed 1 March 2010).

[274] Myndigheternas försiktighetsprincip om lågfrekventa elektriska och magnetiska fälten vägledning för beslutsfattare, p. 2.

According to SSI, this definition of the precautionary principle corresponds to the one included in Article 2:3 Environmental Code.[275] However, a closer look at the wording suggests that the Joint Statement is significantly weaker than the wording of the precautionary principle in the Environmental Code. By making precautionary measures contingent upon whether they can be taken incurring 'reasonable costs and other consequences', it appears that some form of extended cost-benefit analysis (costs and 'other consequences' are included) is at the centre of the equation. This would be in contrast to the Environmental Code, in which it appears that precautionary measures should always be taken unless manifestly unreasonable.[276] Furthermore, the definition provided by the government agencies holds that precautionary measures should only be triggered when radiation 'deviates significantly' from what is 'normal in the local environment'. This seems to indicate that risk levels are not absolute, but that they may vary between different areas. Such an approach appears a far cry from the strict and absolute precautionary principle in the Environmental Court.

That the precautionary principle is applied very differently when it comes to base stations than is the case in pesticide regulation becomes clear when considering the General Advice in detail. The General Advice is based directly on the ICNIRP recommendations, as accepted by the Council Recommendations. Thus, the precautionary considerations taken when constructing the ICNIRP values are the ones in effect in Sweden. However, the ICNIRP guidelines are based on 'scientific data alone'.[277] This means that only scientifically certain[278] data is taken into account when setting the suggested exposure limits. That this is the case is illustrated by the discussion on potential cancer risks in the ICNIRP Guidelines. It is recognised here that epidemiological research has suggested a potential connection between exposure to low-frequency electromagnetic fields and the development of cancer in humans. This issue is also receiving increasing attention from well-reputed researchers in the

[275] Kommentarer till Statens strålskyddsinstituts allmänna råd (SSI FS 2003:3) om begränsning av allmänhetens exponering för elektromagnetiska fält.

[276] See, above, Section 5.1.

[277] International Commission on Non-Ionizing Radiation Protection, 'Guidelines for Limiting Exposure to Time-Varying Electric, Magnetic, and Electromagnetic Fields (up to 300 GHZ)', *Health Physics*, 74:4 (1998), 494 (hereinafter the ICNIRP Guidelines).

[278] Data that is certain with at least a 95 per cent probability. Compare discussion above in Sections 2.3.1–2.3.3.

area.[279] But since the data is 'insufficient', the ICNIRP does not consider it necessary to adjust the exposure guidelines in the light of this.[280] It furthermore indicated that epidemiological studies alone are not sufficient to warrant stricter guidelines. The results of such studies must be substantiated by laboratory experiments.[281] Scientific uncertainty is recognised by the ICNIRP, but is not allowed to influence the setting of exposure limits. A more recent review by the ICNIRP Standing Committee on Epidemiology found that current epidemiological studies were not only marred by inconsistencies but were also lacking in substantive quality. Thus, they did not warrant any stricter guidelines than the ones already in place.[282] The approach suggested by the ICNIRP, which is implemented in the EU Council Recommendations as well as in the SSI's General Advice, must be considered to be protective against scientifically certain risks, but does not aim to be protective under conditions of uncertainty. They do not 'err on the side of caution', but disregard risks that are not scientifically verified. Thus, the guidelines are more preventive than precautionary in character.

The above conclusion does not stop SSI from holding explicitly that 'base stations do not cause any detrimental health effects and do not pose any risk from the point of view of radiation protection'.[283] This is taking the matter a step further even than the ICNIRP, in which, at least, the uncertainties are emphasised, rather than merely glossed over with categorical statements. In November 2009 the same conclusions were repeated by SSM in a joint statement with the radiation protection agencies of the other Nordic States.[284]

The courts have also proven inclined to rely only on what is scientifically verified in cases relating to base stations. For example, one of the cases discussed above was brought by a woman who claimed to be overly

[279] See, for example, M. Blank, 'The Precautionary Principle Must Be Guided by EMF Research', *Electromagnetic Biology and Medicine*, 25 (2006), 203–8.

[280] ICNIRP Guidelines, pp. 496 and 503.

[281] *Ibid.*, p. 503.

[282] ICNIRP (International Commission on Non-Ionizing Radiation Protection) Standing Committee on Epidemiology. Epidemiology of Health Effects of Radiofrequency Exposure. *Environmental Health Perspectives*, 112:7 (December 2004).

[283] 'Strålsäkerhetsmyndigheten bedömer därför att exponeringen från basstationer inte orsakar några skadliga hälsoeffekter och inte innebär några risker från strålskyddssynpunkt.' Taken from the web page of the Swedish Radiation Safety Agency: www.stralsakerhetsmyndigheten.se/allmanhet/Magnetfalt–tradlos-teknik/Mobiltelefoni/Basstationer/

[284] www.stralsakerhetsmyndigheten.se/Global/Pressmeddelanden/2009/091116norden-emf.pdf (accessed 1 March 2010).

sensitive to electromagnetic fields.[285] She held that due to her condition, base stations in the area she lived caused her significant problems. She based her argument upon a number of scientific reports, which suggested biological effects as a result of exposure to electromagnetic fields.[286] The County Administration, however, ignored the scientific reports introduced by the plaintiff, seemingly on the basis that they were marred by uncertainty. As cited above, the Court even went as far as to state that it would only base decisions on what can be considered scientifically proven at the time. Thus, the Court merely concluded that, since the emission of radiation affecting the plaintiff was within the ICNIRP guidelines, no other measures needed to be considered. The case was upheld in full on appeal to the Government as well as to the Supreme Administrative Court. When this approach is compared to the approach assumed in relation to pesticides, the differences are significant. Where uncertainties in relation to pesticides are emphasised and used as a foundation for precautionary regulation, they are downplayed to the point of ignorance in relation to base stations.

In a more recent case, however, the Environmental Court of Appeal took a more nuanced view of the potential risks stemming from base stations. In its judgment of October 2005, it recognised that thermal effects may not be the only risk in relation to base stations, but that some scientific findings also indicate biological effects which are not accounted for in the ICNIRP values. This the Court found sufficient to support a decision to declare the siting of base stations as an environmentally hazardous activity.[287] However, the practical implications of this finding are unclear, as was touched upon above. It does not appear that, in practice, it would require a more precautionary approach to be assumed. Interestingly, the Court also mentioned that it found the psychological unease and worry of people living near base stations sufficient to consider the activity as environmentally hazardous.[288] In practice, it is unlikely that the findings of the Environmental Appeal Court will entail any concrete change with regard to what needs to be shown regarding the risks associated with base stations.

The reluctance of SSI, SSM and the courts to assume a more precautionary stance have, in some cases, prompted the local government to contemplate taking precautionary measures on its own. The discussion

[285] The Judgment of the Administrative Supreme Court Case no. 1163–2001.
[286] *Ibid.*
[287] Judgment of the Environmental Supreme Court in Case MÖD 2005:55.
[288] *Ibid.*

has been centred on using planning decisions taken under PBL to create so-called 'radiation-free zones'. Planning decisions are not legally binding, but are intended to guide the development at local level. Due to the inherent nature of the policy, it is unlikely, at this point, that a planning decision establishing a radiation-free zone could be successfully challenged before the courts. The cases referred to in this chapter have all been cases in which individuals have challenged planning permission, which have clear and unambiguous legal effects. However, it could be foreseen that a request for a building permit regarding a base station could be denied on the basis that the area in question has been declared radiation-free. In that case, the decision to deny the permit could be challenged before the courts, and they would, in turn, be forced to consider whether the establishment of radiation-free zones are within the ambit of the PBL.

5.3.6 *The burden of proof*

At this point, the Swedish Government has not assumed a particularly precautionary approach with regard to base stations. By relying on international standards and EU recommendations, it has effectively prevented itself from making its own risk assessment of the issue. The burden of showing risks stemming from base stations appears to rest firmly with the party making that claim. The courts have, so far, not found that an applicant has managed to fulfil this obligation, and thus no successful appeals regarding decisions to site base stations on health grounds exist.[289] Courts have made firm demands on the scientific evidence needed for a successful appeal.

As mentioned above, the Environmental Appeal Court has, in one instance, highlighted uncertainties in relation to base stations.[290] However, this case related to whether the local government was obliged to check the compliance of an undertaking with the radiation levels currently recommended in Sweden. By declaring that release of radiation from third-generation mobile phone systems is, indeed, to be considered

[289] Judgment of the Supreme Administrative Court in Case no. 1163–2001; Judgment of the Supreme Administrative Court in Case no. 3635 and Judgment of the Supreme Administrative Court in Case no. 3650–1999. The latter was also reported in a note in the yearbook of the Swedish Supreme Administrative Court (RÅ 1999 note 217). See also judgments from lower environmental courts in, for example, M 390–05 and M 3123–04.

[290] Case M 7485–04.

to be an environmentally hazardous activity, the Court found this to be the case. It is unlikely that the Environmental Court of Appeal would find the same scientific evidence considered in the case to be sufficient to overturn the granting of a building permit of a base station which complied with the said recommendations.[291] Previous case law unequivocally indicates that the guidelines issued by SSI, which are, in turn, based on the ICNIRP recommendations, assure an adequate level of protection. This clearly indicates that the burden of proof in cases relating to the risks stemming from base stations is firmly placed with the party claiming the health risk, at least in situations in which the recommended limits are abided by. This is the direct opposite of the situation in relation to pesticides, where uncertainties with regard to potential health risks explicitly mandate strict regulation.

5.3.7 Cost–benefit analysis

Nothing is mentioned in relation to cost–benefit analysis in either the SSL or in the General Advice. However, the general 'unreasonable test' in Article 2:7 of the Environmental Code must be considered applicable also in the area of radiation, just as in the case of pesticides. As discussed above, the requirement enshrined in Article 2:7 of the Environmental Code is not a traditional cost–benefit or proportionality requirement. Instead, it aims to codify earlier Swedish practice which stated that actions to protect the environment should not be taken only in cases in which this was not 'environmentally justified'. Above, it was argued that it appears that such situations are those situations in which excessive costs are incurred and the result is only limited environmental benefits. Thus, the presumption is that measures should be taken. Only where they would be manifestly unreasonable should they not be taken.

In the Joint Statement, on the other hand, the role of economic considerations is strongly emphasised when it comes to regulating the risks from radiation. In the definition of the precautionary principle found there, it is emphasised that measures should be taken only if they incur 'reasonable costs and other consequences'.[292] Through this, effectiveness considerations have been given a more pronounced place when

[291] *Ibid.*

[292] Myndigheternas försiktighetsprincip om lågfrekventa electriska och magnetiska fälten vägledning för beslutsfattare, p. 2.

contemplating precautionary measures in relation to radiation than is the case in relation to, for example, pesticides. That what is referred to is a bona fide cost–benefit analysis is clearly emphasised in the text of the Joint Statement, in which the suggested considerations are centred on the monetary costs of saving statistical lives.[293] However, even though it clearly advocates a cost–benefit-based approach to precautionary measures, the Joint Statement stops short of defining what would be a 'reasonable' cost for saving one statistical life. This issue it apparently leaves to the decision-makers in individual cases. On the whole, the role of cost–benefit considerations appears wholly different in instances of radiation than is the case with regard to pesticides, where officials at KemI held that the prospective costs of regulation play no role whatsoever.[294]

5.3.8 *Conclusion*

Three main legislative acts are involved at different, but partly overlapping, levels when it comes to the regulation of base stations in Sweden. First, the Environmental Code establishes that all aspects of environment and health protection in Sweden should be guided by the precautionary principle. The content of the strict precautionary principle enshrined there has been discussed extensively in the previous chapters. The Environmental Code also establishes that both ionising and non-ionising radiation are to be considered as 'environmentally hazardous activities' if they 'may cause detriment to the surroundings'. That this is the case for base stations has been established by the Supreme Environmental Court. However, so far, this finding has had little impact on how base stations are regulated in Sweden.

Second, instead of a centralised procedure, the placing of base stations is regulated as a general planning decision under the Planning and Building Act. Such decisions are taken at local level by local governments. The Planning and Building Act does not include a specific precautionary principle, but a study of preparatory work and the literature suggests that the Environmental Code version of the precautionary principle should be guiding planning decisions as well. However, the Swedish Agency for Social Affairs, which is mainly responsible for health issues, has, in no uncertain terms, indicated in a communication to local government that

[293] *Ibid.*, pp 3, 6–7.

[294] Interview with officials at KemI, 9 September 2004.

the precautionary principle should not be applied with regard to base stations at local level. This approach has been upheld by courts in the scarce case law existing in the area.

Third, the Radiation Protection Act lays down the framework for radiation protection in general. The Radiation Protection Act covers both ionising and non-ionising radiation, and, read in the light of the preparatory work, it includes a weak precautionary principle. It allows for the delegation of the issuing of standards and the control of the implementation of the said standards to a government agency. The agency that has been given these tasks is SSI, which, in relation to base stations, has stopped short of issuing binding requirements on the emission of radiation from base stations. Instead, it has chosen to implement the EU Council Recommendation, which, in turn, is a direct implementation of the ICNIRP Recommendation, in the form of a General Advice. This General Advice has been relied on heavily by courts in a few cases. However, even if no strong precedents have, so far, been created, it must be assumed that the General Advice is a direct manifestation of the Swedish policy toward radiation from base stations. Since the ICNIRP Recommendation is based solely upon scientifically verified thermal effects in humans, it can hardly be perceived as incorporating a precautionary approach in the broad sense. Minority medical and epidemiological studies exist, which suggest the presence of risks which were not taken into account when the Recommendation was formulated. The scientific uncertainty in the area must be considered as being relatively great, and is generally acknowledged by scientists and decision-makers. This uncertainty is largely due to the lack of conclusive research, rather than due to conflicting evidence. This, however, is not recognised by SSI, which, in its information material, repeatedly states that base stations are safe from a human health perspective.

The finding that the precautionary principle applied in cases of low-frequency radiation is different and weaker than the remarkably strict approach found in the area of pesticides is supported by a joint communication from a number of Swedish agencies. In the communication, the precautionary principle in the area of low-frequency radiation is understood as relative to the area in which it is being applied and directly contingent upon cost–benefit considerations.

Some facts make the leniency toward mobile telephony understandable. In Sweden the Government estimates that nine out of every ten people in the age group of 16–75 years use mobile phones, and Sweden's largest infrastructure project is the extension of third-generation mobile telephony to the whole country. Information technology and telecommunications has rapidly become the most important sector of Swedish industry, and it currently employs directly almost 200,000 people. Telecommunications accounts for 20 per cent of the overall industrial value of Sweden and 15 per cent of total exports.[295] The mobile phone systems manufacturer Ericsson undoubtedly accounts for a sizeable share of these figures.

When considering the political importance that the Swedish Government has attached to full geographical mobile phone coverage, and the macro-economical importance of the Swedish telecom sector, downplaying the risks stemming from base stations makes political sense. However, if one values a rational agenda-setting based on public health protection, the setting of actual risk levels and consistency in the application of the law, this is an approach which is difficult to defend. When comparing the Swedish use of the precautionary principle in the area of pesticides with how it has been used in the area of base stations, it appears that political and economic considerations are highly influential in determining how comparable risks should be treated. This is a far cry from the strict and absolute precautionary principle advocated in the Environmental Code. And, furthermore, it suggests that the precautionary principle in Sweden in its present form may often function as a political tool rather than as a legal instrument which is equally applicable in all similar situations.

[295] http://web.archive.org/web/20080417041920/http://www.sweden.se/templates/cs/FactSheet____14931.aspx (accessed 1 March 2010).

6

The precautionary principle in the United Kingdom

6.1 The development and status of the precautionary principle in the United Kingdom

6.1.1 Introduction

In Sweden the precautionary principle has, in one way or another, been a basic element of environmental and health protection for more than thirty years. It is firmly embedded in legislation and has been fleshed out, to a considerable degree, in preparatory works, case law and guidance documents. In the UK the story is very different. Even though precautionary decisions have, historically, been taken in various policy sectors in the UK, a more consistent approach, spanning all policy areas, was not considered until the early 1990s. Lacking reference in statutory instruments, the precautionary principle has been dealt with in documents of a policy nature. Following this development, the precautionary principle has tentatively been relied on before the UK courts. Since the precautionary principle has not been included in UK legislation, the sources for this chapter mainly include policy documents and relevant judgments of UK courts.

In this chapter the background, development and status of the precautionary principle in the UK will be analysed and discussed. As in the previous chapters, constituent elements of the precautionary principle, such as risk assessment and management, the role of scientific evidence and cost–benefit analysis are discussed separately.

6.1.2 Background and development

Where the precautionary principle in Sweden can be traced back to at least the late 1960s, a general precautionary mechanism in the UK is a much more recent concept. This does not, however, mean that precautionary concerns have not been influential in shaping policy. As the Australian judge Stein J of the Land and Environment Court of New South Wales has stated:

> [The precautionary principle] is a statement of common sense and has already been applied by decision-makers in appropriate circumstances prior to the principle being spelt out.[1]

Even though the term 'precautionary principle' is a recent concept, UK risk regulation can be traced back to nineteenth century and the first Smoke Abatement Law, which restricted the burning of coal in London.[2] Other notable early pre-precautionary principle actions include the introduction of the Alkali Act 1863, which was concerned with the regulation of heavy chemicals.[3] These acts represent regulation undertaken in areas affected by varying, but significant, degrees of uncertainty. Thus, they can also be considered to have been issued on the basis of some basic precautionary considerations. Early developments such as those described above have little to do with a comprehensive, modern version of the precautionary principle, but are important examples of the fact that regulation in the face of varying degrees of uncertainty is hardly a new issue. The precautionary principle as an independent legal instrument, on the other hand, is.

Following the increasing international concern for environmental issues in the last two decades of the twentieth century, UK Government statements and publications tentatively began to recognise the precautionary principle in a broad sense in the late 1980s.[4] In January 1988 the Secretary of State for the Environment stated that the UK officially accepted the precautionary principle.[5] However, in the UK, as elsewhere, a more precise content of the principle was not elaborated on at that time.

Before the UK courts the precautionary principle was relied on as an independent legal ground in the Duddridge case of 1994.[6] The case

1 *Leatch* v. *Director-General of National Parks & Wildlife Service and Shoalhaven City Council* [1993] NSWLEC 191 (23 November 1993) (Land and Environment Court of New South Wales) (No. 10376 of 1993). Available at: www.elaw.org/resources/text.asp?id=2791 (accessed 1 March 2010).

2 See, for example, C. Flick, 'The Movement for Smoke Abatement in 19th-Century Britain', *Technology and Culture*, 21:1 (1980), 29.

3 See, for example, W. A. Damon, 'The Alkali Act: Its Administration and Relation to Chemical Industry', *Journal of the Society of Chemical Industry*, 54:50 (2007), 1070.

4 For a summary of such examples see Haigh, 'The Introduction of the Precautionary Principle into the UK', pp. 247–51.

5 *Ibid.*, p. 247.

6 *R.* v. *Secretary of State for Trade and Industry ex parte Duddridge and Others*, Queen's Bench Division (Crown Office List), *The Independent*, 4 October 1994.

concerned an application for judicial review of a decision whereby the Secretary of State for Trade and Industry had declined to issue regulations restricting the electromagnetic fields created by electric cables. The application was brought on behalf of three children living in an area in which the new National Grid was to be drawn. In this case the ruling judge found that the precautionary principle was essentially designated for environmental protection, and that the only reference of potentially binding nature in the UK would be Article 191 TFEU (formerly Article 174 EC). However, at EU level, the Court found the precautionary principle to be limited to the interpretation of EU legislation, and it thus has no relevance for the application or interpretation of strictly national legislation.[7] The Court did not rule out the possibility of taking precautionary measures, but stated that it would be for the Secretary of State to use his discretion to decide whether to issue these or not. Thus, the Court did not find the precautionary principle to function independently or to bind the Government positively to act in a precautionary manner.[8] The judgment in Duddridge is discussed below and compared to more recent case law and other developments, including, among other documents, the 2002 Report of the Inter-Departmental Liaison Group on Risk Assessment (hereinafter the ILGRA Report).[9]

6.1.3 The current context and definition of the precautionary principle

Early documents relating to the precautionary principle in the UK are mainly concerned with establishing that the UK accepts the concept, but do not flesh out the actual content of it.[10] The most commonly referred to document in this regard is the 1990 White Paper, 'This Common Inheritance', which includes a passage on precaution.[11] This section states that:

7 *Ibid.* 8 *Ibid.*

9 The Precautionary Principle: Policy and Application, Report of the Interdepartmental Liaison Group on Risk Assessment, 2002. Available at: www.hse.gov.uk/aboutus/meetings/committees/ilgra/minrpt3.pdf (accessed 1 March 2010).

10 For a recent study of the precautionary principle in the UK see A. Patterson, *Risk Regulation and Scientific Expertise in the United Kingdom: The Precautionary Principle in Public Policy* (Mellen Press, 2008).

11 White Paper, 'This Common Inheritance: Britain's Environmental Strategy', presented to Parliament, September 1990, HMSO, Cm 1200.

> Where there are significant risks of damage to the environment, the Government will be prepared to take precautionary action to limit the use of potentially dangerous materials or the spread of potentially dangerous pollutants, even where scientific knowledge is not conclusive, if the balance of likely costs and benefits justifies it. This precautionary principle applies particularly where there are good grounds for judging either that action taken promptly at comparatively low cost may avoid more costly damage later, or that irreversible effects may follow if action is delayed.[12]

In the following section of the White Paper, it is emphasised that precautionary action must be based on 'sound science'.[13] The definition contained in the White Paper appears to make any precautionary measures contingent upon cost–benefit analysis. Thus, precautionary action should only be taken where the benefits of action outweigh the costs. Furthermore, it should be pointed out that the definition does not limit the precautionary principle to situations in which the potential damage associated with an activity are irreversible, but it does emphasise that precautionary measures in such situations may be particularly motivated.

After the adoption of the Rio Declaration in 1992, the UK definitions of the precautionary principle have centred on the version included in its Principle 15.[14] Increasingly, the principle appears to be defined in broad and general terms as an instrument which enables action at an early stage and aims to avoid 'paralysis by analysis'.[15] Such broad statements have led some commentators to consider it 'a rather shambolic concept muddled in policy advice and subject to the whims of international diplomacy and the unpredictable public mood over the true costs of sustainable living'.[16] The case law has also failed to define the principle in a more concrete manner. Instead, it has been pointed out that 'no comprehensive or authoritative definition' exists in the UK.[17] The courts have sometimes held that the definition and practice in relation to the precautionary principle which has developed at EU level should only bind the UK in as much as an EU legal act is being interpreted.[18] However, in *Turnbull* the Court seemed to indicate that the precautionary principle as interpreted in Article 191

[12] *Ibid.*, 1.18. [13] *Ibid.*, 1.19.

[14] See, for example, the ILGRA Report. [15] *Ibid.*, pp. 5–6.

[16] C. O'Riordan and A. Jordan, *The History and Contemporary Significance of the Precautionary Principle, in Interpreting the Precautionary Principle* (Earthscan, 1994), p. 23, and more recently quoted in J. Thornton and S. Beckwith, *Environmental Law* (Sweet & Maxwell, 2004), p. 15.

[17] *R.* v. *Secretary of State for Trade and Industry ex parte Duddridge and Others*, p. 3.

[18] *Ibid.*, and *R.* v. *Derbyshire Country Council*, ex parte Murray, Queen's Bench Division (Administrative Court), 6 October 2000.

TFEU (formerly Article 174 EC) could also have some influence in circumstances governed solely by UK law.[19] It was emphasised that this would have to be done on a case-by-case basis. That the precautionary principle as interpreted at EU level also has relevance in UK domestic affairs is further supported by the ILGRA Report. Here it was held that:

> [a]pplying the precautionary principle is essentially a matter of making assumptions about consequences and likelihoods to establish credible scenarios, and then using standard procedures of risk assessment and management to inform decisions on how to address the hazard or threat.[20]

More recently, The Science and Technology Committee of the House of Commons has expressed concern with regard to both the definition and the application of the precautionary principle in the UK. It stated that, due to the confusion surrounding the application of the principle and the different interpretations available, the principle has been 'devalued and [is] of little practical help'. Pending further clarification with regard to scientific methodology, the Committee finds the principle to be unrealistic and impractical. Furthermore, it expressed its preference for using the term 'precautionary approach' instead of 'precautionary principle'. Future work on the clarification of the principle or approach should be focused on developing adequate 'risk management theories' to be applied under uncertainty.[21]

Similar concerns have been expressed by the Select Committee on Economic Affairs in the House of Lords, which has stated that it finds the precautionary principle 'ill-defined and ambiguous', and that it may lead to the assuming of an excessively cautious attitude to risk.[22]

6.1.4 *Judicial review*

Since the precautionary principle lacks an independent history in the UK, it has been held that a 'true test' of its impact will be how it is received by the UK courts.[23] Traditionally, judicial review in the UK is rather

[19] *R.* v. *Environment Agency ex parte Turnbull*, Queen's Bench Division (Crown Office List), CO/4848/99, 12 January 2000.

[20] The ILGRA Report, p. 2.

[21] House of Commons Select Committee on Science and Technology, 'Scientific Advice, Risk and Evidence Based Policy Making', pp. 82–3.

[22] House of Lords Select Committee on Economic Affairs, 'Government Policy on the Management of Risk', Vol. I, p. 25.

[23] E. Fisher, Is the Precautionary Principle Justiciable?, 315–34.

restrained. Not all measures, or aspects of measures, are subject to review by the courts. A few main issues in relation to judicial review are of interest here. First, high-level policy issues are deemed inappropriate for judicial review. As in the case of Sweden discussed above, the courts are prevented from 'second-guessing' what are eminently political choices made by, or on behalf of, elected officials.[24] As has been discussed in previous chapters, this has great significance for the precautionary principle since its application is generally decided on by the government or on behalf of the government with significant discretion.

Second, courts defer to the discretion of bodies that have had decision-making powers delegated to them. In such cases the courts will review the formal procedure under which the decisions were taken, but will only make a limited review of the substance of the case. This review is centred on the issue of whether the body acting with discretion has acted 'unreasonably', and, if this is found to be the case, the measure can be quashed. The requirements for this to be the case are set very high.[25] It has been held that this review is comparable to the limited review performed by the European Court of Justice.[26] Such review is significantly more restrictive than the review performed by Swedish courts.[27]

Since the application of the precautionary principle is often highly political, and is used in cases of substantive uncertainty, often by bodies which have had decision-making powers delegated to them, it is not surprising that the UK courts have, to date, trod carefully in relying on the principle as an independent ground. In *Alfred McAlpine Homes* the Secretary of State for the Environment applied the precautionary principle as a statement of policy, and found explicit support for this before the Queen's Bench, and, even if it was not as explicit, on appeal before the Appeals Court.[28] In *Duddridge* the Court refrained from testing the substance of the case against the precautionary principle based on the fact that it considered its application to be within the scope of the discretion of the Secretary of State for the Environment.[29] A similar approach was

[24] Fisher, 'Is the Precautionary Principle Justiciable?', 315–34.

[25] P. Cane, *Administrative Law* (Oxford University Press, 2004), pp. 250 *et seq.*

[26] Fisher, 'Is the Precautionary Principle Justiciable?', 315–34.

[27] In this regard see, above, Section 5.1.4.

[28] *Alfred McAlpine Homes (North) Ltd.* v. *Secretary of State for the Environment and Another*, Queen's Bench Division (Crown Office List), 5 April 1993 and *Alfred McAlpine Homes (North) Ltd.* v. *Secretary of State for the Environment and Another*, Court of Appeal (Civil Division), 3 November 1994.

[29] *R.* v. *Secretary of State for Trade and Industry ex parte Duddridge and Others.*

assumed by the Queen's Bench in *Re Cullen*.[30] Similar for these cases is that they all uphold the decisions taken by the Secretary of State using his discretion. This is the case in *Alfred McAlpine* and *Re Cullen* in which the precautionary principle was used to justify measures, but also in *Duddridge*, in which the principle was disregarded. Thus, it appears that the courts are reluctant to review measures taken with reference to the precautionary principle as well as measures taken disregarding it. Despite this, the courts have been inclined to perform a surprisingly material review of the scientific evidence placed before them. This issue is returned to in the following section.

6.1.5 *Risk assessment and risk management*

No general explicit reference to formal requirements on decision-making in relation to risks is found in UK legislation. However, during the 1980s and 1990s the understanding and use of analytical risk assessment increased markedly in the UK.[31] The relevance of risk assessment in the respective sectors relevant here is returned to in the following sections.

In 1995 the Department of the Environment published its first Guidelines outlining how it intended to make use of risk assessment, risk management and risk communication.[32] This guide has since been updated and is available in its present form on the Department of the Environment webpage.[33] In the Guidelines the separation of risk assessment, risk management and risk communication is maintained, and each is elaborated on in a rather sophisticated fashion. When compared to, for example, the Commission Communication on the Precautionary Principle,[34] the Guidelines appear nuanced and more overtly concerned with the process of decision-making. However, where one main feature of the Communication on the Precautionary Principle is the issue of how to obtain independent scientific advice and the fundamental importance of the separation of risk assessment and risk management, the Guidelines

[30] *Re Cullen*, Queen's Bench Division, NIQB 9, 14 February 2005.

[31] E. Fisher, 'The Rise of the Risk Commonwealth and the Challenge for Administrative Law', *Public Law*, 2003, 455.

[32] Department for Environment, Food and Rural Affairs, *A Guide to Risk Assessment and Risk Management for Environmental Protection* (HMSO, 1995).

[33] Department for Environment, Food and Rural Affairs, *Guidelines for Environmental Risk Assessment and Management*. Available at: www.defra.gov.uk/environment/risk/eramguide/ (accessed 1 March 2010) (hereinafter the Guidelines).

[34] Communication from the Commission on the Precautionary Principle, COM 2000/1/Final.

centres mainly on procedural issues of risk assessment and substantive issues of relevance for the risk management decision. This is exemplified by the organisation chart included, which prescribes rather detailed theoretical models for risk assessment,[35] as well as the concern with risk trade-off analysis and cost–benefit analysis in the risk management phase. Less effort is spent on illuminating the border between what are risk assessment issues and what are management choices. Furthermore, little attention is given to the issue of how to make sure that as objective and independent scientific advice as possible is obtained.

In the ILGRA Report on the Precautionary Principle it is not explicitly stated where in the process the precautionary measures should actually be decided. However, considering that emphasis in the Report is put on the fact that the principle should be applied after a risk assessment has taken place with consideration of cost–benefit analysis, it appears that the decision to issue precautionary measures is intended to be taken in the risk management phase.[36]

Given the lack of a clearer distinction between risk assessment and risk management in either legislation or in other government documents, it is hardly surprising that the courts have not touched upon the issue. However, considering what was stated above in relation to the deference of the courts to the discretion of the decision-maker, it could be presumed that they consider the application of the precautionary principle a management choice. Since management decisions are taken using the discretion that the relevant body has had delegated to it, the courts show significant deference in reviewing their substance.

6.1.6 Scientific evidence and the introduction of precautionary measures

So far, this chapter has shown that the role of the precautionary principle as a generally applicable principle in the UK is rather weakly defined. The courts have proven as reluctant as the legislator to flesh out its content and application. Despite this, there is evidence that precautionary measures are applied in the UK in specific situations. The question, then, arises as to when and how such measures are invoked.

[35] Figure 2.1. Available at: www.defra.gov.uk/environment/risk/eramguide/03.htm (accessed 1 March 2010).

[36] ILGRA Report, pp. 6 and 10.

As mentioned above, scientific analysis and scientific proof were instrumental in setting the UK risk agenda during the 1980s and 1990s. It has been held that failure to act before full scientific proof was available contributed to, for example, the BSE crisis.[37] The importance of scientific evidence also in cases of uncertainty is supported in UK Government documents relating to the precautionary principle. The UK Government's White Paper on its Environmental Strategy of 1990 held that precautionary measures should be considered either 'where there are good grounds for judging either that action taken promptly at comparatively low cost may avoid more costly damage later, or that irreversible effects may follow if action is delayed'.[38] 'Good grounds', here, appears to allude to the fact that some scientific basis needs to support the measure. Similar sentiments were voiced almost ten years later in the UK Government's White Paper on Environmental Strategy of 1999. Although it was emphasised that action to protect the environment should not only be considered in situations of 100 per cent certainty, it was stressed that 'significant risk' of damage to the environment would have to be present.[39] This indicates that, even if full scientific proof is not required, some degree of scientific evidence needs to support the decision.

In the ILGRA Report scientific evidence is further emphasised and elaborated on. Here, it is held that 'good reason' to believe that harmful effects would occur needs to be:

> demonstrated by empirical evidence; by analogy with another activity, product or situation which has been shown to carry a substantial adverse risk or by showing that there is a sound theoretical explanation (tested as necessary by peer review) as to how harm might be caused.[40]

Thus, it appears that 'good reason' can be established by empirical studies as well as by theoretical modelling or through analogy with similar activities. Nothing is stated in relation to whether the scientific evidence relied upon must represent the majority of scientific opinion.

In *Duddridge* the Court appears to have made a relatively material review of the scientific evidence relevant for the case, only to conclude that a 'significant risk' of cancer had not been shown to emanate from the electromagnetic fields at issue in the case. However, the Court did not rule out the fact that electromagnetic fields may still constitute a risk. In

37 Fisher, 'The Rise of the Risk Commonwealth and the Challenge for Administrative Law', 7 (in LEXIS NEXIS version).

38 'This Common Inheritance, Britain's Environmental Strategy', Section 1.18.

39 *Ibid.*, Section 11.

40 ILGRA Report, p. 6.

fact, it appears that the Court considered the EU definition of the precautionary principle to be wider than the more circumscribed version which it considers being in effect in the UK. It even conceded that, if the Court were to consider the EU version to be in effect in the UK (which it did not), then the Secretary of State would have been under an obligation to apply the precautionary principle.[41] This is interesting for three reasons. First, the Court considers the UK version of the precautionary principle to be different and less onerous on the regulator than that in effect at EU level. The Secretary of State in this case was explicitly allowed to determine the scope of the principle using his discretion and the courts would not challenge this. Second, the UK courts, if they follow the *Duddridge* model, will materially examine scientific evidence put before them in order to establish the level of risk. Third, if a 'significant risk' is found to be present, the regulator appears under an obligation to act in a precautionary manner provided that a cost–benefit analysis supports the measures.[42] It is important to observe that the term 'significant risk' is not equivalent to any risk, but something needs to qualify it as significant.

In a more recent case in relation to base stations, which will be returned to in more detail below, the Queen's Bench generally upheld the approach established in *Duddridge*. The issue in *Al-Fayed* centred on the risks posed by base stations for mobile telephones.[43] The Court held that the uncertainty surrounding the health implications of radiation from base stations was not sufficient to justify a precautionary approach. The judge evaluated the available scientific evidence and failed to find any scientifically supported allegation of risk. Thus, the Court saw no reason to issue any type of precautionary measures on the side of the regulator. The judgment was upheld in this part on appeal.[44]

6.1.7 *The burden of proof*

In the ILGRA Report on the Precautionary Principle the burden of proof is dealt with under a separate sub-heading.[45] Here, the issue of

[41] *R. v. Secretary of State for Trade and Industry ex parte Duddridge and Others*, p. 7 (in LEXIS NEXIS version).

[42] *Ibid.*

[43] *R. v. Tandridge D.C. and another, ex parte Al-Fayed*, Queen's Bench Division Crown Office List) 79 P / CR 227, [1999] JPL 825, [1999] 1 PLR 104, *The Times*, 28 January 1999, CO/3665/97, 14 January 1999.

[44] *R. v. Tandridge D.C. and another, ex parte Al-Fayed.*

[45] ILGRA Report, p. 11.

pre-authorisation regimes is primarily dealt with. It is in this area, the Report implies, that the reversed burden of proof plays its largest role. Pesticide regulation and the permitting regime for nuclear plants are given as examples. The Report points out that the burden of proof may be fashioned differently in different areas in which pre-authorisation regimes are in place. In pesticide regulation the level of acceptable risk is decided on by the Government. It is then the responsibility of the manufacturer of a pesticide to convince the decision-maker that its product meets this level in a concrete case. This is done by producing a risk assessment in favour of allowing the pesticide in question. As mentioned in previous chapters, the burden of proof may, in such cases, be conceived as being reversed. However, the initial burden of proof generally rests with the Government to indicate a potential risk with a substance. This level of proof may, indeed, be low. Once it has been met, the onus is shifted on to the applicant to show that the assumption of risk is unfounded.[46]

Furthermore, it is made very clear in the ILGRA Report that the Government wants to retain wide latitude to prescribe in greater detail where the burden of proof should be placed in different regimes.[47] Thus, it appears that the burden of proof is not intended to be static in relation to the precautionary principle. Instead, it should be allowed to fluctuate according to the intent of the legislator.

The above, in combination with the limited scope for judicial review that the courts have allowed themselves, makes it difficult to foresee a situation in which a challenge of a precautionary measure could be successful. In *Duddridge* it was held that only when 'significant proof' of harm is present would it be possible to demand regulation on the part of the regulator.[48] However, following the *Duddridge* reasoning, nothing excludes the regulator from using its discretion and regulating a risk before 'significant proof' exists. Thus, 'significant proof' does not represent a level of risk which the regulator needs to meet in order to act in a precautionary manner, but rather a level of proof in which it can be expected to act in a precautionary manner.

The limited review, or ultra vires, doctrine employed in the UK prescribes that, only if the regulator has acted in a formally incorrect manner, outside of its competences, or has acted 'unreasonably', can a measure be quashed on substantive grounds. In relation to the substantive review of precautionary measures, the last of these options is most relevant.

[46] *Ibid.* [47] *Ibid.*, p. 12.

[48] *R.* v. *Secretary of State for Trade and Industry ex parte Duddridge and Others.*

The unreasonableness test used in the UK emanates from the famous *Associated Provincial Picture Houses Ltd.* v. *Wednesbury Corporation* case.[49] In this case it was stated that a decision or rule must be 'so unreasonable that no reasonable authority could ever have come to it' for it to be quashed by the courts.[50] The bar for so-called 'Wednesbury unreasonableness' is set high. It is not sufficient that the Court does not agree with the decision of a regulator, but it must be apparent that the measure is unreasonable.[51]

With regard to challenges to decisions involving the precautionary principle, this may have great relevance for the possibilities for effective redress. In areas surrounded by varying degrees of uncertainty, it is exceedingly difficult for a plaintiff to convince a court that the actions of a regulator were 'Wednesbury unreasonable'. Since high levels of uncertainty per definition are situations in which no clear course of action can be prescribed, it is nearly impossible even to meet the manifest error test employed by the ECJ.[52] As was concluded in the chapter on EU law, a successful challenge of a precautionary measure before the ECJ would have to end the scientific uncertainty in order to prove the decision taken to be vitiated by a manifest error. In the UK legal order this may not even be enough. What needs to be shown is that no reasonable person would have taken the decision at the time. However, if it is shown that the decision was objectively contrary to scientific evidence, but not unreasonable, the courts will not quash it. Thus, the burden of proof, once reversed, appears to be even higher in the UK than at EU level. This is further emphasised in a statement by in *R.* v. *Ministry of Defence, ex parte Smith*, in which it was held that:

> the greater the policy content of a decision, and the more remote the subject matter of the decision from ordinary judicial experience, the more hesitant the court must necessarily be in holding a decision to be irrational.[53]

Since this adequately describes the circumstances under which precautionary measures are likely to be taken, the already minuscule scope for effective redress appears even smaller.

[49] *Associated Provincial Picture Houses Ltd.* v. *Wednesbury Corporation* [1948] 1 KB 223.

[50] *Ibid.* [51] Cane, *Administrative Law*, p. 250.

[52] For a discussion on the judicial review before the EC courts see, in particular, Section 4.3.3 above.

[53] *R.* v. *Secretary of State for Defence, ex parte Smith* (Queens Bench Division) (1996) 2 WLR 305 B 517.

6.1.8 *Cost–benefit analysis*

That cost–benefit analysis can be one of the decisive elements for precautionary regulation was already indicated, but not elaborated on, in 'This Common Inheritance' in 1990.[54] In *Duddridge* the Court placed great weight on the performance of cost–benefit analyses. In the case, it was indicated that the decision to introduce precautionary measures would have to depend on two major elements, the first of which was the risk assessment, and the second a cost–benefit analysis. It was repeatedly stressed that a cost–benefit analysis would have to be carried out before precautionary measures could be issued.[55] In the ILGRA Report on the Precautionary Principle the use of cost–benefit analysis was also stressed, but not elaborated on in any detail.[56]

Since the Environment Act of 1995, it has been explicitly established that the Environment Agency must consider the costs and benefits of its actions whenever it exercises its powers.[57] How the Agency is to go about this has been elaborated on in a Consultation Paper of 2003,[58] and more recently in the document 'Managing Risks to the Public: Appraisal Guidance' (hereinafter the Risk Appraisal Guide).[59] However, since the scope of the Environment Agency is limited to issues not dealt with in any detail in this book (for example, pollution control and waste), the Risk Appraisal Guide will be dealt with in greater detail since its scope is not limited to certain areas, but applied generally to health as well as to the environment.

The UK approach to costs and benefits analysis is nuanced and detailed, even though it is summarised in a brief form in the available documents. Distinction is made between cost–benefit analysis and cost-effectiveness analysis, the former being a decision-making tool used to inform the decision-maker as to the desirability of a certain type of regulation, and the latter being an instrument to find the most effective way of carrying out a predetermined piece of regulation.[60] It is emphasised that in the absence of a market, surveys measuring the public's 'willingness to pay'

[54] 'This Common Inheritance, Britain's Environmental Strategy', p. 11.

[55] *R. v. Secretary of State for Trade and Industry ex parte Duddridge and Others.*

[56] ILGRA Report, p 10.

[57] Environment Act 1995, Article 39.

[58] HM Treasury, *The Green Book, Appraisal and Evaluation in Central Government* (2003). Available at: www.hm-treasury.gov.uk/d/green_book_complete.pdf (accessed 1 March 2010).

[59] HM Treasury, *Managing Risks to the Public: Appraisal Guidance* (2005). Available at: www.hm-treasury.gov.uk/consult_greenbook_index.htm (accessed 1 March 2010).

[60] *Ibid.*, p. 21.

(WTP) and 'willingness to accept' (WTA) can be used for input in the analysis.[61] Benefits to health can also be measured in quality adjusted life years (QUALYs). This basic abstraction builds on the fact that one life year in full health equals 1.0 QUALY. This value is then adjusted downward depending on perceptions of the decrease in the quality of life with age. Perceptions are measured through the use of statistical surveys. Thus, the older a person becomes, the fewer QUALYs equal one life in real time.[62] The values that the cost–benefit analysis generate should be adjusted for ethical or moral considerations connected to, for example, the willingness to provide higher levels of protection for children, or against particularly undesirable causes of loss in QUALYs.[63] In general, it is emphasised that all considerations put into a cost–benefit analysis or cost-effectiveness analysis should be made explicit so that the sources of the equation can be readily understood.[64] Uncertainties should be expressed and highlighted.[65]

The Risk Appraisal Guide provides an example of the importance that the UK Government attaches to the use of cost–benefit analysis (CBA) and cost-effectiveness analysis (CEA) in risk regulation. It is also an unusually detailed outline of how the Government intends to make use of these instruments.

6.1.9 The targeting of risks

The order in which to target risks, or 'risk prioritisation', is dealt with in a number of UK policy documents. In 'Guidelines for Environmental Risk Assessment and Management' it is stressed that ranking, rating and weighing risks objectively is fundamental to effective risk regulation. It is held that:

> All good policy decisions rely on the effective analysis of alternative options. Therefore, a systematic appraisal is important to ensure that the decision-maker is clear about the objectives and how to decide where the balance lies between the benefits from the reduction of the risk and the implications for society of introducing potential control measures.[66]

[61] *Ibid.* For a general discussion on this in the context of the precautionary principle see Sunstein, *The Laws of Fear.*

[62] HM Treasury, *Managing Risks to the Public: Appraisal Guidance*, pp. 22–3.

[63] *Ibid.*, p. 28. [64] *Ibid.*, pp. 24–5.

[65] *Ibid.*, p. 29.

[66] Guidelines for Environmental Risk Assessment and Management, HMSO 2000, Section 8.2. Now updated and available at: www.defra.gov.uk/environment/quality/risk/eramguide/ (accessed 1 March 2010).

The basis for tackling environmental risks in this fashion should be environmental impact assessments, strategic environmental assessments and cost–benefit analyses.[67]

In the Risk Appraisal Guide this approach is reinforced more generally in the area of risk regulation. The approach advocated here is more overtly of an economic nature. It is held that WTP analysis can be used as an input in the ranking of risks in order to find a balance between the risk and the public's willingness to protect themselves from the risk.[68] The ethical problems involved in comparing the saving of lives are emphasised.[69] Interestingly, the issue of risk trade-offs is not discussed in any detail, even though the literature on the subject is included in the bibliography to the document.[70]

The commitment to a rational ranking of risks by using economic models such as cost–benefit analysis and cost-effectiveness analysis seems firm in the UK. However, the lack of the inclusion of the concepts in any binding documents may make it difficult for the UK courts to test decisions against this commitment due to the limited review generally performed. Thus, it would have to be considered that what has been discussed above in this section shows that an elaborated and rational approach is intended by UK decision-makers, but they can hardly be interpreted as legally binding requirements.

6.1.10 Public opinion

As has been shown above, public opinion is taken into account in a sophisticated manner through the use of WTP, WTA and QUALYs. All of these mechanisms essentially build on public perception of how much one is willing to pay to avoid a risk, how big a risk one is willing to accept or how one values decreasing life quality with age. Thus, public opinion in an abstract form is made an integral part of the essentially economic approach suggested by the UK Government in the Risk Appraisal Guidance.

However, the issue is also dealt with as a separate topic under Appendix A, in which the emphasis is on understanding the causes and levels of public concern regarding a given risk. The public concern should be identified

67 *Ibid.*

68 HM Treasury, *Managing Risks to the Public: Appraisal Guidance*, pp. 25–6.

69 *Ibid.*, p. 28.

70 For example, Graham and Wiener, *Risk versus Risk*. For a further discussion on the issue of risk trade-offs see, above, Section 2.3.3.

and given a number on a five-point scale, where one represents the lowest level of concern, and five the highest. This scale is then used as input for the decision-maker when choosing between regulatory choices, communication strategies, and so on.[71]

The fact that the above-described methods of allowing public perception to influence economic considerations and decision-making are not exclusive is suggested in the Guidelines for Environmental Risk Assessment and Management. Here, it is first emphasised that allowing the public to influence the procedure is important in order to enhance the legitimacy and trust in the process.[72] Second, it is held that this can, perhaps, best be done through the involvement of stakeholders in risk regulation.[73] Similar concerns are expressed in the UK Government's Strategy Report of 2002,[74] although no concrete proposals are made as to how public participation is to be effectively carried out in particular policy areas. The involvement of the public, as well as stakeholder participation, appears to be entirely up to the discretion of the decision-maker in the individual cases.

In practice, it has been predicted by Haigh that public opinion may count much more than any principle of precaution.[75] In relation to this, the courts stated relatively clearly in *R.* v. *Tandridge District Council, ex parte Al Fayed* that public concern can, possibly, be relied upon to compel the authorities to examine the matter further through a risk assessment. Whether the Government allows public opinion to be decisive for the action it takes seems to lie entirely within its discretion. There appears to be little in the way of documents, binding or not, which describe in any detail how the Government should go about this. The only thing which appears to be clear is that continued public opinion in opposition to a measure which was issued after a risk assessment, abiding by formal requirements, and not affected by Wednesbury unreasonableness, is not sufficient to override it.[76]

In conclusion, it is clear that the involvement of the public in risk regulation is an issue that has received a fair amount of attention from the UK

[71] HM Treasury, *Managing Risks to the Public: Appraisal Guidance*, pp. 34–5.

[72] Guidelines for Environmental Risk Assessment and Management, Sections 3.2–3.5.

[73] *Ibid.*, Section 3.8.

[74] Strategy Unit Report, 'Risk: Improving Government's Capability to Handle Risk and Uncertainty', November 2002, pp. 74–92. Available at: www.cabinetoffice.gov.uk/media/cabinetoffice/strategy/assets/su%20risk%20summary.pdf (accessed 1 March 2010).

[75] Haigh, 'The Introduction of the Precautionary Principle into the UK', p. 240.

[76] *R.* v. *Tandridge District Council, ex parte Al Fayed*, p. 7 (the page number follows the version printed from LEXIS NEXIS).

Government, and is perceived to be of importance when deciding on issues involving uncertainty and precaution. However, involvement is carried out on an ad hoc basis and failure to take public opinion into account does not lead to a measure being challengeable before the courts. Despite this, it must be held that the emphasis on public opinion and rational communication is far greater in the UK than is the case in Sweden, where this issue is barely touched upon at all.

6.1.11 Conclusion

Precautionary decisions in areas of environmental and health protection in the UK are not a new feature of the UK legal order. It has been shown that such decisions have been a part of the UK legal order for centuries. However, the precautionary principle as an overarching policy that is adopted consistently by the UK Government is a very recent feature. Not included in any statutory instruments, the principle has mainly been referred to in statements by UK ministers and policy documents, the most detailed of which being the ILGRA Report of 2002. In general, policy documents do little to flesh out the content of the precautionary principle and even less to discuss how it is intended to apply in the UK. During the 1990s discussions rarely went further than reiterating Principle 15 of the Rio Declaration. In the ILGRA Report it was made clear that the precautionary principle should apply after a risk assessment has been performed and in conjunction with a cost–benefit analysis. This indicates that the principle is considered to be a tool to be applied by the risk manager. However, no clear separation between risk assessment and risk management is maintained in the UK, so it is often unclear to which category a decision belongs. Furthermore, in the last couple of years both the House of Commons and the House of Lords have expressed their concern with the unclear nature of the precautionary principle, and have recommended that the concept be better defined.

With regard to when the precautionary principle is to apply, not much guidance is given. In the ILGRA Report, as well as in other earlier documents, it has been held that the principle should come into effect where there is 'good reason' to believe that harmful effects may occur. Such good reason would be demonstrated by empirical studies, modelling or analogy with a similar activity. This indicates that some form of scientific evidence lies at the heart of the application of the principle.

The UK courts have not proven any more helpful in interpreting the precautionary principle than the UK Government. Usually considering themselves prevented from material review, they have allowed the UK Government near total freedom in deciding on when and how to apply the precautionary principle. Due to the severely limited material review that the courts are allowed in administrative cases, they are content to review whether measures taken by, or on behalf of, the UK Government are reasonable. This is in stark contrast to the Swedish courts which do, at least in some cases, take an active part in also reviewing detailed scientific evidence involved in a case. In *Duddridge* the Court considered the precautionary principle in force at EU level through Article 191 TFEU (formerly Article 174 EC) to be restricted to the application of EU rules and found that it could not be used to challenge national rules or decisions. The Court clearly indicated that the UK version of the precautionary principle was less onerous than the EU version, and that the UK Government was afforded full discretion in deciding on how to make use of it. In *Al Fayed* it was held that the Government was under an obligation to act in a precautionary manner in cases concerning base stations for mobile telephones, but the Court could not find any clear scientific evidence with which to support such a claim.

The lack of a definition or a more defined content of the precautionary principle, in combination with the extreme reluctance of the courts to review precautionary decisions, has led to a situation in which the decision-maker enjoys unbridled discretion. Consequently, it is excessively difficult to foresee a situation in which a precautionary decision could be successfully challenged by the UK courts. In the chapter on the EU it was shown that a successful challenge of a precautionary decision by the Council or the Commission before the European Court of Justice would effectively have to end the scientific uncertainty in order to meet the high level of proof created through the manifest error doctrine. However, in the UK even this may not be enough. Since what needs to be shown is that the decision taken is 'unreasonable', it may, in fact, be wrong, as long as a reasonable person could potentially have taken the decision. What needs to be shown, therefore, appears to border on showing that a decision was taken by negligence or was intentional. It is difficult to foresee how a more onerous burden for the plaintiff could be envisaged. The conclusion is that it is virtually impossible to challenge any decisions taken under the pretext of the precautionary principle in the UK.

It has been indicated in, for example, the ILGRA Report that the application of the precautionary principle is contingent upon the carrying-out of a cost–benefit analysis. This was also supported by the reasoning by the judge in *Duddridge*. In a government Consultation Paper of 1993, as well as in the Risk Appraisal Guidance, the UK Government has, in considerable detail, fleshed out how all government decisions in the areas of health and environmental protection should be preceded by the performance of cost–benefit analyses and risk trade-off analyses. Advanced models are suggested for making such comparisons. These include willingness to pay and willingness to accept calculations, as well as quality adjusted life years. Whether, and if so how, the UK Government has made use of these instruments in relation to the precautionary principle is unclear. The documents in which they are suggested are not binding, and no other rules indicate that the performance of a cost–benefit analysis would be a formal requirement. The courts have not tested the application of the principle against any formal requirements.

The role of public opinion is also emphasised in UK policy documents. Once again, no formal requirement has been created to safeguard this. However, as is discussed in the section on base stations, perhaps public opinion in the end plays a larger role when opting for the issuing of precautionary measures than that of an independent and legally binding precautionary principle.

In conclusion, it should once again be pointed out that the precautionary principle lacks independent legal status in the UK, at least outside the application of EU law. The decision to apply precautionary measures in individual cases is entirely in the hands of the UK Government. The failure to issue any binding documents regarding its potential definition or application in combination with the reluctance of the courts to review matters substantively means that no formal rules for the application of the principle exist. Thus, if the UK G overnment decides to use its discretion to issue precautionary measures, the scope for effective redress is very limited.

6.2 UK pesticides regulation

6.2.1 Introduction

As in other EU countries, the regulation of pesticides in the UK is a bundle of national rules and EU directives, which is not always easy to straighten out. Unlike the situation in Sweden, where the Chemicals Agency reigns more or less supreme, a number of national bodies are involved in the regulation and oversight of pesticides in the UK. Their competences and scopes of power are not always readily discerned and their functions sometimes overlap. Risk assessments are largely performed by the

Advisory Committee on Pesticides but other standing scientific committees are involved on an ad hoc basis.

In Sweden the precautionary principle in relation to pesticides underlies all policy and has led to the adoption of the policy goal of a 'Non-Toxic Environment', which effectively means that all man-made substances are to be eliminated from the environment by 2020. In the UK, on the other hand, the precautionary principle is not codified anywhere in statutes and has a significantly lesser impact on policy in relation to pesticides.

This section aims to explain how the precautionary principle has been expressed in the area of pesticides regulation, and how precautionary considerations have been used in practice in the UK. The chapter begins with an outline of the regulatory and institutional framework in the area, before the topic of the precautionary principle is discussed in detail. As in previous chapters, the precautionary principle is disseminated and dealt with in its constituent parts. Of particular importance, here, are the issues of risk assessment, the role of scientific evidence and the burden of proof, as well as cost–benefit analysis, risk trade-off analysis and the role of public opinion.

6.2.2 The legal framework

In the UK the legal basis for the control of pesticides consists of a number of provisions. The basic framework is made up of the Food and Environment Protection Act 1985 Part III (FEPA)[77] and the eight Control of Pesticides Regulations of 1986 (COPR).[78] The aims of FEPA are to protect health, safeguard the environment, secure safe and effective methods of controlling pests and to make information on pesticides available to the public.[79] The Act is mainly of enabling nature, with the aim of authorising ministers to issue more detailed legislation and set up bodies for advising on the use of pesticides.[80]

In COPR, which was issued the year after FEPA, it is established that advertising or trade in pesticides is prohibited unless authorised by the ministers.[81] In the Regulations following the prohibition, it is prescribed how ministers may issue approvals of pesticides and grant permission for the sales and advertising of the products.[82] In schedules attached to

[77] Food and Environment Protection Act 1985, Chapter 48, s. 16.

[78] Control of Pesticides Regulations, SI 1986/1510, s. 1–8.

[79] FEPA, Article 1. [80] *Ibid.*, Articles 2 and 7.

[81] COPR, s. 4, Articles 1 and 2. [82] *Ibid.*, s. 5 and 6.

the Regulations more specific requirements for approval and consent are established.

Since the early 1990s pesticide regulation has been mainly subjected to harmonised rules at EU level under Directive 91/414 (hereinafter the Directive). In the UK the Directive was implemented in the Plant Protection Products Regulation (PPPR) of 1995. Anyone applying for a permit to use a new active substance has to request permission from the Secretary of State, and, furthermore, must submit the application to the Commission and the competent authorities of the other Member States of the EU.[83] The decision regarding whether to include a substance in Annex I of the Directive is then taken using a comitology procedure.[84]

If the active substances in a pesticide have been duly included in Annex I of the Directive, the product still has to be approved by the Secretary of State.[85] In practice, this task has been delegated to the Pesticide Safety Directorate, which functions as an Executive Agency of the Ministry of Agriculture, Fisheries and Food.[86]

6.2.3 *The institutional framework*

The task of regulating agricultural pesticides in the UK falls, in principle, to the Department for Environment, Food and Rural Affairs (DEFRA). Since 1993 this task has been delegated by DEFRA to the Pesticide Safety Directorate (PSD).[87] The PSD was established as an Executive Agency in The Pesticides Safety Directorate's Framework Document (the Framework Document) in 1993.[88] With regard to non-agricultural pesticides, the Health and Safety Executive, and its counterparts in Scotland, Wales and Northern Ireland, are the relevant bodies.[89]

The system of executive agencies in the UK was conceived in the late 1980s and came as a response to the realisation that government ministries were being increasingly overburdened with implementation tasks.[90]

[83] Plant Protection Products Regulations 2005, SI 1435, s. 4.

[84] See, above, Section 5.2.2. [85] PPPR, s. 5, Article 1.

[86] The Pesticides Safety Directorate's Framework Document, Article 2.3. Available at: www.pesticides.gov.uk/corporate.asp?id=232 (accessed 1 March 2010).

[87] *Ibid.*, Article 1.1.

[88] *Ibid.* The above reference is to the Framework Document of 1996, which is an updated version of the Framework Document of 1993 and it is the one presently in force.

[89] Advisory Committee on Pesticides, *A Guide to Pesticide Regulation in the UK and the Role of The Advisory Committee On Pesticides*, ACP 19 (311/2005), 18. Available at: www.pesticides.gov.uk/acp.asp (accessed 1 March 2010).

[90] 'Improving Management in Government: The Next Steps', A Report to the Prime Minister (Ibbs Report), HMSO 1988.

In the case of PSD, it has effectively taken over the overseeing and application of the legislative framework from DEFRA. Its principal objectives thus include maintaining and administering approval, reviewing and monitoring systems in place for pesticides, providing policy advice to DEFRA and the promotion of the minimisation of pesticide use.[91]

The PSD is headed by a Chief Executive who is appointed by the Minister for Environment, Food and Rural Affairs.[92] The Ownership Board functions as a liaison between the Chief Executive and the Minister since its tasks are to advise the Minister on issues pertaining to the PSD, monitor the work of the PSD and, at the same time, provide advice and support to the Chief Executive. Furthermore, the Board functions as an arbiter between the PSD and the Ministry on certain issues.[93] Its members are appointed by the Minister and the Chief Executive has to be one of them.[94]

The general guidelines for the work of the PSD are drawn up by the Minister,[95] but the Chief Executive may make changes to the organisational structure to ensure its efficiency and overall performance.[96] In principle, the day-to-day management of affairs is performed by the PSD without interference from the Government. However, no constitutional guarantee granting the agency independence in deciding on cases before it, such as the one in place in Sweden, is to be found in the UK.[97] Instead, it is explicitly held in the framework document that the PSD functions under the DEFRA, and that the Minister may, in exceptional circumstances, issue direct instructions to the Chief Executive of the PSD.[98] The budget for the PSD is granted by Parliament for each budgetary year. However, the bids for the budget to be put before Parliament have to be agreed by the Chief Executive and the Minister. Parts of the budget are financed through the collection of fees.[99]

The Advisory Committee on Pesticides (ACP) was introduced in 1977 to give scientific advice to the Minister in relation to pesticides.[100] In its present form it functions under section 16(7) of the FEPA.[101] The ACP

[91] The Pesticides Safety Directorate's Framework Document, Articles 2.1–2.3. Available at: www.pesticides.gov.uk/corporate.asp?id=232 (accessed 1 March 2010).

[92] *Ibid.*, Articles 3.2 and 4.4. [93] *Ibid.*, Article 4.7.

[94] *Ibid.*, Article 4.6. [95] *Ibid.*, Article 4.1.

[96] *Ibid.*, Article 3.3. [97] For a Swedish comparison see, above, Section 5.2.3.

[98] The Pesticides Safety Directorate's Framework Document, Article 4.3.

[99] *Ibid.*, Article 6.3.

[100] Advisory Committee on Pesticides, *A Guide to Pesticide Regulation in the UK and the Role of the Advisory Committee on Pesticides*, p. 20.

[101] Control of Pesticides (Advisory Committee on Pesticides) Order 1985, SI No. 1985 No. 1516.

consists of as many members as the Minister sees fit.[102] Currently there are eighteen members, including the Chairman and the Deputy Chairman.[103] The committee consists of eminent scientists and the importance of their independence from government as well as industry is strongly emphasised.[104] The role of the committee is strictly advisory, and it does not have a regulatory function.[105] Since the Minister is bound by FEPA to consult the relevant advisory body before taking any decision, the amount of work before the ACP is significant.[106] In order to deal with this, the ACP applies a simplified procedure which allows it to review only novel products in detail. A system is in place to let the PSD and the Health and Safety Executive (HSE) review issues which have already been evaluated by ACP.[107]

It should also be noted that other scientific committees are sometimes consulted by the PSD on an ad hoc basis. There is no statutory basis for this procedure but it takes place where the expertise of other committees are required for 'obvious reasons'.[108]

In conclusion, the only truly independent organ involved in UK pesticide regulation is the ACP, which has only an advisory function. Even though formally independent when it comes to day-to-day issues, the PSD is closely tied to the Ministry through its organisational structure, its financing and the possibility of the Minister taking decisions actively in individual cases under exceptional circumstances.

6.2.4 Risk assessment and risk management

No formal separation of risk assessment and risk management is established in UK legislation. However, as demonstrated in the previous

[102] Control of Pesticides (Advisory Committee on Pesticides) Order 1985, SI No. 1985 No. 1517, s. 2.

[103] Advisory Committee on Pesticides, *A Guide to Pesticide Regulation in the UK and the Role of the Advisory Committee on Pesticides*, p. 20.

[104] *Ibid* and Control of Pesticides (Advisory Committee on Pesticides) Order 1985, SI No. 1985 No. 1517, s. 2.

[105] Advisory Committee on Pesticides, *A Guide to Pesticide Regulation in the UK and the Role of the Advisory Committee on Pesticides*, p. 22.

[106] See FEPA, Section 16, Article 9.

[107] Advisory Committee on Pesticides, *A Guide to Pesticide Regulation in the UK and the Role of the Advisory Committee on Pesticides*, p. 21.

[108] *The Queen on the Application of Amvac Chemical UK Limited* v. *Secretary of State for Environment, Food and Rural Affairs, Secretary of State for Transport, Local Government and Regions, Food Standards Agency*, High Court of Justice, Queen's Bench Division (Administrative Court) 3 December 2001, 2001 WL 1479891.

section, in most cases risk management decisions are ultimately taken by the Minister, after preparation by the authority to which this task has been delegated (in the case of pesticides, this means the PSD). This arrangement is supported in UK soft law instruments.[109] The risk assessment is also performed by the PSD together with the Advisory Committee on Pesticides.

When requesting an authorisation, companies are requested to complete and submit an extensive package of scientific data, which forms the basis for the risk assessment.[110] In this package the applicant needs to account for the physico-chemical properties of the substance, the product's potential toxicity in humans, the tolerance of humans towards residues of the pesticide, how operators and bystanders may be affected by it and the environmental and ecotoxicological behaviour of the product.[111] If the active substance has been included in Annex I to Directive 91/414, it is held that 'the government would be expected to draw upon the scientific assessment that had already been agreed [on the EU level]'.[112] Despite this fact, it is emphasised that the assessment and approval of pesticides is a decision to be taken at national level.[113] It could be assumed that what is meant by this is that, if an active substance is included in Annex I, the risk assessment with regard to the active substance has been carried out at EU level. However, the properties of the pesticide in which the active substance is included is to be assessed at national level. Since the coming into force of Annex I of Directive 91/414/EC, the UK authorities, in practice, rely on the risk assessment performed at EU level. That this is the case was illustrated in, for example, *British Agrochemicals Association Limited* which concerned the question of parallel importations of pesticides with similar characteristics to the substances included in Annex I.[114] It does not appear that the UK Government pursues its own risk assessment and management agenda in the way that Sweden does in the area of pesticides through the setting of far-reaching environmental quality objectives, the strict application of the precautionary principle and the setting of cut-off criteria.

[109] Compare with the discussion above in Section 6.2.5.

[110] Advisory Committee on Pesticides, *A Guide to Pesticide Regulation in the UK and the Role of the Advisory Committee on Pesticides*, p. 8.

[111] *Ibid.*, pp. 8–14. [112] *Ibid.*, p. 6.

[113] *Ibid.*

[114] *R.* v. *Ministry of Agriculture, Fisheries and Food ex parte British Agrochemicals Association Ltd.*, English Court of Appeal (Civil Division), 9 November 2001, [2002] 1 C.M.L.R. 8.

In cases before the UK judiciary, courts have proven reluctant to allow substantive review of the scientific evidence underlying a decision.[115] With regard to the formal procedure, however, the Court in *Inquinosa International* clarified that the PSD is at the centre of the risk assessment procedure, and that it takes the decision on which scientific evidence is to be passed onto the ACP for evaluation. In this case, the claimant argued that the advice given by the ACP was not complete due to the fact that the ACP had not had access to the full scientific study carried out by the applicants. Parts of this study had been 'filtered' by the PSD. The ACP report proved decisive for the decision by the ministers to revoke the authorisation of the pesticide lindane. The claimants held that since the report was not based on all the scientific evidence available, but rather a selection of it, the report was flawed. The Court did not agree. Instead it held that there is 'nothing unusual, and nothing objectionable in the PSD acting as a filter in this way'.[116] It is interesting to note that the PSD, which is a branch subordinate to the Secretary of State for Environment, is explicitly allowed to control the material that is to be assessed by the only independent body in the UK pesticide regulatory regime, the ACP. Thus the risk manager can maintain control over the scientific material put before the risk assessor and thus it can potentially exert a very direct control of the outcome of the risk assessment.

6.2.5 *Scientific evidence and the introduction of precautionary measures*

As shown in the previous section, the precautionary principle in the UK is not defined, and the decision of whether and how to apply precautionary measures lies entirely within the discretion of the risk manager. It is not even entirely clear that health risks could trigger precautionary action. In a paper for the ACP Open Meeting of 2001, the former ACP Chairman David Coggon held that the precautionary principle 'focuses only on environmental degradation'.[117] This statement was quoted by the Court in *Amvac Chemical UK Limited* in order to limit the scope of the principle

[115] See, for example, *R.* v. *Minister for Agriculture ex parte Inquinosa International*, Queen's Bench Division (Crown Office List), 17 January 2000.

[116] *R.* v. *Minister for Agriculture ex parte Inquinosa International.*

[117] D. Coggon, 'The Precautionary Principle and Pesticide Regulation', Chairman's Paper for ACP Open Meeting 2001. Available at: www.pesticides.gov.uk/acp.asp?id=490 (accessed 1 March 2010).

to issues of environmental protection.[118] However, the Guide to Pesticide Regulation, issued jointly by DEFRA and the HSE in 2005, held that:

> Effectively, the precautionary principle is applied as routine in the regulation of new pesticides. As has already been explained, new products are not authorised unless there is adequate scientific evidence that their use will not pose unacceptable risks to health and to the environment.[119]

This indicates clearly that not only environmental concerns are considered when deciding on whether to issue precautionary measures. Instead the precautionary principle is applied to ensure that 'adequate scientific evidence' supports the fact that a product does not pose unacceptable risks. So far this construction has not been clarified further by the ministries or elaborated on by the courts, but it seems to indicate that the UK authorities concerned with pesticide regulation do not intend to restrict the use of the precautionary principle to solely environmental issues.[120]

Having stated that, the issue of what lies in the concept 'adequate scientific evidence' arises. First, it should be noted that what needs to be shown is that a product does not pose unacceptable risks. Thus, the presumption with regard to pesticides seems to have changed, moving away from the regulator showing the risk inherent in a product, towards the applicant showing that it does not bring unacceptable risks. This has to be backed up with 'adequate scientific evidence'. What is meant by this is further qualified in the Guide:

> In other words, uncertainties in the risk assessment must have been sufficiently eliminated before a new product is approved.[121]

If the two statements are combined, the conclusion is can be drawn that 'adequate scientific evidence' should have 'sufficiently eliminated' uncertainties before a product is authorised. The fact that the phrase 'sufficiently eliminated' does not mean that full scientific certainty regarding the harmlessness of a product is required appears relatively clear. It has been stated that requiring this 'would be as unrealistic as requiring full

[118] *The Queen on the Application of Amvac Chemical UK Limited* v. *Secretary of State for the Environment, Food and Rural Affairs, Secretary of State for Transport, Local Government and Regions, Food Standards Agency*, p. 79.

[119] Advisory Committee on Pesticides, *A Guide to Pesticide Regulation in the UK and the Role of The Advisory Committee on Pesticides*, p. 15.

[120] This issue has also been discussed in the area of base stations for mobile telephones, and is returned to in relation to this below.

[121] Advisory Committee on Pesticides, *A Guide to Pesticide Regulation in the UK and the Role of The Advisory Committee on Pesticides*, p. 15.

scientific certainty of the existence of a hazard'.[122] Building on this explicit recognition of uncertainties, the precautionary principle in pesticide regulation has been elaborated on as a specific example of a general principle of risk management, which holds that uncertainties must be appropriately taken into account. However, very little is stated in relation to the actual level of risk required for precautionary action.[123] It can be assumed that new products will be approved in the UK following the result of risk assessments taken at EU level. In relation to this, it has been emphasised that the trend is progressively moving towards a more precautionary approach towards chemicals in general.[124] A more complicated area, in which national risk management decisions still play a large role, is that of pesticides that were originally approved before the EU regime came into effect.

One of the few cases, if not the only one, that deals in some detail with the scientific evidence underlying a regulatory decision on a pesticide is *Amvac Chemical UK Limited*, in which the High Court of Justice was faced with a challenge regarding a revoked authorisation for the pesticide dichlorvos. Dichlorvos had not been included in Annex I of Directive 91/414, but had been authorised in the UK before the coming into force of Annex I.[125] The Court did not grant permission to challenge the substance of the decision, but only the issue of whether the decision had been taken according to formal requirements. Furthermore, the issue of whether the precautionary principle had been properly regarded was considered.[126]

Even if the substance of the case was not tried before the Court, the judgment proved to at least give some indication as to how the ACP and other scientific bodies advise the PSD and the impact their advice has on the outcome of decisions. In this case the Committee on Mutagenicity of Chemicals in Food, Consumer Products and the Environment (COM) was consulted on an ad hoc basis, beside the mandatory consultation of the ACP. Interestingly enough, in its advice the ACP deferred to the

[122] Coggon, 'The Precautionary Principle and Pesticide Regulation'.

[123] *Ibid.* [124] *Ibid.*

[125] The substance was originally approved in 1994, thus before Annex I of Directive 91/414/EC of 15 July 1991 concerning the placing of plant protection products on the market, OJ L 230/1 had come into effect. According to provisional measures in Article 8.2 of Directive 91/414/EC during a transitional period national administrations were allowed to authorise pesticides pending review of the Commission.

[126] *The Queen on the Application of Amvac Chemical UK Limited v. Secretary of State for the Environment, Food and Rural Affairs, Secretary of State for Transport, Local Government and Regions, Food Standards Agency*, p. 2.

expertise of the COM. Instead of giving a material advice on the facts, it held that if the COM found dichlorvos to possess *in vivo* mutagenic properties and it could not be excluded that the occurrence of tumours in animal tests were due to a genotoxic mechanism, the authorisation for the substance should be revoked.[127] This was the advice given, the ACP stated, despite the fact that the risk of carcinogenicity in humans must be regarded as very small. The COM subsequently declared that it found it 'prudent to assume a genotoxic mechanism on the basis of the available data', and it concluded that:

> The Committee agreed that in the absence of appropriate mechanistic data a precautionary approach should be adopted and no threshold could be assumed for the mutagenic and carcinogenic effects of Dichlorvos.[128]

The COM conclusions were based on a rather extensive study of the literature, which was summarised in a statement dated January 2002.[129] In the statement it appears that the risk of mutagenicity perceived by the COM is rather low, and it explicitly stated that 'there was no evidence for systemic mutagenic effects'.[130] The uncertainty surrounding the issue seems to have been due to the lack of adequate scientific evidence, rather than conflicting opinions.[131] Facing this situation, the COM advised that 'until evidence was provided to the contrary and the absence of appropriate mechanistic data, a precautionary approach should be adopted'.[132] The ACP also chose to recommend a suspension of the authorisation for dichlorvos rather than the revocation of the substance, pending the resolution of the scientific uncertainty.[133]

It appears clear that the decision taken in the case of dichlorvos was based exclusively on scientific advice provided by the COM.[134] No mention is made of other concerns such as a weighing of the costs and benefits of regulation. As shown, it appears that the level of proof for the triggering

[127] *Ibid.*, p. 47. [128] *Ibid.*, p. 49.

[129] Mutagenicity of dichlorvos, COM statement COM/02/S2 – January 2002. Available at: www.advisorybodies.doh.gov.uk/com/dichlorvos.htm (accessed 1 March 2010).

[130] *The Queen on the Application of Amvac Chemical UK Limited* v. *Secretary of State for the Environment, Food and Rural Affairs, Secretary of State for Transport, Local Government and Regions, Food Standards Agency*, p. 23.

[131] *Ibid.*, pp. 20 and 21. [132] *Ibid.*, p. 23.

[133] Minutes of the 290th Meeting of the Advisory Committee on Pesticides (ACP) on 4 March 2002. Available at: www.pesticides.gov.uk/acp.asp?id=269 (accessed 1 March 2010).

[134] DEFRA News Release, 19 April 2002. Available at: www.hse.gov.uk/press/2002/e02076.htm (accessed 1 March 2010).

of measures is low here. Little evidence indicated that the substance might lead to mutagenicity, but even less evidence indicated that it would not. Thus, in the absence of convincing evidence in either direction, the UK authorities opted for a textbook precautionary approach and suspended the application of the product. It should be noted that, due to the high level of uncertainty, the UK authorities settled for an interim measure rather than a permanent one, that is, it suspended the authorisation rather than revoking it.

If the regulation of dichlorvos as highlighted in *Amvac Chemical UK Limited* is indicative of the process in the UK, it tells us a number of important things about UK pesticide regulation. First, the decision-maker, in these cases the Secretary of State for the Environment and Health, relies heavily on the advice of scientific expertise. The advice in this case was provided by the COM and the ACP, which are both scientific agencies, formally independent from the Government. Second, the threshold for invoking precautionary measures appears low here. The risk of cancer is, of course, serious, but the probability of the substance actually causing it, even if mutagenic effects are proven, is generally recognised as very low. Third, the scientific agencies will, in some cases of scientific uncertainty, actively endorse the precautionary principle and base their advice on it. Fourth, and connected to this, in the judgment the Court once again reinforces the approach that the precautionary principle is a tool to be used at the full discretion of the UK Government. It does not view it as a legal principle to be consistently applied in all cases, and no formal rules of risk assessment accompany it.

That the precautionary principle in the UK is not approached similarly in other instances of pesticide regulation in which the risks are similar or greater than in the case of dichlorvos can be illustrated by the issue of paraquat. As discussed in the chapter on Swedish pesticide regulation, the pesticide paraquat is acutely toxic even in small doses when ingested and possesses a number of other health risks of a more uncertain nature.[135] As with dichlorvos, paraquat mainly poses a risk when handled. The risk from residues in food can be considered negligible at the doses normally employed. Paraquat has been described as 'one of the absolutely most hazardous pesticides' there is.[136] Despite this fact, the UK Government

[135] See, above, Section 5.2.5.

[136] Statement of the Swedish Chemicals Agency (KemI), 16 September 2003. Available at: www.kemi.se/templates/News____3090.aspx (accessed 1 March 2010).

chose to defer to the risk assessment at EU level.[137] Interestingly, the UK was the rapporteur Member State for paraquat within the framework of Regulation 91/414, and its recommendation was that paraquat was sufficiently safe for professional use in the EU.[138] Thus, the UK considered paraquat, a pesticide with more well-documented risks than dichlorvos, to be sufficiently safe to be allowed on the UK and the EU markets, whereas dichlorvos, a pesticide for which the magnitude of risk is low, and a great degree of uncertainty pertains to actual existence of that risk, is forbidden, based on the 'precautionary principle'. From the documentation available for this research, it has been impossible to discern whether facts other than scientific facts have influenced the UK stance on the pesticides discussed above, and what such concerns may have entailed. Officials at the PSD declined the opportunity to be interviewed for this book.

6.2.6 *The burden of proof*

In *Amvac Chemical UK Limited* the claimant argued that, since the precautionary principle was relied upon by DEFRA, it would be bound by some form of formal risk assessment. It held that, if this were not the case, the conclusion would be that the claimant would have to prove the absence of risk to successfully challenge the decision. 'Rarely', it held, 'is it possible to prove a negative.'[139] The Court, however, chose not to deal with this plea, hiding behind the fact that a substantive review had not been granted in the case. Instead, it seemed to misunderstand slightly the plea put forth by Amvac as seeking to connect the precautionary principle to a formalistic type of risk assessment.[140] The Court concluded:

> [M]y very firm conclusion is that there is – at least so far – no settled, specific or identifiable mechanism of risk assessment in the field of pesticide approval that the Claimant is entitled to rely on as part of the 'precautionary principle'.[141]

137 Short Report of the Meeting of the Standing Committee on the Food Chain and Animal Health (Phytopharmaceuticals Section) held on 3 October 2003 in Brussels. SCFA 4/2003, p. 2. Available at: http://ec.europa.eu/food/committees/regulatory/scfcah/phytopharmaceuticals/summary16_en.pdf (accessed 1 March 2010).

138 Review report for the active substance paraquat, 3 October 2003, SANCO/10382/2002–final. Available at: http://ec.europa.eu/food/plant/protection/evaluation/existactive/list_paraquat.pdf (accessed 1 March 2010).

139 *The Queen on the Application of Amvac Chemical UK Limited* v. *Secretary of State for the Environment, Food and Rural Affairs, Secretary of State for Transport, Local Government and Regions, Food Standards Agency*, p. 67.

140 *Ibid.*, p. 68. 141 *Ibid.*, p. 84.

This is a rather peculiar statement since, as was shown in the previous sections, an institutionalised system for risk assessment involving agencies is foreseen in UK legislation and was made use of in the case. What does not seem to exist are substantive guidelines for how the risk assessment is to be performed, against which a plea could be formally tried. However, this discussion seems to be missing the point made by the claimant. What was actually argued was a repetition of the classic criticism of precautionary decisions taken under uncertainty, namely, that they force the parties challenging them to prove a negative, and, thus, render effective redress impossible.[142] This issue was not addressed by the Court at all.

As discussed above, when it comes to pesticide regulation, the presumption is reversed, and, instead of the regulator having to demonstrate harm, the applicant needs to demonstrate that 'adequate scientific evidence' 'sufficiently eliminates' uncertainties before a product is allowed. The fact that scientific certainty with regard to the harmlessness of the product 'would be just as unrealistic as requiring full scientific certainty of the existence of hazard' has been emphasised by the ACP.[143] The keyword is thus 'sufficiently'. It has previously been discussed that a change of presumption may, at first, be interpreted as a reversal of the burden of proof onto the applicant, but that this, in reality, is not the case. Instead, it has been held that this situation is one of a 'false' burden of proof placed on the applicant. The real, initial, burden of proof still rests with the regulator, who must indicate what kind of harm it expects from the product, and this then needs to be refuted by the regulated party. The level of proof that needs to be met by the regulator can, in these instances, be very low. However, once it has been met, the burden transfers to the regulated party and, subsequently, the level of proof that the regulated party needs to meet can be set very high.[144]

It is, at present, unclear what the UK authorities would consider a sufficient elimination of uncertainties. Since the courts make use of the unreasonable test for substantive review of government decisions, not much help can be found in the case law. In *Amvac Chemical UK Limited* Amvac presented scientific evidence to the ACP and the COM during the regulatory process. Some of the studies criticised the methodology and the conclusions drawn from the evidence initially relied on by the COM.[145]

[142] *Ibid.*, p. 67. [143] Coggon, 'The Precautionary Principle and Pesticide Regulation'.

[144] See, above, Section 5.1.7, in particular, footnote 95.

[145] *The Queen on the Application of Amvac Chemical UK Limited* v. *Secretary of State for the Environment, Food and Rural Affairs, Secretary of State for Transport, Local Government and Regions, Food Standards Agency*, pp. 15–17.

This evidence was considered by COM, but it concluded that when viewing the available evidence as a whole, the remaining uncertainties were sufficient to warrant precautionary advice. However, when bearing the issue of paraquat in mind, this conclusion appears inconsistent at best. Far more scientific evidence of risk exists in this case, but no action to counter these risks has been proposed.

6.2.7 *Cost–benefit analysis*

In the previous section it was established that the issue of cost–benefit regulation in the area of UK risk regulation has received a fair amount of attention and has been assigned a great deal of importance.[146] Statutory support for the carrying-out of cost–benefit analyses in relation to environmental and health decision-making is, so far, only to be found with regard to the Environment Agencies set up in the Environment Act 1995.[147] How such analyses should be carried out was further clarified in 2003.[148] That not only the Environment Agencies should be subject to a requirement to carry out cost–benefit analysis when carrying out regulation was indicated in the document 'Managing Risks to the Public: Appraisal Guidance' (hereinafter the Risk Appraisal Guide).[149] The Risk Appraisal Guide gives a detailed account of how the costs and benefits of a regulation can be compared using monetary units and so called quality adjusted life years (QUALYs). In *Duddridge* the Court placed great emphasis on the two constituent parts of the application of the precautionary principle, the first of which being the risk assessment, and the second being the performance of a cost–benefit analysis.[150] The relevance of analysis of costs and benefits was further emphasised in *R.* v. *Tandridge District Council, ex parte Al Fayed.*[151]

In the light of the lip service that has been paid to the importance of cost–benefit analysis in risk regulation in general, one would expect to find ample examples of it in the area of pesticide regulation. The bodies involved in the regulation of pesticides should be considered subject to, if not the Environment Act, then at least the Risk Appraisal Guide. However, if cost–benefit analyses are being performed in UK pesticide regulation,

146 See, above, Section 6.1.8. 147 Environment Act 1995, Article 39.

148 HM Treasury, *The Green Book, Appraisal and Evaluation in Central Government.*

149 HM Treasury, *Managing Risks to the Public: Appraisal Guidance.*

150 *R.* v. *Secretary of State for Trade and Industry ex parte Duddridge and Others.*

151 *R.* v. *Tandridge District Council*, ex parte *Al Fayed*, p. 7 (of the LEXIS NEXIS version).

this is being carried out at an informal level. No formal requirements have been made on the precautionary principle being contingent upon the comparison of costs and benefits. On the contrary, as was discussed above, the Court in *Amvac Chemical UK Limited* held that there was no settled mechanism in place for the application of the precautionary principle against which a risk management choice could be tried.[152] Thus, it appears once again that use of cost–benefit analysis that is made in the risk management phase lies solely within the discretion of the UK Government. It appears that since no mechanism is in place for the use of the precautionary principle, the UK Government can use the it as it sees fit.

In the previous section it was discussed that the issue of rational targeting of risks and risk trade-off analysis are concepts that have been elaborated on to a certain degree by the UK Government,[153] but that no firm commitment has been made to such mechanisms in binding documents. In the area of pesticides regulation, the ACP has stated that it aims to take, among other things, risk off-sets and risk substitutions into account when scientifically examining pesticides before it. When preparing advice, it will highlight such issues and emphasise the remaining uncertainties.[154] In the case law, once again, the courts have left such considerations up to the exclusive discretion of the Government.

6.2.8 *Conclusion*

Three main pieces of legislation lay down the legislative framework for pesticide regulation in the UK. The FEPA and the COPR were both issued before Directive 91/414/EC, which effectively harmonised the approval of active substances in pesticides at EU level. The Directive has been implemented in the UK in the Plant Protection Products Regulations. The regulation of pesticides primarily falls under the competence of the Department for Environment, Food and Rural Affairs, which is headed by a government minister. Non-agricultural pesticides are regulated by the Health and Safety Executive. In practice, many tasks have been delegated to the Pesticide Safety Directorate, which functions as an executive agency

152 *The Queen on the Application of Amvac Chemical UK Limited* v. *Secretary of State for the Environment, Food and Rural Affairs, Secretary of State for Transport, Local Government and Regions, Food Standards Agency*, pp. 84–6.

153 See, for example, Guidelines for Environmental Risk Assessment and Management, Section 8.2.

154 Advisory Committee on Pesticides, *A Guide to Pesticide Regulation in the UK and the role of The Advisory Committee on Pesticides*, p. 16.

under DEFRA. The PSD maintains strong ties with DEFRA and cannot be said to function as an independent agency. The Advisory Committee on Pesticides mainly performs risk assessments in relation to pesticides. The role of the ACP is to function as an independent advisory group and its members consist of eminent experts. Even though the ACP is formally independent, the case law has illustrated that the PSD, in practice, can exert control with regard to what information is actually put before the ACP.[155] A strict separation of risk assessment and risk management is not upheld.

As discussed extensively in the previous section, UK courts have, in general, avoided pronouncing judgment on the merits of cases pertaining to the precautionary principle. This has led to a situation in which the UK Government has a de facto unlimited discretion in deciding on issues involving scientific uncertainty. In the few cases involving the regulation of pesticides, this trend has been maintained. In *Amvac Chemical UK Limited* the Court refused to investigate the facts. Despite this, a few interesting conclusions can be drawn from the case. First, the recommendations of scientific bodies are instrumental for the decisions taken by the UK Government in cases of uncertainty. Second, the level of potential risk does not have to be particularly high for measures to be invoked if they are recommended by scientific bodies. In *Amvac Chemical UK Limited* all the parties agreed that the risk resulting from the use of the pesticide lindane was very low even though it was demonstrated that it had mutagenic properties. Despite this, a precautionary approach was taken. Third, even though the PSD did not refer to the precautionary principle in this case, it passively endorsed it, since its decisions were based on the opinions of the scientific bodies which were explicitly precautionary. Fourth, and most importantly, the three previous points do not establish any form of trend that can be relied upon for future reference. The use of the precautionary principle is entirely in the hands of the Government. In the case of lindane, a very small risk of cancer was allowed to justify precautionary action, but in the case of paraquat, a pesticide with well-documented toxic properties, the UK Government has taken no action. Instead it has, in the risk assessment carried out at the EU level, held that it considers it safe for normal use.

As discussed in the previous section, this apparently leads to a situation in which, in order to challenge a precautionary decision, one

[155] See *R. v. Minister for Agriculture ex parte Inquinosa International* where the Court let the PSD 'filter' information be considered by the ACP.

would first have to 'eliminate sufficiently' the scientific uncertainty, and, second, show that it was unreasonable for the Government to take the decision it has taken. Despite this, the ACP has stated that asking someone to show the harmlessness of a product is just as unrealistic as asking someone to prove the hazardousness of a product. However, this is exactly what the courts are doing under the current regime, and it may not even be sufficient to show that the product is harmless. One also has to show that no reasonable person would have taken the erroneous decision.

In the previous section it was discussed extensively how the UK Government emphasises the importance of cost–benefit analysis and, to some degree, risk trade-off analysis. However, in relation to pesticides no formal regime for taking such considerations into account is in place. If costs and benefits are compared and trade-offs considered, this is done in an informal manner. Here, too, the UK Government has avoided laying down formal requirements against which its decisions can be checked. The same is true for the role of public opinion.

The lack of formal requirements on the application of the precautionary principle and the ineffective judicial review in the UK system of pesticide regulation undoubtedly leaves the Government much room to manoeuvre. The price of this leeway is limited judicial protection for undertakings and individuals. Possibilities of redress and foreseeability are severely limited under the UK regime. Furthermore, it is difficult to obtain background material and reports on the basis of which decisions have been taken by the UK Government. It should also be noted that requests to interview UK civil servants involved in the regulation of pesticides were turned down.

The fact that pesticide regulation nowadays is mainly carried out at EU level within the framework of Regulation 91/414 does little to remedy the situation. The risk assessment under the Regulation is, as shown above, Member State-driven. It is Member States that are appointed rapporteurs for the risk assessments carried out and it is the Member States that recommend the actions to be taken. The fact that the Member States treat the issue of the precautionary principle inconsistently leads to problematical situations. One example of this is the evaluation of the pesticide paraquat, in which the strict Swedish precautionary approach stands out in stark contrast to the more toned-down approach assumed by other Member States, in this case notably the UK. Different approaches to this issue will continue to lead to different assessments being made and a continuation of the friction and imbalance in the treatment of the same products in different Member States.

6.3 UK regulation of base stations

6.3.1 *Introduction*

Unlike the Swedish example above, the UK does not employ a multitude of overlapping laws and regulations in the field of radiation in relation to mobile telephony. However, this does not render the UK situation any less complicated or ambiguous. In the UK no general statutory provision aiming to address the issue of radiation risks to the public exists. The only legal instruments that do exist in the area concern workers who come in contact with radiation when carrying out their duties.[156] The issue of base stations siting is regulated through the use of general planning rules, where it is stipulated that base stations, in most cases, are exempt from planning permission. However, non-binding recommendations relating to non-ionising radiation issued at national level have proven important.

In contrast to the area of pesticide regulation discussed above, the regulation of base stations has yielded an increasing body of case law. Primarily, issues before the courts have dealt with how public opinion should be taken into account, and how big a role it should be allowed to play in the outcome of decisions. Connected to this is the issue of whether local authorities are allowed to be more precautionary when deciding on the siting of base stations than national guidelines would recommend.

This section, like sections above, deals with the constituent parts of the precautionary principle separately. First, an overview of the legislative and institutional framework is given. Then, the attention is turned towards issues of risk assessment and management, the role of scientific evidence, the burden of proof and the role of cost–benefit and risk trade-off analyses. Finally, the issue of how public opinion at local level is allowed to influence the application of precautionary measures is addressed.

6.3.2 *The legal framework*

In Sweden the Radiation Protection Agency has been given the explicit statutory right to issue regulations and recommendations to ensure effective protection from radiation. In the UK no such framework is in effect. Instead, the only recommendations in place with regard to base stations for mobile phones have been issued by the National Radiation

[156] Health and Safety at Work, etc. Act 1974 and the Management of Health and Safety at Work Regulations 1999.

Protection Board (NRPB).[157] The institutional framework and the content of the recommendations are returned to below. The lack of a legislative framework regarding radiation from base stations means that guidance with regard to the precautionary principle needs to be sought in planning law and in judgments from the courts.

Like other UK statutory instruments, the relevant planning law is silent on the issue of the precautionary principle.[158] Instead, it explicitly states that the establishment of 'telecommunications systems' is a permitted development, provided that a number of aesthetical requirements are met.[159] These requirements are primarily concerned with the height of the proposed masts. Development of base stations is subject to a procedure in which the local planning authority must decide if the project needs prior approval with regard to the siting and the appearance of the development.[160] If the development does not exceed two cubic meters in size, there is no need to consult the local planning authority.[161] The request for prior approval has to be made in writing and has to be accompanied by plans and a description of the proposed development.[162]

6.3.3 *The institutional framework*

As was seen in the Swedish example, at least two institutional structures are involved in the regulation of radiation from base stations.[163] The issue of setting safety standards in relation to the radiation from base stations is usually taken care of at national level by a radiation protection agency. In the UK the body responsible for this was the NRPB, but its authority was transferred to the new Health Protection Agency (HPA) in the spring of 2005.[164] The siting of base stations falls under the authority of local governments and is subject to normal planning rules.[165] Special rules for workers are the prime responsibility of the Health and Safety Executive, which will not be dealt with in any detail here.[166]

157 Available at: www.hse.gov.uk/radiation/nonionising/ (accessed 1 March 2010).

158 The Town and Country Planning Order 1995, Statutory Instrument 1995 No. 418.

159 *Ibid.*, Part 24, A and A.1.

160 *Ibid.*, Part 24, A.2 (4) (b) (ii).

161 *Ibid.*, Part 24, A.2 (4) (b)

162 *Ibid.*, Part 24, A.2 (4) (b) (iii).

163 See, above, Section 5.3.3.

164 Health Protection Agency Act 2004, Article 3. Available at: www.opsi.gov.uk/acts/acts2004/20040017.htm (accessed 1 March 2010).

165 The Town and Country Planning Order 1995.

166 www.hse.gov.uk/radiation/nonionising/index.htm (accessed 1 March 2010).

Without going into too much detail with regard to the NRPB, it should be pointed out that it was assisted by a group of experts which was set up by its director in 1990. The group of experts still exists and reports to a subcommittee of the HPA. Its advice is not binding but is regularly referred to by the HPA.[167] The position of the HPA is that of an independent agency. Unlike the Pesticide Safety Directorate discussed in the previous section, the HPA is formally independent from direct influence by the Government.[168] In this sense it is more similar to the agencies discussed in the Swedish section. A chairman, who is appointed by the Secretary of State, heads the agency. The chairman, in turn, is assisted by a number of non-executive members, who also are appointed by the Secretary of State and by the representatives of the home nations.[169] These serve on the board of the HPA together with executive members and the chief executive, who are all employed by the HPA or are seconded staff.[170] The chief executive leads the day-to-day work of the agency.

To assist it in its scientific work, the agency may appoint committees and subcommittees, which may consist of people not employed by the agency.[171] In 1990 the NRPB set up an advisory group, the Advisory Group on Non-Ionising Radiation (AGNIR), whose aim is to 'to review work on the biological effects of non-ionising radiation relevant to human health and to advise on research priorities'.[172] The AGNIR consists of a number of scientists who were appointed by the NRPB. The Advisory Group now fulfils the same tasks, but answers to the HPA. It has no formal decision-making role, but functions as a strictly advisory body.[173] Formal documents pertaining to its internal workings, as well as to its actual influence on UK policy, are difficult to come by. However, it appears that, in practice, its advice has been referred to as the official UK policy towards non-ionising radiation.[174]

Following widespread public concern regarding the use of mobile phones, an independent expert group was formed in 1999 following the

[167] Available at: www.hpa.org.uk/webw/HPAweb&Page&HPAwebAutoListName/Page/1207821636407?p=1207821636407 (accessed 1 March 2010).

[168] Health and Protection Agency Act 2004, Schedule 1, Articles 5 and 6.

[169] *Ibid.*, Article 1 (1)–(3). [170] *Ibid.*, Article 1 (4).

[171] *Ibid.*, Article 11 (1)–(3).

[172] Health Protection Agency, AGNIR Reports. Available at: www.hpa.org.uk/webw/HPAweb&Page&HPAwebAutoListName/Page/1207821636407?p=1207821636407 (accessed 1 March 2010).

[173] *Ibid.*

[174] *Ibid.* The reports so far issued by AGNIR can be downloaded from this website.

recommendation of the House of Commons' Science and Technology Committee.[175] Its official terms of reference were:

> To consider present concerns about the possible health effects from the use of mobile phones, base stations and transmitters, to conduct a rigorous assessment of existing research and to give advice based on the present state of knowledge. To make recommendations of further work that should be carried out to improve the basis for sound advice.[176]

Professor Sir William Stewart was appointed to lead the work of the Independent Expert Group on Mobile Phones (IEGMP) as the group was called. Its conclusions were published in the form of a report on 11 May 2000 and have proven to be highly influential as a basis for UK policy in relation to mobile phones.[177] The report, which recommended a precautionary approach to radiation from mobile phones and base stations, is widely referred to as 'the Stewart Report' and is returned to in more detail below.

As mentioned above, the actual siting of base stations follows planning rules. Local Planning Authorities (LPAs) are responsible for reviewing planning requests and issuing planning permission where necessary. In the section which deals specifically with telecommunications, no reference to health concerns is made. Instead, the main focus is on aesthetic considerations. As already stated, for masts below the height of 15 meters no building permit is required. The Stewart Report made far-reaching recommendations to change UK planning rules in a more precautionary direction. These recommendations included the revocation and re-examination of all planning decisions for base stations, the introduction of clear templates of protocols to be used by LPAs, and the establishment of exclusion zones around base stations.[178] Furthermore, it recommended that the ICNIRP standards, which also form the basis for the EU Council Recommendations in the area,[179] be accepted as the UK standard for non-ionising radiation as well.[180]

175 House of Commons, Science and Technology Committee Publications, Third Report, printed on 2 September 1999. Available at: www.parliament.the-stationery-office.co.uk/pa/cm199899/cmselect/cmsctech/489/48902.htm (accessed 1 March 2010).

176 *Ibid.*, Annex 2.

177 For a background to the work of the IEGMP and the report in full see the homepage of IEGMP at: www.iegmp.org.uk/ (accessed 1 March 2010).

178 Independent Expert Group on Mobile Phones, 'Mobile Phones and Health', 11 May 2000, 1.30–1.48. Available at: www.iegmp.org.uk/report/text.htm (accessed 1 March 2010).

179 Council Recommendation 1999/519/EC of 12 July 1999 on the limitation of exposure of the general public to electromagnetic fields (0 Hz to 300 GHz). OJ L 199/59.

180 Independent Expert Group on Mobile Phones, 'Mobile Phones and Health', Section 1.27.

The UK Government chose to adopt a significantly more moderate approach than that advocated in the Stewart Report. The only modifications it chose to make to the planning rules were to strengthen public consultation mechanisms and increase time limits and fees to allow for this.[181]

The issue of whether it is possible for LPAs to adopt a more precautionary approach than the ICNIRP standards allow, that is, the establishment of radiation-free zones and lower acceptable levels of radiation, was discussed by the UK courts in *Al Fayed*. In that case, the Queen's Bench Division ruled that the LPAs should be bound by safety standards developed nationally, and should not be allowed to set their own, individual risk levels.[182] However, on appeal, the Court of Appeal indicated that the issue might not be as clear-cut as this. Instead, it held that LPAs could be allowed to be more restrictive than the national policy would allow in certain cases.[183] A similar stance has been taken in some later cases.[184] The details of these judgments are returned to below.

6.3.4 *Risk assessment and risk management*

As has been shown in previous chapters, no formal separation between risk assessment and risk management is in place in UK legislation.[185] However, in practice, in the area of base stations, risk assessment is separated from risk management. In this area the Health and Safety Executive (HSE), the HPA and the Local Planning Authorities share decision-making powers. The HSE and the HPA are the main risk managers since they are responsible for commissioning scientific research into issues of radiation, and have the formal powers to issue more specific guidance. The risk assessment is carried out by expert groups, for example the AGNIR, which was set up by, and answers to, the HPA.

At present the authoritative document relating to risks from base stations in the UK is the Stewart Report. The Report is, in effect, a risk assessment performed by a group of experts, and not a risk management

[181] R. Ford, 'Planning For a New Telecommunications Network in the UK', *J.P.L.*, September 2001, 1035.

[182] *R.* v. *Tandridge District Council, ex parte Al Fayed*, p. 7 (of the LEXIS NEXIS version).

[183] *R.* v. *Tandridge District Council and another, ex parte Al Fayed*, Court of Appeal, p. 2 (of the LEXIS NEXIS version).

[184] Most notably in *T-Mobile (UK) Ltd* v. *First Secretary of State*, Court of Appeal (Civil Division), [2005] Env. L.R. 18, C2 and C3.

[185] See, above, Section 6.1.5.

decision based on scientific evidence. However, in the absence of binding rules, the Stewart Report has become the authoritative source in the field. Guidelines and recommendations by the HPA and other institutions endorse most of its scientific conclusions, if not all of its policy recommendations. Courts have proven willing to rely on the conclusions of the Stewart Report, at least as far as they have been reflected in government recommendations.[186]

The experts who took part in the Independent Expert Group for Mobile Telephones and in the writing of the Stewart Report were appointed by the NRPB at the request of the Minister for Public Health.[187] The Report principally comprised a study of the literature in the area of non-ionising radiation. Preceding the commencement of the work of the group, a call for scientific evidence was made. Oral evidence regarded as valid by the committee is to be found on its website.[188]

The conclusions and recommendations of the expert group were far more precautionary in nature than had previously been the case in the UK.[189] The basis for this is discussed in detail in the following sections. It is, however, important to point out, in relation to risk assessment and risk management, that the Stewart Report was the first government-commissioned research to recommend an adoption of the ICNIRP values in the UK.

6.3.5 *Scientific evidence and the introduction of precautionary measures*

The risk assessment in the UK in relation to radiation from mobile telephony and base stations has traditionally been performed by AGNIR and the HPA. However, as shown in the previous section, the most authoritative risk assessment since 1999 has been the Stewart Report. A closer look at this report is, therefore, warranted.

186 See, for example, *T-Mobile (UK) Ltd* v. *First Secretary of State*.

187 Independent Export Group on Mobile Phones, 'Mobile Phones and Health', 11 May 2000 (The Stewart Report). Available at www.iegmp.org.uk/report/text.htm (accessed 1 March 2010).

188 www.iegmp.org.uk/evidence/index.htm (accessed 1 March 2010).

189 Compare with the reports published by AGNIR, which preceded the publishing of the Stewart Report. Apart from the obvious references to precaution in the Stewart Report, the exposure levels recommended in the AGNIR reports were significantly higher. AGNIR reports are available at: www.hpa.org.uk/web/HPAweb&HPAwebStandard/HPAweb_C/1195733801791 (accessed 1 March 2010).

Up until 1999 AGNIR and the HPA had recommended maximum radiation levels from base stations that were considerably higher than those advised by the ICNIRP and in use in other countries.[190] The reason for this recommendation was that the AGNIR considered that ordinary citizens would not be exposed to radiation from base stations on a scale that would warrant the stricter approach reflected in the ICNIRP standards. For workers who would routinely be exposed to radiation, the AGNIR considered the UK approach to be in line with the ICNIRP Guidelines.[191]

The Stewart Report, on the other hand, found that the ICNIRP Guidelines would be better suited to take into account the special needs of those in the general public who might be 'unusually susceptible' to the heating effects of radiation.[192] Thus, it recommended that:

> as a precautionary approach, the ICNIRP guidelines for public exposure be adopted for use in the UK rather than the NRPB guidelines.[193]

It is interesting to note that the Stewart Report explicitly views the implementation of the ICNIRP values as a precautionary measure. However, as has been discussed earlier, the standards are not necessarily precautionary in nature. The foundation for the ICNIRP values is the level at which thermal effects on humans are scientifically certain. It was discussed above that they are based on 'scientific data alone',[194] and, thus, can best be described as preventive rather than precautionary measures.[195] It seems that what the Stewart Report finds to be precautionary in the ICNIRP recommendations is the fact that its values incorporate a reduction factor of five to ensure that the scientifically certain effects do not occur.[196] This factor had previously been considered arbitrary by the AGNIR.[197] However, such reduction factors are the rule, rather than the exception, in risk regulation. It could be argued that they are precautionary in nature since they recommend a safety margin that is not directly scientifically justifiable above or below a scientifically verified level in which harm ensues. If

[190] National Radiation Protection Board, *Restrictions on Human Exposure to Static and Time Varying Electromagnetic Fields and Radiation*, Doc. NRPB, 4(5) 1993.

[191] National Radiation Protection Board, *Advice on the 1998 ICNIRP Guidelines for Limiting Exposure to Time-varying Electric, Magnetic and Electromagnetic Fields (up to 300 GHz)*, Doc. NRPB, 10(2), 1, 1999.

[192] The Stewart Report, 6.34. [193] *Ibid.*, 6.35.

[194] The ICNIRP Guidelines, p. 495. [195] See above, Section 5.3.5.

[196] The Stewart Report, Sections 6.32–6.35.

[197] National Radiation Protection Board, *Advice on the 1998 ICNIRP guidelines for limiting exposure to time-varying electric, magnetic and electromagnetic fields (up to 300 GHz)*.

the exposure is set at X, a reduction factor of five may be used. This means that, at X, harmful effects are certain; consequently, the approved emission level is set at X divided by five in order to ensure that the harmful level is not approached. If the level was set exactly at the level of scientifically proven harm, the margin of error would be very small. By adding a safety margin around the exposure level, the margin of error is widened. This does not, however, have to be viewed as a strictly precautionary decision, in that it does not necessarily aim to counter uncertainties in the data. Instead, it can be seen as a prudent, albeit arbitrary, way to establish a 'fence' around the scientifically certain harm. In relation to the ICNIRP values, the aim of the reduction factor is thus not to take any of the significant and previously discussed uncertainties in relation to exposure from radiation from base stations into account.[198] Instead, the purpose is to ensure that a sufficient safety margin is maintained between the exposure levels at which harm undoubtedly ensues and actual exposure levels. The Stewart Report recommended that the ICNIRP values should be incorporated into UK statutes, the reason for this being that it considered it likely that the emergence of more scientific information would make them liable to change over time.[199]

However, the Stewart Report did not limit itself to recommending the implementation of the ICNIRP values as a precautionary measure in the UK; it went on to state that scientific evidence suggests that thermal effects may not be the only hazard of mobile telephone systems. It also held that radiation may result in biological effects in humans.[200] Since the ICNIRP Guidelines focus solely on thermal effects, biological effects may present themselves even though the Guidelines are abided by. Therefore, the Report concluded that:

> it is not possible at present to say that exposure to RF radiation, even at levels below national guidelines, is totally without potential adverse health effects, and that the gaps in knowledge are sufficient to justify a precautionary approach.[201]

The Group of Experts expressed the view that it seemed 'highly unlikely' that the level of radiation normally emitted from base stations would lead to 'significant, direct adverse effects on health' but that this 'could not be ruled out with confidence'.[202] To address these uncertainties, a number of recommendations were made, for example, the establishment of a national

[198] See, above, Section 5.3.5.
[199] The Stewart Report, Section 6.36.
[200] *Ibid.*, Sections 1.18 and 6.38.
[201] *Ibid.*, Sections 1.19 and 6.39.
[202] *Ibid.*, Section 6.43.

register for base stations,[203] the establishment of a protocol to be followed when establishing base stations[204] and the mandatory establishment of clearly marked exclusion zones around every base station.[205] It also made the far-reaching suggestion that all planning decisions for base stations issued at the time of the issuing of the Report should be revoked and re-examined.[206] Furthermore, base stations near schools should be designed in a manner to minimise the radiation affecting the actual school grounds.[207]

If compared with the Swedish risk assessment of radiation from base stations, it is clear that the Stewart Report chose to emphasise uncertainties to a much larger extent than the Swedish Radiation Protection Agency.[208] Where the Swedish Agency glossed over uncertainties and focused upon what is scientifically verifiable, the Stewart Report emphasised the existing uncertainties and chose to recommend a far-reaching precautionary approach. It is particularly interesting that this is the case, given the fact that both the Swedish and the UK risk assessments are based upon essentially the same set of data.

The response from the UK Government to the Stewart Report was widely positive. The ICNIRP values were not incorporated into statutes, but they have become the standard advocated by the Government in policy documents.[209] Following the publishing of the Report, a review of the UK planning system was put in place. The far-reaching proposal in the Report to revoke all planning decisions and re-evaluate them was not followed in practice. However, Northern Ireland and Scotland have essentially implemented the recommendation to require full planning permission for all base station sites. The same rule has not been implemented in England and Wales.[210] In addition, following the recommendations of the Stewart Report, a website was established where the general public can find information about where base stations are located.[211] Furthermore, two documents with recommendations for how the siting process of base stations is to be carried out have been issued.[212] The primary focus of these

[203] *Ibid.*, Section 6.48. [204] *Ibid.*, Sections 6.54, 6.58 and 6.59.

[205] *Ibid.*, Section 6.50. [206] *Ibid.*, Section 6.57.

[207] *Ibid.*, Section 6.68.

[208] For a discussion of the Swedish approach see, above, Section 5.3.5.

[209] See, for example, Planning Policy Guidance 8, p. 95.

[210] Report by the Board of NRPB, 'Mobile Phones and Health', Doc. of the NRPB 15:5 (2004). Interestingly, it thus appears that different precautionary decisions have also been implemented within the UK based on the same data.

[211] www.sitefinder.ofcom.org.uk/ (accessed 1 March 2010).

[212] Planning Policy Guidance 8 and code of Best Practice on Mobile Phone Network Development, Office of the Deputy Prime Minister, 2002.

documents is to provide an easy-to-access version of the relevant planning rules, but they also emphasise the importance of public participation. This is returned to below.

The courts have generally accepted the ICNIRP values as the standard to be used in assessing radiation from base stations. However, in some cases it has been indicated that it is not impossible for local planning authorities to be allowed to act in a more precautionary manner when deciding on the siting of base stations than the national rules stipulate.[213] This is in stark contrast to the UK Government's own opinion that the general rules relating to the expansion of the telecommunications system is best decided at national level.[214] The courts have pointed out that the guidance document only has limited value when interpreting legislation, and is not to function as a rule to 'fetter' the judgment of the decision-maker.[215] Thus, it would appear that the Local Planning Authorities do, in fact, enjoy some discretion to apply the precautionary principle, at least in theory. When the issue of a more precautionary approach at local level has been brought before the courts, the justification for the more restrictive policy has generally been negative public opinion in relation to the siting of the base station in question. It, thus, seems that, in some cases in the UK, the precautionary principle is closely related to public opinion. The issue is returned below.

6.3.6 *The burden of proof*

As in the case of Sweden discussed above, the UK has not adopted a particularly precautionary approach at national level. The ICNIRP standards which have been incorporated through policy documents and case law primarily offer protection against scientifically established risks and cannot be said to reflect the precautionary principle in the sense discussed here. The Stewart Report included frequent references to the precautionary principle or approach, but stopped short of actually suggesting more restrictive measures than those which can be justified by preventive considerations. It is, therefore, difficult to say what would have to be shown

213 *R.* v. *Tandridge District Council and another, ex parte Al Fayed*, Court of Appeal, and *T-Mobile (UK) Ltd* v. *First Secretary of State*, Court of Appeal.

214 Planning Policy Guidance 8, pp. 8, 29 and 30.

215 *R. (on the application of S)* v. *Brent LBC* [2002] E.W.C.A Civ 693; [2002] E.L.R 556, para. 15. As quoted in relation to Planning Policy Guidance 8 in *Trevett* v. *Secretary of State for Transport Local Government and the Regions and Others* [2002] E.W.H.C. Admin 2696, 25 November 2002, paras. 23–4.

in order to challenge a precautionary measure at national level in the UK, simply because a precautionary approach is not really in effect at the moment.

However, at local level the issue is slightly different in the UK than in Sweden. Where the Swedish Government and courts have made it clear that they do not consider the precautionary principle to be in effect when local authorities decide on planning matters, the matter is more complicated in the UK. As discussed briefly above, Planning Policy Guidance 8, issued by the office of the Deputy Prime Minister, holds that, in relation to base stations, the issue of health safeguards is best dealt with at a centralised level.[216] However, the document is not binding, and courts have, albeit inconsistently, held that Local Planning Authorities may actually be allowed to act in a more precautionary manner than the relevant policy documents would seem to permit in certain cases. In practice, such precautionary measures are primarily based on public opinion that is opposed to the siting of base stations.

6.3.7 *Cost–benefit analysis*

As has been held in previous chapters, the official commitment to cost–benefit analysis in the UK is firm. Different departments of the central UK Government, as well as agencies, have issued non-binding guidelines and reports reaffirming both the importance and the usefulness of cost–benefit analysis. In official documents pertaining to the establishment of base stations, this issue is not elaborated on in any detail. The relevant planning regulations do not require a cost–benefit analysis to be carried out, and no proportionality requirement appears to override the issue. The same can safely be said with regard to the analysis of risk trade-offs.

In Planning Policy Guidance 8 the emphasis is clearly on the benefits of mobile telephony and the importance of consumer choice.[217] As has been discussed extensively above, the UK Government appears to consider the adherence to the ICNIRP Guidelines to address the issue of potential risks of mobile phone technology effectively.

However, as is shown in the following section, the courts are increasingly allowing local authorities to act in a more precautionary manner than the ICNIRP standards stipulate when such action is supported by public opinion. It appears that the rudimentary mechanism that the

[216] Planning Policy Guidance 8, para. 30. [217] Planning Policy Guidance 8, paras. 1–6.

courts have applied include a weighing of the public opposition towards base stations against the current scientific evidence and any other potential costs of not having adequate mobile phone coverage.[218] What may develop out of this is a requirement of an informal cost–benefit analysis in the broad sense, without a firm theoretical basis. What such an analysis should entail and how it should be performed appears to have been adjudicated upon by courts on an ad hoc basis so far. That courts would also consider the proportionality of measures and risk trade-off analysis when conducting such a balancing seems likely

6.3.8 Public opinion

There is no formal, statutory requirement in place to safeguard the consideration of public opinion in the UK. Generally, base stations are considered Class A developments and do not require a building permit.[219] Only if the masts are above 15 meters in height or in any other way exceed the limits established in Section A.1 of Part 24 of the planning code must a formal building permit be applied for. In such cases normal planning rules include a formal consultation procedure.

However, even though statutory obligations make relatively weak requirements on public involvement, the UK Government stresses the importance of involving 'interested parties' in the process in policy documents.[220] Furthermore, a system of 'traffic lights' has been developed, whereby the public attitude can be gauged.[221] The mobile phone operator carries out the assessment and is responsible for choosing the appropriate communication strategy based on its results.[222] If the result of the survey is to be classified as green, public opinion is deemed to be benign to the development and normal planning rules can be relied on. If the result of the survey is red, public opinion is considered hostile to the development and a more thorough consultation procedure must take place. Such a procedure includes increased publicity around the planned development and a fourteen-day consultation period.[223]

218 *Trevett* v. *Secretary of State for Transport Local Government and the Regions and Others*, para. 25 and *T-Mobile (UK) Ltd* v. *First Secretary of State*, Court of Appeal, para. 29.

219 The Town and Country Planning Order 1995, Part 24, A.

220 See, for example, Planning Policy Guidance 8, para. 61.

221 Code of Best Practice on Mobile Phone Network Development, Office of the Deputy Prime Minister, paras. 38–48.

222 *Ibid.*, para. 44. 223 *Ibid.*, para. 45.

Even though a robust, albeit non-binding, public consultation procedure is foreseen and advocated by the UK Government, it remains unclear what effects a consistently negative public opinion at local level may have on actual planning decisions. Planning Policy Guidance 8 stresses both that 'health considerations and public concern can in principle be material considerations in determining application for planning permission and prior approval',[224] and that 'it is the government's firm view that the planning system is not the place for determining health safeguards'.[225] It has been left to the courts to interpret this somewhat schizophrenic approach.

Unfortunately, the UK courts have not dealt with the issue in a particularly consistent manner either. In two planning cases stemming from the mid- and late-1990s, which respectively concerned dioxin emissions from a proposed development[226] and the perceived risks from a chemical waste treatment plant,[227] the Court of Appeal reached what were, essentially, contradictory conclusions. In *Gateshead* the Court held that public opinion without the firm support of scientific evidence cannot be allowed to function as a justification for disallowing development. If this were allowed, very little development at all would take place.[228] In *Newport*, on the other hand, the Court stated, in diametrical opposition to the above, that 'public perception can [...] warrant refusal, even though the factual basis for that fear has no scientific or logical reason'.[229]

In successive case law directly concerning the siting of base stations, this inconsistency has been maintained. In *Al Fayed* the Court of Appeal held that it was 'common ground' between the parties that 'objectively unjustified fears in the locality' are, in some cases, legitimate factors for the planning authority to consider.[230] It has been pointed out that the fact that the point was conceded by both parties reduces the precedential value of this issue.[231] In *Phillips* the Queen's Bench ruled that if two alternative sites with equal properties exist, 'as a matter of principle' the option to refuse approval for the site where public opposition to the development was strong could be justified.[232] In *Trevett* the Queen's Bench made an attempt

224 Planning Policy Guidance 8, para. 97. 225 *Ibid.*, para. 98.

226 *Gateshead MBC* v. *Secretary of State for the Environment*, 1 P.L.R., 85. 1994.

227 *Newport CBC* v. *Secretary of State for Wales*, J.P.L, 377, 1998.

228 *Gateshead MBC* v. *Secretary of State for the Environment*, para. 95.

229 *Newport CBC* v. *Secretary of State for Wales*, 385.

230 *R.* v. *Tandridge District Council and another, ex parte Al Fayed.*

231 C. Hilson, 'Planning Law and Public Perceptions of Risk: Evidence of Concern or Concern Based on Evidence', *J.P.L*, 2004, 1641.

232 *Phillips* v. *First Secretary of State and others*, Queen's Bench Division (Administrative Court) [2003] EWHC 2415, para. 41.

to link the two contradictory statements in Planning Policy Guidance 8 after it had stated that the Policy Guidance could not be considered to 'fetter' the judgment of the Court.[233] The judgment in this case related to the refusal of a local council to issue planning permission for TETRA (terrestrial trunked radio) communications masts mainly due to public opposition to the establishment of the masts. Here, the Court held that public opposition to a base station was not necessarily sufficient to justify a refusal. However, it also held that it would be equally erroneous to ignore pubic opinion, even if public fears were not objectively justified. What the Court suggests is that public opinion must be carefully weighed against the current scientific evidence on a case-by-case basis.[234] The inevitable conclusion of this must be that the Court in *Trevett* did not find the ICNIRP Guidelines to reign supreme, but that public opinion could, under certain circumstances, justify the refusal to grant planning permission even though the ICNIRP standards were met. In the case at hand, the Court did not find reason to override the ICNIRP standards and thus it considered the local council's decision to have been flawed when it refused to grant planning permission. The *Trevett* judgment does little to resolve the tension built in to Planning Policy Guidance 8. Instead, it reinforces its ambiguity. The same can be said for the more recent *T-Mobile* judgment, in which the Court of Appeal held that:

> Health considerations and public concern can in principle be material considerations in determining applications for planning permission and prior approval. Whether such matters are material in a particular case is ultimately a matter for the courts.[235]

Once again, the Court leaves the door open for the local authorities to take public opinion into account, and, interestingly, a more precautionary reading of potential health effects than would be allowed were the ICNIRP values considered binding.

The role of public opinion in decision-making in relation to base stations is ambiguous at best. In recent case law it appears that courts are increasingly becoming willing to accept public opposition to the establishment of base stations as a justification to deny planning permission at local level, even when such base stations are abiding by the national rules

[233] *Trevett* v. *Secretary of State for Transport Local Government and the Regions and Others*, paras. 23 and 24.

[234] *Ibid.*, para. 25.

[235] *T-Mobile (UK) Ltd* v. *First Secretary of State*, Court of Appeal, para. 29.

and standards. However, it appears that a weighing of the interests must precede such a decision.

6.3.9 *Conclusion*

The UK legal system lacks a direct statutory instrument which covers the area of non-ionising radiation. Instead, the National Radiation Protection Board was originally given the task of issuing recommendations in the field. Now this task has been transferred to the Health Protection Agency, an independent agency that does not answer directly to the Ministry in the way that the PSD does in the field of pesticide regulation. The degree of autonomy is thus greater on the part of the HPA. The Advisory Group on Non-Ionising Radiation was set up to assist the HPA by performing the de facto risk assessments. However, with regard to mobile telephony, the most authoritative report to date was published by the Independent Expert Group on Mobile Phones, a temporary advisory group charged with the task of providing scientific advice on mobile phones. The conclusions of the group were published in the Stewart Report.

The siting of base stations is subject to planning rules like all other installations. For base stations that do not exceed certain limitations in size, no planning permission is required. However, a request for establishment must be made to the relevant Local Planning Authority, which has to reply that planning permission is not needed before construction can begin.

As has been stated repeatedly above, no formal separation of risk assessment and risk management is in effect in UK legislation. The UK Government has not turned the risk assessment included in the Stewart Report into binding regulations, but has broadly embraced its conclusions in policy documents. Thus, the Stewart Report's recommendation that the ICNIRP standards be adopted in the UK has been welcomed by the UK authorities. However, it has stopped short of implementing some of the more far-reaching planning recommendations made in the Report. The Stewart Report explicitly endorsed a precautionary approach to the regulation of base stations. Instead of glossing over the existing uncertainties, as, for example, the Swedish Radiation Protection Agency has done in its assessments, uncertainties were highlighted and discussed. It is thus clear from the Stewart Report that the health effects of radiation from mobile phones are far from certain.

When adopting the ICNIRP standards, the UK Government stated that it found health effects stemming from base stations to be regulated most effectively at national level. However, in Planning Policy Guidance 8 it left a door open for public opinion to be considered at local level. UK courts seem to have expanded this leeway for local governments. In *Brent* and *Trevett* it was pointed out that the government documents in the area of planning for base stations are not binding and therefore cannot be considered to 'fetter' the judgment of the courts. This would seem to allow some room for manoeuvre on the side of the local authorities when applying the precautionary principle and taking public opinion into account. The issue of whether, and, if so, how much, scientific evidence must underlie a precautionary decision by the local authorities has received very contradictory treatment by the courts. In *Gateshead* the Court of Appeal held that public opinion could be relied on, but only if strong support for the views of public opinion were to be found in science. In *Newport* the Court held that public opinion can warrant a more restrictive policy even if its factual basis 'has no scientific or logical reason'. Successive case law has done little to resolve this conflict. The Court of Appeal concluded in *Al Fayed* that it can sometimes be legitimate for 'objectively justified fears' to be taken into consideration by planning authorities when deciding on matters relating to base stations. However, the precedential value of the case was undermined by the fact that both parties had apparently conceded this point at the outset of the proceedings. The Court in *Phillips* ruled that in cases where two alternative but equal sites were available for a base station, the one with the lesser public resistance could be chosen. In *Trevett* the Queen's Bench aimed to negotiate the contradictory statements in Planning Policy Guidance 8. It essentially concluded that public opinion without a scientific basis to justify its fears was not sufficient to refuse the establishment of the site in question. Simultaneously, it held that the fact that no scientific basis for fears existed, but that public opinion was still negative, did not necessarily mean that a base station should be allowed. It is difficult to imagine how the Court could have offered any less guidance on how actually to carry out this balancing act. The only thing that it does is reinforce the ambiguity of Planning Policy Guidance 8 and the previous case law. The same thing can be said for *T-Mobile*, where the Court held that 'public concern can in principle be material considerations' when deciding on planning issues relating to base stations.

In conclusion, it can be said that health and safety concerns stemming from radiation from base stations are, in principle, only addressed by the recommended ICNIRP standards in the UK. Expert groups, the UK Government, Local Planning Authorities and courts have generally agreed upon this. So far, regulation in the UK resembles that of Sweden.

However, whereas in Sweden the Government and courts have clearly established that local governments are prevented from taking a more precautionary stance, the UK situation is open to this, provided that public opinion is opposed to the establishment of base stations. Since Local Planning Authorities are not directly bound by regulations in the area of non-ionising radiation, they should not be fettered by the recommendations made by the UK Government. In *Duddridge* the Court pointed out that local authorities were not the bodies to consider precautionary measures. However, as can be seen from the case law in this chapter, public opinion clearly can play an, as of yet, undefined role in influencing planning decisions. In the case law public opinion is often treated as an independent ground not included in the precautionary principle. This development indicates that, in the area of base stations, public opinion may be as important as that of any legally binding precautionary principle. This would be in line with what Haigh had already stated in 1994 in relation to the general implementation of the precautionary principle in the UK:

> When it comes to political action, public pressure is as likely to count as much as, or more than, principle.[236]

[236] Haigh, 'The Introduction of the Precautionary Principle into the UK'.

7

The precautionary principle in the United States

7.1 The development and status of the precautionary principle in the United States

7.1.1 Introduction

That the US view of the precautionary principle differs significantly from the view advocated by the EU is apparent from recent international trade conflicts,[1] as well as from the official rhetoric from both sides of the Atlantic.[2] The American authorities tend to argue that the EU version[3] of the precautionary principle slows down scientific and technological development and hinders international trade. Where the EU is seen as risk averse, the USA appears to require firm scientific evidence of harm before recognising the need for regulation. Babich has held that the American mentality is not predisposed to precautionary considerations:

> US culture embraces change and risk. Our heroes look forward – entrepreneurs and inventors that create what we have barely imagined, athletes who break physical boundaries, and stars of music and film who transcend behavioral conventions. The precautionary principle is not at the core of our national character.[4]

In the same vein, the *New York Times* held in 2003 that 'precaution is for Europeans'.[5]

[1] See, above, Section 3.3.2.

[2] See, for example, J. D. Graham, 'The Perils of the Precautionary Principle: Lessons from the American and European Experience', Speech before the Heritage Foundation, Washington DC, 20 October 2003. Available at: www.heritage.org/Research/Regulation/hl818.cfm (accessed 1 March 2010).

[3] For a discussion on the problems of considering one EU version of the precautionary principle to exist see previous sections of this book.

[4] A. Babich, 'Too Much Science in Environmental Law', *Colum. J. Envl. L.*, 28 (2003), 119.

[5] S. Loewenberg, 'Precaution is for Europeans', *New York Times*, 18 May 2003, Section 4, p. 14.

However, the facts do not seem to match the rhetoric. In fact, historically, the USA has acted in as precautionary a manner as the EU, if not more so.[6] US environmental legislation of the 1970s went a long way to establishing a form of the precautionary principle in many areas. Agencies and courts have interpreted statutes in order to afford precautionary considerations even greater weight. Applegate has held that:

> [p]recaution is clearly one of the principal goals of congress in drafting and of the Environmental Protection Agency (EPA) in administering environmental legislation.[7]

This chapter aims to look beyond the rhetoric and give a brief overview of the development of what some have called a 'precautionary preference' in US legislation.[8] Even if this preference does not amount to an all-encompassing precautionary principle, like the one found, for example, in Sweden, it is argued here that it is definitely comparable to how precaution is used in the EU. In order to facilitate this comparison, the precautionary principle is broken down into its constituent parts and dealt with separately as in the previous chapters. In the following sections it is discussed whether, and if so, how, the precautionary principle, or something resembling the precautionary principle, has influenced regulation of base stations for mobile telephones and pesticides.

7.1.2 Background and development

The precautionary principle as an independent legal concept is not applied in the United States. The US Constitution does not refer to the principle, and neither do representative state constitutions.[9] The US Supreme Court has never referred to the concept, and only one federal appellate court has

[6] See, for example, R. öfstedt, 'The Changing Character of Regulation: A Comparison of Europe and the United States', *Risk Analysis*, 21 (2002), 399.

[7] J. S. Applegate, 'The Precautionary Preference: An American Perspective on the Precautionary Principle', *Human and Ecological Risk Assessment*, 6:3 (2000), 420.

[8] Applegate, 'The Precautionary Preference'.

[9] This conclusion is based on the study of the state constitutions of California, Florida, Iowa, Illinois, Massachusetts, Michigan, New York, Oregon, Texas and Utah, which was carried out by Wood *et al.*, and presented in S. G. Wood, S. Q. Wood and R. A. Wood, 'The Precautionary Principle? An American Assessment', in M. Paques, *Le Principe de precaution en droit administratif/Precautionary Principle and Administrative Law, National and International Reports, International Academy of Comparative Law* (Bruylant, 2007), p. 247.

ever referred to it.[10] Moreover, reference to the precautionary principle in legal doctrine is scarce.[11]

However, even though the precautionary principle as a general, legal principle certainly does not feature prominently in the US legal system, there is little doubt that precautionary considerations have played a significant part in US regulatory history.[12] One commentator has even stated that 'the precautionary principle has found its way into United States environmental laws, regulations, and policies'.[13] Vogel has pointed out that, over the last thirty years, it appears that the EU and the USA have traded places when it comes to being risk averse and precautionary.[14] Together with Löfstedt, Vogel has also argued that some of the regulatory policies in place in the EU were originally borrowed from the USA, and vice versa.[15] Already from the outset it can thus be noted that historically the regulatory similarities between the EU and the USA are greater than they are often perceived to be.

One example of early precautionary legislative instruments in the USA is the Endangered Species Act of 1973. The Act states that its purpose is the 'conservation'[16] of 'endangered',[17] 'threatened'[18] and 'jeopardised'[19] species. The fact that this not only includes scientifically certain risks of extinction is clarified by the fact that species 'likely to become so within the foreseeable future' should also be protected.[20] In the highly commented-on *Snail Darter* case of 1978, the US Supreme Court, citing the legislative history, held that the Endangered Species Act constituted

[10] *Ibid.*, p. 248. However, the case in which the federal court in question made the reference to the precautionary principle related to international torts, and not to US domestic affairs. It should also be mentioned that the 5th Circuit in the case did not find that the precautionary principle rose 'to the level of an international tort'. *Banal* v. *Freeport-McMoran, Inc.*, 197 F 3d 161 (5th Cir. 1999).

[11] Wood *et al.*, 'The Precautionary Principle?', pp. 248–9.

[12] See, for example, Wiener, 'Precaution in a Multi-Risk World', pp. 1–2;Applegate, 'The Precautionary Preference'; B. D. Goldstein and R. S. Carruth, 'Science in the Regulatory Process: Implications of the Precautionary Principle for Environmental Regulation in the United States: Examples from the Control of Hazardous Air Pollutants in the 1990 Clean Air Act Amendments', *Law & Contemp. Prob.*, 66 (2003), 247; and Wood *et al.*, 'The Precautionary Principle?', p. 250.

[13] P. M. Kannan, 'The Precautionary Principle: More than a Cameo Appearance in United States Environmental Law', *Wm. & Mary Envtl. L & Pol'y Rev.*, 31 (2007), 409.

[14] Vogel, 'Ships Passing in the Night'.

[15] Löfstedt, 'The Changing Character of Regulation'.

[16] The term is used throughout the Act; see, for example, Endangered Species Act 1973, 16 U.S.C. 1531–44, Sect. 4 (B) and (D).

[17] *Ibid.*, Section 4. [18] *Ibid.*, Section 4.

[19] *Ibid.*, Section 7 (A) (4). [20] *Ibid.*, Section 4 (B) (1) (a) ii.

an expression of 'institutionalized caution'.[21] Interestingly, the Court here seems to have acknowledged and afforded weight to the existence of uncertainty in relation to endangered species. However, in *Snail Darter* the Court did not primarily consider uncertainty in relation to whether the species may be extinct, but, instead, considered how it may benefit humanity in the future. The Court concluded that, since the benefits of the continued survival of the Snail Darter in the future are uncertain, it should be prevented from becoming extinct. Thus, potential benefits from the species for humanity will be safeguarded. On the face of it, this does not seem directly similar to the precautionary principle, as it has been discussed in the previous chapters in this book. Despite this, Applegate has held that the expression of the Supreme Court in *Snail Darter* is a 'fair paraphrase of the precautionary principle'.[22]

Other early legislative instruments in the USA that were based on basic precautionary assumptions include The Clean Air Act 1970[23,24] and the Clean Water Act 1972.[25,26] The precautionary nature of The Clean Air Act was reinforced and elaborated on in quite some detail in the influential *Ethyl Corp.* case.[27] Furthermore, in *Reserve Mining Co.* v. *EPA* the US Supreme Court once again stated that the term 'endangering', this time in the context of the Clean Water Act, signifies that not only scientifically certain hazards may be regulated, but also that precautionary action may be taken by administrators.[28] This was also established in *Lead Indus. Ass'n, Inc.* v. *EPA*, in which the DC Circuit Court held that, in the Clean Air Act, Congress had 'directed the Administrator to err on the side of caution in making the necessary decisions'.[29]

The most overt example of the use of a precautionary principle or approach in the USA is found in food regulation. The (in)famous Delaney Clause imposed zero tolerance for residues which may cause cancer in food, regardless of how small the actual risk would be.[30] In *Public Citizen* v. *Young* the DC District Court established that the apparent absolutism in the clause was intended by Congress, and should, therefore, be

[21] *Tennessee Valley Authority (TVA)* v. *Hill*, 437 US 153 (1978).
[22] Applegate, 'The Precautionary Preference', 420.
[23] The Clean Air Act 1970, 42 U.S.C. 7401. [24] Vogel, 'Ships Passing in the Night', 3.
[25] Federal Water Pollution Control Act (Clean Water Act), 33 U.S.C. 1251.
[26] Tickner and Raffensberger, 'The Precautionary Principle in Action'.
[27] *Ethyl Corp.* v. *EPA*, 598 F.22d 91 (D.C. Cir. 1979). The case is returned to in detail below.
[28] *Reserve Mining Co.* v. *EPA*, 514 F.2d 492 (8th Cir. 1975), 528.
[29] *Lead Indus. Ass'n, Inc.* v. *EPA*, 647 F.2d 1130 (D.C. Cir. 1976), 1155.
[30] Federal Food, Drug and Cosmetic Act 1958, 21 U.S.C.. 348, Section 408 and 409.

upheld in the application of the statute.[31] The Delaney Clause has since been revoked, but its influence on US pesticide regulation is returned to in detail in the section on pesticides below.[32]

It is no coincidence that most of the above examples of precautionary action by US Congress and courts relate to legislation passed in the 1970s, a time when the environmental movement was growing strong. During the presidency of Ronald Reagan (1981–8) the impetus for environmental regulation in general was weakened. This trend was continued during the Republican Control of Congress in the 1990s and 2000s (1994–2007). It has been argued that, during this time, environmental groups have been largely concerned with (successfully) maintaining the status quo of environmental regulation, rather than trying to break new ground.[33] Since the coming into force of the Data Quality Act in October of 2002, this balance might finally be conclusively shifted away from precautionary considerations and towards the strict requirements of scientific evidence for regulation.[34] The concrete implications of this issue is returned to in detail below in this chapter.

7.1.3 *The current context and definition of the precautionary principle*

In the absence of any explicit, or even implied, reference to a generally applicable precautionary principle in US legislation, the content of precautionary measures, and the circumstances under which they may be taken, have been dealt with primarily by the US courts.[35] Following the introduction of the National Environmental Policy Act 1969 (NEPA), US courts became a battleground for 'competing environmental ideologies'.[36] On one side, the newly established environmental movement argued for a wide introduction of environmental concerns in US regulation, and, on the other side, industry fought to maintain the status quo. In this environment, even though not relying particularly on the NEPA, activist US courts started carefully to define a precautionary principle of sorts in certain policy areas.

31 *Public Citizen* v. *Young*, 831 F.2d 1008 (D.C. Cir. 1987).
32 See also Kannan, 'The Precautionary Principle', 438–40.
33 Vogel, 'Ships Passing in the Night', 32.
34 For an interesting account of this development see C. Mooney, *The Republican War on Science* (Basic Books, 2005), pp. 102–20.
35 Jasanoff, *Science at the Bar*, pp. 36 *et seq.* 36 *Ibid.*, p. 37.

As outlined briefly above, two cases in the mid-1970s became central for the inclusion of precautionary considerations in the United States. The Clean Air Act and the Federal Water Pollution Control Act, which were the statutes at issue in the respective cases, lacked any direct reference to precautionary measures. However, this did not stop the courts in the mid-1970s from allowing precautionary measures taken by the Environmental Protection Agency (EPA) under these acts. The vessel used to facilitate this development was the term 'endanger', which was included in both statutes.[37]

In *Reserve Mining* the US Court of Appeals for the 9th Circuit needed to assess if low-grade iron ore emissions from a mining operation was a sufficient threat to public health to warrant regulatory intervention. The evidence regarding the emissions was not conclusive and the issue was surrounded by a great deal of uncertainty. As mentioned above, the relevant statute, the Federal Water Pollution Control Act, held that substances that 'endanger' human health should be regulated. No more guidance as to how to interpret the term 'endanger' was to be found in the statute, nor had the US Supreme Court at that time made any rulings on how the term was to be interpreted.[38] Thus, the Court had some discretion with regard to whether to interpret the term narrowly and demand firm scientific evidence of harm before action,[39] or to expand the term to incorporate situations fraught with scientific uncertainty. Relying on a dissenting opinion in the first *Ethyl Corp.* v. *EPA* case,[40] the 8th Circuit decided to opt for the latter.[41] It stated:

> In the context of this environmental legislation, we believe that Congress used the term 'endangering' in a precautionary or preventive sense, and,

[37] As regards the Clean Air Act see Section 211(c)(1)(A) which authorises the Administrator of the EPA to regulate gasoline additives whose emission products 'will endanger the public health or welfare'. With regard to the Federal Water Pollution Control Act see Section 11, which holds that regulation should be enacted with regard to substances 'endangering the health or welfare of persons'.

[38] *Reserve Mining* v. *EPA*, 90–1.

[39] This was the traditional common law approach in situations of scientific uncertainty up until the 1970s, as exemplified by for example in *Fletcher* v. *Bealey* 28 Ch.D. pp. 688–700. For a more detailed discussion on this see Jasanoff, *Science at the Bar*, p. 36.

[40] The Court here referred to the first of two identical cases brought by Ethyl Corp. before the US Court of Appeals for the DC Circuit. In this case, the majority found in favour of the plaintiff and quashed the precautionary action by the EPA. However, as is shown below, the case was referred back to the Court for an *en banc* decision before long and the verdict was reversed.

[41] *Reserve Mining* v. *EPA*, 89–90.

> therefore, evidence of potential harm as well as actual harm comes within the purview of that term.[42]

Interestingly, it found further support for this reading of the statute in a 1966 case from the US Supreme Court in which the importance of 'common sense' interpretations of environmental statutes was emphasised.[43] Thus, it appears that the issuing of precautionary measures in this case was considered to be common sense by the Court. This interpretation of the term 'endangering' must be considered a relatively strong endorsement of a form of the precautionary principle. This is particularly the case when it is considered that the Federal Water Pollution Control Act and the Clean Air Act 'together constitute the bulk of this nation's substantive environmental protection legislation' at the time of the *Reserve Mining* and *Ethyl Corp.* rulings.[44]

As was mentioned above, the majority of the Court in the first *Ethyl Corp.* case had rejected the expansive reading of the term 'will endanger' in the Clean Air Act.[45] Instead, the *Reserve Mining* case was constructed on the basis of the minority opinion of Judge Wright in the same case. However, the *Ethyl Corp.* case was referred back to the DC Circuit Court, this time for an *en banc* decision. And this time, Judge Wright's stance was supported by the majority of the Court, which stated in conclusion:

> In sum, based on the plain meaning of the statute, the juxtaposition of Section 211 with Sections 108 and 202, and the Reserve Mining precedent, we conclude that the 'will endanger' standard is precautionary in nature and does not require proof of actual harm before regulation is appropriate.[46]

In a few other cases stemming from around the same time as *Reserve Mining* and *Ethyl Corp.*, courts continued to hold that, in areas that are 'particularly sensitive and fright laden',[47] regulation based on 'lower standards of proof than otherwise applicable'[48] could be allowed.[49]

[42] *Ibid.*, 89. [43] *Ibid.*, 91.

[44] *Ethyl Corp.* v. *EPA*, 37.

[45] The relevant provision, Section 211 (c) (1) (a) of the Clean Air Act 1970, holds that the regulator may 'control or prohibit the manufacture, introduction into commerce, offering for sale, or sale of any fuel or fuel additive for use in a motor vehicle or motor vehicle engine (A) if any emission products of such fuel or fuel additive *will endanger* the public health or welfare'. *Highlighting* added by the author.

[46] *Ethyl Corp.* v. *EPA*, 39.

[47] *Environmental Defence Fund* v. *EPA*, 465 F.2d 528 (D.C. Cir. 1972).

[48] *Certified Color Manufacturers Association et al.*, v. *David Mathews, Secretary of the Department of HEW et al.*, 465 F.2d 528 (D.C. Cir. 1976), 297–8.

[49] For a similar discussion on this see Jasanoff, *The Fifth Branch*, p. 51.

If the judgments of the appellate courts in *Reserve Mining* and *Ethyl Corp.* seemed to establish a precautionary principle of sorts in the US, as well as going some way towards defining it, subsequent case law was to put a serious dent in this development. The National Environmental Protection Act 1969 (NEPA) established the use of Environmental Impact Statements in the USA and has been considered by some to be an embodiment of the precautionary principle in US legislation.[50] Others have held that, even though it has precautionary elements, it lacks 'the substantive teeth' that can be expected from an instrument which truly reflects the precautionary principle.[51] This is supported by the fact that the US Supreme Court explicitly stated that the NEPA does not assign any particular weight to environmental considerations.[52] Furthermore, there is no effective obligation under the NEPA that mandates the EPA to mitigate harms which may be identified in the Environmental Impact Statement.[53] Even if it is contemporary to the Clean Air Act and the Federal Water Pollution Control Act, the precautionary element in the NEPA must be considered to be significantly less pronounced.

In the *Benzene* case of 1980, the US Supreme Court confirmed that precautionary regulatory action could be taken under the Occupational Health and Safety Act, but that such action would have to be backed up by a certain amount of scientific evidence.[54] The case marked the beginning of a new era in US risk regulation under the Reagan administration. Regulatory decisions would, from now on, have to be based on detailed risk assessments, rather than on more sweeping precautionary considerations.[55] The fact that the increased scope of courts' judicial review was instrumental to this development is discussed in the following section.

7.1.4 *Judicial review*

While the judicial review in the national legal orders previously discussed here can best be described as narrowly confined,[56] the situation in the USA

[50] See, for example, the position of the Science and Environmental Health Network (SEHN), *Precautionary Principle FAQs*. Available at: www.sehn.org/ppfaqs.html (accessed 1 March 2010).

[51] Applegate, 'The Precautionary Preference', 429.

[52] *Marsh* v. *Oregon Natural Resources Council* 490 U.S. 332 (1989).

[53] Applegate, 'The Precautionary Preference', 429.

[54] *Industrial Union Department, AFL-CIO* v. *American Petroleum Institute*, 448 U.S. 607 (1980).

[55] Wiener, 'Precaution in a Multi-Risk World', p. 2.

[56] This, in practice, is also the case in Sweden, even if, as was shown above, the actual legal scope for judicial review leaves a rather wide leeway for the courts. However, traditionally

is rather different. The scope of judicial review of administrative action in the USA is partly defined in a statutory instrument.[57] The Administrative Procedure Act (APA) holds that administrative decisions should be overturned if they are 'arbitrary, capricious, an abuse of discretion, or otherwise not in accordance with law',[58] or if they are not based on 'reliable, probative, and substantial evidence'.[59] However, the interpretation of this provision has not been consistently upheld in the case law, as is the case in the UK, where the 'reasonableness' test, as developed by the courts, is a veritable judicial axiom.[60] Instead, the US courts' willingness to review administrative decisions concerning scientific assessments has varied over time, seemingly in unison with the reigning political paradigm.[61]

The fact that agency actions in the USA are subjected to judicial review is thus established by federal statute. The APA allows for judicial review of all actions for which there is no other adequate remedy, where such review has not been precluded by statute, or where agency discretion is commanded by law.[62] The exceptions to judicial review have been narrowly defined in the case law, and the prevailing legislative intent of the APA is that agency action should be open to review by the courts.[63]

Jasanoff has discussed a number of different schools of judicial review which have been applied by the courts in relation to scientific evidence since the 1970s.[64] The first school, which has to a certain extent been discussed above, occurred in the 1970s, when activist federal courts went a long way to use expansive readings of environmental legislation in order to usher through precautionary decisions. A famous example of this stance of the courts is the *Ethyl Corp.* decision, which was touched upon above. A methodology central to this development was the so-called 'hard

courts have made very limited use of this possibility. Potentially, this is changing with the expanded use of scientific expertise in the courts' composition within the framework of the newly established environmental courts. For a detailed discussion see, above, Section 5.1.4.

[57] Section 704 of the Administrative Procedure Act 1974, 5 U.S.C. 551 (APA).

[58] *Ibid.*, Section 706(2)(A). For a concise discussion on judicial review under the APA see F. Anderson *et al.*, 'Regulatory Improvement Legislation: Risk Assessment, Cost-Benefit Analysis, and Judicial Review', *Duke Env. L. and Pol'y F.*, 11 (2000), 89.

[59] Section 706(2)(E). [60] See, above, Section 6.1.6.

[61] Jasanoff, *Science at the Bar*, pp. 70–1.

[62] Sections 701(a)(1) and 701(a)(2) respectively, Administrative Procedure Act 1974.

[63] See, for example, M. Pak, 'Note & Comment: An IQ Test for Federal Agencies? Judicial Review of the Information and Quality Act Under the APA', *Wash. L. Rev.*, 80 (2005), 731.

[64] Jasanoff, *Science at the Bar*, pp. 69–72.

look doctrine', developed primarily by the DC Circuit Court. The hard look doctrine aimed to ensure that agencies had, indeed, reviewed the relevant scientific evidence underpinning an issue and had understood it. The method has been defined in the following manner:

> Assumptions must be spelled out, inconsistencies explained, methodologies disclosed, contradictory evidence rebutted, record references solidly grounded, guesswork eliminated and conclusions solidly supported 'in a manner capable of judicial understanding'.[65]

However reasonable the review described appears to be in theory, in practice it proved very difficult to strike the required balance.[66] These difficulties led to a resort to proceduralism by the courts, and, in order to facilitate this development, they even began to invent formal procedures against which agency decisions were measured.[67] However, in *Vermont Yankee* the US Supreme Court curtailed this practice and held that court-prescribed procedures would only exaggerate the uncertainties inherent in the judicial process.[68] At the same time, *Vermont Yankee* explicitly allowed courts to substantially review the data on which decisions were made.

In the 1980s the courts, following a number of precedents from the US Supreme Court, began to offer more deference to the interpretation of scientific evidence by agencies.[69] However, the hard look doctrine was, in principle, maintained and solidified by the Supreme Court in *State Farm*. In that case, the Court famously held that:

> the agency has relied on factors which Congress has not intended it to consider, entirely failed to consider an important aspect of the problem, offered an explanation for its decision that runs counter to the evidence before the agency, or is so implausible that it could not be ascribed to a difference in view or the product of agency expertise.

However, in *Baltimore Gas* the US Supreme Court elaborated on the limits to the hard look doctrine and held that a court 'must generally be at its most deferential' when it is reviewing assessments made at the 'frontiers of science', thus indicating that the scope of the courts' review should be

65 *EI DuPont de Nemours and Co.* v. *Train*, 541 F.2d 1018 (4th Cir. 1976).

66 Jasanoff, *Science at the Bar*, p. 76.

67 *Ibid.*, p. 77.

68 *Vermont Yankee Nuclear Power Corp.* v. *Natural Resources Defense Council*, 435 U.S. 519 (1978), 547.

69 Jasanoff, *The Fifth Branch*, p. 56.

more limited in cases of scientific uncertainty.[70] In *Sierra Club* v. *Costle* the DC Circuit Court concluded that agency 'informal rule-making' was not only a technocratic process, but that it must also be allowed to include political considerations. Thereby, the Court seemed to indicate that a certain leeway should be afforded to agencies to enable them to consider aspects of regulation other than those strictly based in science.[71]

A few other cases, dating from the early 1980s, indicate that the change in the evolving character of judicial review perhaps had more to do with curtailing the substantive environmental activism on the part of some US federal courts than any genuine procedural issues of judicial review. Where the courts in cases such as *Ethyl Corp.* and *Reserve Mining* had effectively reversed the burden of proof onto the regulated party, the *Benzene* case showed that the Supreme Court wished it to remain with the agency. In order to do so, it substantively reviewed the data which underlay the decision to regulate benzene emissions and found it insufficient to warrant regulation.[72] This trend is also visible in *Gulf South Insulation*.[73] The issue of the allocation of the burden of proof is returned to in greater detail below. Here, it is sufficient to point out that courts in the 1980s started to break up the activist partnership that had formed between agencies and courts in the 1970s. This task was further facilitated by the fact that Congress began to issue narrower and more detailed rules in the field of the environment, thus further limiting the possibilities of wide interpretation by agencies that previous legislation had allowed for.[74]

In *Chevron* the Supreme Court attempted to clarify the scope of agency discretion by introducing a two-pronged test. This test meant that courts should first ask themselves whether 'Congress has directly spoken to the precise question at issue' in the case. If so, the Court had to 'give effect to the unambiguously expressed intent of Congress'.[75] Thus, the discretion of the agency in question was strictly limited. However, if Congress has not directly addressed the issue, courts can only consider whether the agency's interpretation is 'is based on a permissible construction of

[70] *Baltimore Gas and Electric Co.* v. *Natural Resources Defence Council*, 462 U.S. 87 (1983), 103.

[71] *Sierra Club* v. *Costle*, 657 F.2d 298 (D.C. Cir. 1981), 408.

[72] *Industrial Union Department* v. *American Petroleum Institute*, 448 U.S. 607 (1980).

[73] *Gulf South Insulation* v. *Consumer Product Safety Commission*, 701 F.2d 1137 (5th Cir. 1983).

[74] S. Shapiro and R. L. Glicksman, 'Congress, The Supreme Court and the Quiet Revolution in Administrative Law', *Duke L. J.* (1988), 819.

[75] *Chevron USA. Inc.* v. *Natural Resources Defence Council*, 467 U.S. 87 (1984), 842–3.

the statute'.[76] If this is the case, the Court can do nothing but defer to the interpretation of the agency.[77] In short, the Court's reasoning meant that, unless Congress has addressed an issue unambiguously, it is for the relevant agency to interpret the statute, and this interpretation is only subject to a limited review by the courts. It has been held that this doctrine distorts the system of checks and balances at the expense of the judiciary.[78]

In the more recent case *Texas Oil & Gas Ass'n* v. *EPA*, which dealt with discharge limits established by the EPA, the 5th Circuit Court reconfirmed that the judicial review of agency decisions must respect a great margin of deference when 'the agency's decision rests on an evaluation of complex scientific data within the agency's technical expertise'.[79]

Following the highly publicised *Daubert* v. *Merrel Dow Pharmaceuticals*[80] in 1993, in which the US Supreme Court explicitly allowed federal judges to reject irrelevant or unreliable scientific evidence, voices were raised in favour of courts implementing the *Daubert* approach in the review of agency decisions.[81] In *Daubert* the Supreme Court established a number of methods or principles regarding how courts should treat scientific evidence. Among other things, it held that courts must exercise an element of screening when deciding on what scientific evidence to consider. The aim of this would be to ensure that only 'relevant' and 'reliable' evidence is relied upon in court proceedings.[82] However, the Supreme Court also prohibited judges from drawing their own conclusions on the basis of the scientific evidence presented. Instead, they should evaluate and criticise the 'principles and methodology, not [...] the conclusions they generate'.[83] Since this judgment, courts have expressly rejected the extension of the *Daubert* approach to agency decisions.[84] However, some commentators have vigorously supported its application in administrative cases.[85] Other commentators have found that deep substantive review

[76] *Ibid.*, 843. [77] *Ibid.*

[78] Shapiro and Glicksman, 'Congress, The Supreme Court, and the Quiet Revolution in Administrative Law', 862.

[79] *Texas Oil & Gas Ass'n* v. *EPA*, 161 F.3d 923 (5th Cir. 1998), 933–4.

[80] *Daubert* v. *Merrel Dow Pharmaceuticals*, 509 U.S. 579 (1993).

[81] See, for example, A. C. Raul and J. Z. Dwyer, 'Science in the Regulatory Process: "Regulatory Daubert": A Proposal to Enhance Judicial Review of Agency Science by Incorporating Daubert Principles into Administrative Law', *Law & Contemp. Prob.*, 66 (2003), 7.

[82] *Daubert* v. *Merrel Dow Pharmaceuticals*, 589.

[83] *Ibid.*, 595.

[84] See *Sierra Club* v. *Marita*, 46 F.3d 606 (7th Cir. 1995), 622.

[85] See, for example, Raul and Dwyer, 'Science in the Regulatory Process', 7.

possibilities, including the hard look doctrine, have led to courts splitting according to partisan lines when deciding on the outcome of cases.[86]

On the whole, judicial review by US courts has remained as it was developed by the US Supreme Court during the 1980s. This means that the activist approach by courts in *Ethyl Corp.* and *Reserve Mining* has, to a large extent, been abandoned in favour of a more deferential review, although the hard look doctrine still remains in force. However, stronger demands for scientific evidence and the reasoning of agencies before regulatory action is taken have progressively been made.[87] What this means for the application of the precautionary principle with regard to scientific evidence, burden of proof and cost–benefit analysis is discussed below.

Finally, it should be mentioned that in the recent case *Massachusetts* v. *EPA*, a split Supreme Court found that Massachusetts should be awarded standing to challenge the decision of the EPA not to regulate greenhouse gas emissions from mobile sources.[88] Even though the Court treated the link between greenhouse gas emissions and global warming as, more or less, scientifically certain, the case provides an interesting move towards opening a theoretical possibility for claimants to gain access to the court system for challenges of agency decisions based on the precautionary principle. Some commentators have held that the Supreme Court missed an opportunity to make a strong endorsement of 'precautionary standing' in US courts in the case.[89] However, others have stated that the case was an example of undue judicial activism on the part of the Supreme Court.[90]

7.1.5 Risk assessment and risk management

As has been elaborated on in previous sections, US risk regulation during the 1970s was often precautionary, although strict requirements on risk assessment before issuing regulation were not always made. However, in 1980 a divided US Supreme Court in the *Benzene* decision stated that the Occupational Safety and Health Administration (OSHA) needed to base its decision to regulate benzene emissions on scientific evidence. The

[86] T. J. Miles and C. R. Sunstein, 'The Real World of Arbitrariness Review', John M. Olin Law & Economics Working Paper No. 368, November 2007.

[87] Raul and Dwyer, 'Science in the Regulatory Process'.

[88] *Massachusetts* v. *EPA*, 127 U.S. 1438 (2007).

[89] J. R. Nash, 'Standing and the Precautionary Principle', *Colum. L. Rev.*, 108 (2008), 494.

[90] Editorial, 'Jolly Green Justices', *Wall Street Journal*, 3 April 2007, at A14.

judgment marked a clear shift away from the environmentalist 'partnership'[91] between government agencies and courts that had emerged during the 1970s, which had allowed for precautionary interpretations of environmental statutes. In only a few years after *Benzene*, the use of risk assessment in the US regulatory regime became 'almost routine' when taking decisions on chemicals and physical agents.[92]

In 1983 the US National Research Council issued an influential report which set out to define some of the technical aspects of risk assessment.[93] It also included a section which recommended that the assessment of risks be kept separate from the decision of how to manage the risk.[94] During the 1990s the Republican majority in the US Congress sought to pass legislative measures which would lay down detailed requirements on risk assessment, including, in some cases, methods and procedural requirements. However, this effort was not successful and the use of risk assessment in environmental and health regulation has not been effectively included in binding legislation.[95] It is thus largely within the discretion of the respective agencies entrusted with risk management tasks to decide on how to carry out risk assessments in the relevant cases, unless more detailed formal requirements on the use of scientific committees, and so on, have been laid down by the US Congress.[96] However, the performance of risk assessments can be described as a 'quasi-procedural requirement',[97] and the courts have, at least in some cases, proven willing to check in which manner the risk assessment has been performed.

The issue of the separation of risk assessment and risk management has received surprisingly little attention in the USA.[98] In a seminal report issued by The Presidential/Congressional Commission on Risk Assessment and Risk Management in 1997, the issue was not dealt with in any greater detail, but it can be concluded both from the graphs

[91] Jasanoff, *Science at the Bar*, pp. 88–9.

[92] B. D. Goldstein, 'The Interface Between Science and Law', *Colum. J. Envtl. L.*, 14 (1989), pp. 343, 355.

[93] National Research Council, *Risk Assessment in the Federal Government*.

[94] *Ibid.*, p. 152.

[95] L.-J. Schierow, 'Environmental Risk and Cost-Benefit Analysis: A Review of Proposed Legislative Mandates, 1993–1998', CRS Report for Congress, 22 January 1999.

[96] This issue is returned to in detail in the respective case studies below.

[97] J. S. Applegate, 'A Beginning and Not an End in Itself: The Role of Risk Assessment in Environmental Decision-Making', *U. Cin. L. Rev.*, 63 (1994/1995), 1643.

[98] R. M. Sedman and P. W. Hadley, 'Risk Assessment and Risk Management: Mending the Schism', *Risk – Issues in Health & Safety*, 3 1992, 189.

and the general discussion that a separation of risk assessment and risk management is, in fact, fundamental.[99] It also concluded that:

> Failure to achieve a clear separation of scientific evidence from policy considerations has negative implications beyond impaired credibility: it inevitably obscures the overall level of uncertainty in the assessment and will influence consideration of societal trade-offs and ultimate decision making.[100]

However, the EPA has stated that, in practice, a formal, institutional separation of the risk assessment and risk management is not upheld. On the contrary, the EPA considers it important that the two processes take place side by side, in order for the risk manager to be adequately informed on how the process of risk assessment has been conducted. It is, however, emphasised that the respective roles of risk assessor and managers are different. Risk assessors analyse the scientific material, and risk managers take the decisions based on this scientific evaluation.[101]

Recently, it would seem that the importance of risk assessments is increasing even further. With the passing of the Data Quality Act in 2000, all government decisions can be subject to 'correction' if they do not meet strict requirements established in relevant guidelines.[102] In 2002 the Office of Management and Budget issued detailed guidance on the scientific data to be used for regulatory purposes by federal authorities.[103]

The increased demand for decisions to be based on scientific evidence from the early 1980s onwards means that risk assessments are at the core of US risk regulation. Courts will mandate the performance of a risk assessment and may penalise a government body that has failed to perform it. Furthermore, it appears that risk assessment is, in principle, considered a scientific process which is to be kept separate from risk management, which is the process of taking decisions upon the basis of

[99] The Presidential/Congressional Commission on Risk Assessment and Risk Management, Risk Assessment and Risk Management in Regulatory Decision-making, Final Report Vol. 2, 1997, 2 and 3.

[100] *Ibid.*, p. 192.

[101] US Environmental Protection Agency, 'An Evaluation of EPA Risk Assessment Principles and Practices', Staff Paper Prepared for the US Environmental Protection Agency by members of the Risk Assessment Task Force, EPA/100/B-04/001, pp. 22–3.

[102] Data Quality Act, Amendment to the Paperwork Reduction Act, 4 U.S.C. 3501.

[103] Office of Management and Budget, Guidelines for Ensuring and Maximizing the Quality, Objectivity, Utility, and Integrity of Information Disseminated by Federal Agencies; Notice; Republication, Federal Register / Vol. 67, No. 36 / 22 February 2002. This is discussed in more detail below in the section on scientific evidence.

the risk assessment. It has been emphasised that these processes, even if separated, should take place side by side in order for the best decisions to be taken. How this is done in practice, and how it influences the decisions taken, is dealt with in the following section.

7.1.6 Scientific evidence and the introduction of precautionary measures

With the ruling in the *Benzene* case, and the subsequent development of a strict reliance on risk assessment during the 1980s, the importance of scientific evidence when regulating risks in the USA has been strongly emphasised. As has been shown above, the demands on scientific evidence required for regulation during the 1970s had been low. Agencies had often used scientific uncertainty as sufficient proof of risk to warrant regulation,[104] and the courts, content with the fact that the agencies had had a 'hard look' at the scientific evidence brought before them, would generally allow for this. However, in the *Benzene* ruling the US Supreme Court mandated, with the thinnest margin possible, that agencies base their actions on some form of scientific evidence of risk. The majority held that before 'any permanent health or safety standard' can be required, 'a threshold finding [...] that significant risks are present and can be eliminated or lessened by a change in practices' must be established.[105] However, it was not possible for the judges in *Benzene* to agree on a more defined and generally applicable allocation of the scientific evidence which would be required to establish the 'significant risk'. Three of the judges in the case thought it prudent to call for regulatory action in situations in which it would be 'more likely than not' that the relevant activity would lead to the realisation of the suspected risk.[106] However, this assertion was not supported by the majority.

The majority in the *Benzene* case established that a 'significant risk' triggers regulatory action, but proved hesitant to lay down any more precise indications of how much, or what kind of, scientific evidence this would require. At the same time, it did make it clear that the scientific evidence would not need to represent 'anything approaching scientific certainty'.[107] Thus, it appears that:

[104] See the discussion on *Ethyl Corp.* and *Reserve Mining*, above.

[105] *Industrial Union Department, AFL-CIO* v. *American Petroleum Institute et al.*, 448 U.S. 607 (1980), 408.

[106] *Ibid.*, 653. [107] *Ibid.*, 656.

> [the agency] may use conservative assumptions in interpreting data, thus erring on the side of overprotection, so long as its decision is supported by a body of reputable scientific thought.[108]

However, the *Benzene* case did not put an end to precautionary considerations by US authorities. On the contrary, it explicitly stated that 'policy considerations' play an important role when deciding on regulatory action under conditions of uncertainty.[109] However, it did make it clear that scientific uncertainty alone cannot justify regulatory action. The absence of scientifically certain evidence in support of the harmlessness of a product or an activity is not sufficient to warrant regulation. Some, but not necessarily conclusive, evidence of a risk is required for a measure to stand.

'Significant risk', as it was dealt with in the *Benzene* ruling, has not been defined in any detail. The general deference of the courts, when it comes to dealing with scientific uncertainty, allows the regulator considerable freedom in deciding on what measures to take based on the scientific evidence in each individual case.[110] However, since the passing of the controversial Data Quality Act in 2000, the requirements on scientific evidence have been elevated further. The purpose of the Act is to ensure that government agencies rely on 'good science' when issuing regulations. The Act mandates the Office of Management and Budget (OMB) to issue guidelines on how data quality can be ensured,[111] and more detailed and agency-specific requirements have been issued by individual agencies.[112] In short, four principles are identified as fundamental for the information relied upon by regulators: quality, utility, objectivity and integrity.[113] If data held by an agency does not meet the standards laid down in the OMB or agency-specific guidelines, a 'correction' of this data can be sought by application to the relevant agency holding the data.[114]

[108] J. C. McElveen and C. Amatea, 'Risk Symposium: Legislating Risk Assessment', *U. Cin. L. Rev.*, 63 (1995), 1553.

[109] *Industrial Union Department, AFL-CIO* v. *American Petroleum Institute et al.*, 448 U.S. 607 (1980), 656.

[110] See, for example, *Chevron* and *Texas Oil & Gas Ass'n* v. *EPA*, discussed above.

[111] Office of Management and Budget, Guidelines for Ensuring and Maximizing the Quality, Objectivity, Utility, and Integrity of Information Disseminated by Federal Agencies; Notice; Republication, 8452.

[112] In relation to the EPA see United States Environmental Protection Agency, *Guidelines for Ensuring and Maximizing the Quality, Objectivity, Utility, and Integrity of Information Disseminated by the Environmental Protection Agency*, EPA/260R-02-008. Available at: www.epa.gov/quality/informationguidelines/documents/EPA_InfoQualityGuidelines.pdf (accessed 1 March 2010).

[113] *Ibid.* [114] *Ibid.*, p. 12.

The aim of the Act is to force agencies to rely on accurate and established information, when issuing regulation. However, many commentators have held that the Act leads to a situation in which all science-based regulations can be effectively challenged, since few scientific findings would meet a strict application of the requirements. Some level of uncertainty always remains, and this can be exploited by parties who wish to challenge legislation by 'manufacturing uncertainty'.[115] Even though frequent use appears to be made of the possibility of having agencies 'correct' the data held by them, primarily by industry, courts have so far proven reluctant to afford any type of judicial review of the Act or the guidelines. The fact remains that a strict application of the Data Quality Act, and the guidelines based on it, could potentially lead to a situation in which the regulator can base decisions solely on established scientific facts. If this is the case, the possibility of regulatory action under uncertainty would effectively disappear, and only scientifically proven risks could be regulated.

7.1.7 The burden of proof

As has been seen in the previous chapters, precautionary decision-making rarely, if ever, entails an initial reversal of the burden of proof.[116] Instead, the initial burden of proof generally rests with the regulator.[117] However, the initial level of proof may, in some cases, be very low, and thereby create the illusion that the burden of proof is originally placed on the regulated party. In administrative cases in the USA, the burden of proof explicitly rests with the proponent of a rule or an order. Thus, it is for an agency wishing to regulate a product to demonstrate that the risk associated with it warrants regulation. In the previous section, it was shown that the level of proof needed to be met for regulation seems to have changed over time. For courts in the 1970s, mere scientific uncertainty was sufficient for regulation. However, after the *Benzene* case some evidence of risk is required, even if it does not have to meet the 50 per cent likelihood requirement that is in effect in civil litigation.[118] Once this level of proof

115 See, for example, R. Weiss, 'Law is Nemesis of Regulation', *Washington Post*, 16 August 2004, p. A01 and Mooney, *The Republican War on Science*.

116 See above in relation to Sweden, Section 5.1.4, and in relation to the UK, Section 6.1.4.

117 An exception to this may be preauthorisation systems, for example for pesticides, where a risk level has been established and must be met by a party wishing to have its product authorised.

118 Jasanoff, *Science at the Bar*, p. 10.

has been met, the burden of proving the agency's conclusions wrong shifts onto the regulated party.

Successful challenges against agency decisions under varying degrees of uncertainty appear mainly to relate to situations in which the relevant agency has failed to explain adequately how and why it has made use of scientific evidence.[119] In *State Farm* the US Supreme Court found that the National Highway Traffic Safety Administration had failed to explain why it had made a requirement for new vehicles to be equipped with passive restraints, and therefore the decision was vacated.[120] Similarly, in *National Mining Association* v. *Babbitt* the DC Circuit Court found that the Office of Surface Mining Reclamation and Enforcement had failed to justify the presumptions underlying its decision and, consequently, it came to the conclusion that the decision of the agency was not based upon scientific evidence.[121] Furthermore, in *Midwater Trawlers Cooperative* v. *Department of Commerce* the 9th Circuit Court struck down a measure that it found to be 'a product of pure political compromise, not reasoned scientific endeavour'.[122] In these cases the courts have generally considered the measures to be arbitrary and capricious in accordance with the APA.[123]

None of the above cases concerned a situation in which an agency had acted flawlessly and a challenging party had successfully challenged the decision on its merits. Instead, they were cases in which the challenging party had successfully shown arbitrariness and capriciousness on the side of the regulator, a bar that is, perhaps, comparable to the 'manifest error' doctrine employed by the European Court of Justice (ECJ). As in cases before the ECJ, US courts, to a great extent, defer to the judgments of the agencies. However, the 'hard look' doctrine still shines through before the US courts, and they often appear to assess the validity of the scientific evidence put before them, if only to assess whether the agencies have taken it into account in an appropriate fashion when issuing decisions.[124]

[119] Raul and Dwyer, 'Science in the Regulatory Process', 22–4.

[120] *Motor Vehicle Manufacturers Association* v. *State Farm Mutual Auto Insurance, Co.*, 463 U.S. 29 (1983).

[121] *National Mining Association* v. *Babbitt*, 172 F.3d 906 (D.C. Cir. 1999).

[122] *Midwater Trawlers Cooperative* v. *Department of Commerce*, 282 F.3d 710 (9th Cir. 2002), p. 720.

[123] See, above, Section 7.1.4.

[124] However, as was shown in the section on the precautionary principle in the EU above, the ECJ does also assess scientific evidence in quite some detail in order to establish that scientific uncertainty is at hand. Compare, for example, with the ECJ in the *Pfizer* case; see above, Sections 4.3.3–4.3.5.

With the implementation of the Data Quality Act, the requirements on scientific data have been raised further in the USA. This may lead to an increased scope for challenges made against regulatory decisions on the basis that the scientific data used by the regulator is not adequate. However, so far US courts have proven reluctant to test decisions against the Act.

7.1.8 *Cost–benefit analysis*

The history of institutional cost–benefit analysis in the US regulatory system dates back to the time of the Quality of Life Reviews introduced by President Nixon (1969–74) in the early 1970s.[125] Since then, the use of cost–benefit analysis has continued to be mandated in Executive Orders issued by US presidents, Republicans as well as Democrats.[126] It has been held by some influential commentators that 'cost–benefit balancing is now the official creed of the executive branch'.[127]

Recent Executive Orders instruct agencies to perform regulatory impact assessments (RIA) before issuing regulations.[128] Such analyses include the carrying-out of cost–benefit analysis, but also include a statement relating to the need for the proposed action by the agency, as well as an examination of alternative approaches.[129] Through the use of RIAs, the executive branch obliges agencies to consider the wider implications of issuing regulation. However, it is important to note that this also creates a mechanism whereby the independence of agencies is made contingent on control by the executive branch. The actual executive oversight is carried out by the Office of Management and Budget (OMB), which is itself

[125] C. Sunstein, 'The Cost-Benefit State', *Chicago Working Paper in Law & Economics*, No. 039, 05/1996, 24. Available at: www.law.uchicago.edu/files/files/39.sunstein.pdf (accessed 1 March 2010).

[126] The famous Executive Order No. 12,291, 3 C.F.R. 127 (1982) issued by President Reagan in 1982, which became a strong incentive for making agency action contingent upon the performance of cost–benefit balancing, was heavily criticised by Democrats, but its general content has been upheld by consecutive presidents, including the Democratic President Clinton (Executive Order 12,866, 3 C.F.R. 638 (1994)).

[127] R. W. Hahn and C. R. Sunstein, 'A New Executive Order for Improving Federal Regulation? Deeper and Wider Cost-Benefit Analysis', *Chicago Working Paper in Law & Economics*, No. 150, 04/2002, p. 1. Available at: www.law.uchicago.edu/files/files/ISO.CRS_Cost-Benefit.pdf (accessed 1 March 2010).

[128] See, for example, Executive Order 12, 866.

[129] Office of Management and Budget, Regulatory Analysis, Circular A-4 To the Heads of Executive Agencies and Establishments, 17 September 2003, p. 2.

an entity within the Executive Office of the President (EOP).[130] Within the OMB, the Office of Information and Regulatory Affairs (OIRA) is responsible for the oversight of agency regulations.[131]

The rules prescribed in Executive Orders are not formally subject to judicial review, and, thus, the failure of agencies to perform a cost–benefit analysis adequately should, for the most part, be beyond the reach of the courts, unless statutes passed by Congress prescribe their performance.[132] It has been pointed out that some statutes do include independent requirements of some kind of balancing.[133] On the other hand, other statutes actually prohibit the regulator from comparing costs and benefits.[134]

Even though, in most cases, the courts do not uphold requirements to perform cost–benefit analyses, the influence of the OIRA in upholding the meaning of the Executive Orders should not be underestimated. In 2001 and 2002 the United States General Accounting Office investigated how a number of draft regulations had faired in review by the OIRA, and it concluded that, out of a total number of eighty-five measures investigated, twenty-five had been significantly modified. Often, the modifications related to how costs and benefits had been compared by the relevant agency.[135] It should be noted that, in consecutive Executive Orders, it has been held that the OIRA's review is to focus on 'particularly important' agency rule-making.[136] Furthermore, it should be pointed out that the OIRA's influence is not limited to the formal review. Informal contacts

[130] See, for example, E. Kagan, 'Presidential Administration', *Harv. L. Rev.*, 114 (2001), 2245. For a recent discussion on the influence of OMB in a number of agency decisions, see J. D. Graham, 'The Evolving Regulatory Role of the U.S. Office of Management and Budget', *Review of Environmental Economics and Policy*, 1:2 (2007), 171–91.

[131] For a detailed summary of how the OMB and the OIRA survey agency rule-making see United Stated General Accounting Office, *OMB's Role in Reviews of Agency's Draft Rules and the Transparency of Those Reviews*, GAO-03–929, 2003. Available at: www.gao.gov/new.items/d03929.pdf (accessed 1 March 2010). For a critical analysis of the impact of OMB review on Agencies' performance of cost–benefit analysis see Johnston, 'A Game Theoretic Analysis of Alternative Institutions for Regulatory Cost-Benefit Analysis', 1343–411.

[132] Sunstein, for one, has held that the courts have played a 'limited role' in making agencies implement cost–benefit analysis; see Sunstein, 'The Cost-Benefit State', p. 23.

[133] For example the Toxic Substances Control Act, 7 U.S.C., 136–136y and the Fungicide, Insecticide, and Pesticide Act, 15 U.S.C., 2601–92. See Sunstein, 'The Cost-Benefit State', p. 3. These provisions are dealt with in more detail below in the section on pesticides.

[134] For example the Clear Air Act, the Clean Water Act and the Delaney Clause.

[135] United States General Accounting Office, *OMB's Role in Reviews of Agency's Draft Rules and the Transparency of Those Reviews*, p. 5.

[136] S. Croley, 'White House Review of Agency Rulemaking: An Empirical Investigation', *U. Chi. L. Rev.*, 70 (2003), 821.

between agencies and the OIRA during the drafting process appears to be a common occurrence.[137]

In its Circular A-4 of 2003 the OMB went to considerable lengths to define how regulatory impact analysis in general, and cost–benefit (or benefit–cost analysis as it is called in the document) in particular, is to be carried out by agencies.[138] Unlike the cases of Sweden and the UK, where the performance of cost–benefit analysis is at the discretion of the regulator, the OMB gives a detailed account for how balancing is expected to be carried out by agencies. It also discusses the usefulness of cost-effectiveness analysis (CEA), which is defined as being 'designed to compare a set of regulatory actions with the same primary outcome [...] or multiple outcomes that can be integrated into a single numerical index'.[139]

The OMB makes it clear that costs as well as benefits are to be measured in 'opportunity costs', using calculations based upon willingness to pay (WTP) or willingness to accept (WTA) assumptions.[140] All values should, as far as possible, be quantified in monetary terms.[141] However, the Circular recognises that it may not be possible to express all the relevant factors in quantitative terms, and that, in such cases, qualitative considerations may be included, as long as they are clearly identified and motivated.[142] Thus, the OMB does not necessarily translate all benefits of regulation to numerical values, but may also accept rules that are 'proposed on philosophic grounds that have nothing to do with economic efficiency'.[143] Furthermore, it is explicitly stated that issues surrounded by varying degrees of uncertainty should also be quantified and assigned economic values. It is stressed that transparency and openness with regard to the models underlying the assumptions is essential. Worst-case scenarios and conservative estimates should be avoided when analysing the material in question, and the regulator should consider whether regulation could be postponed until a clearer picture has emerged.[144] This would imply that precautionary considerations do not play a prominent

[137] For detailed accounts of the influence of the executive branch on agency rule making see United States General Accounting Office, *OMB's Role in Reviews of Agency's Draft Rules and the Transparency of Those Reviews*, p. 5; Kagan, 'Presidential Administration'; and Croley, 'White House Review of Agency Rulemaking'.

[138] Office of Management and Budget, Regulatory Analysis, Circular A-4 To the Heads of Executive Agencies and Establishments.

[139] *Ibid.*, p. 11. [140] *Ibid.*, p. 18.

[141] *Ibid.*, p. 10. [142] *Ibid.*, p. 27.

[143] Graham, 'The Evolving Regulatory Role of the U.S. Office of Management and Budget', 172–3.

[144] *Ibid.*, 28–9.

role when carrying out the cost–benefit analysis. This is understandable when one considers that the result of the regulatory impact assessment does not necessarily mandate the degree of regulation, but, instead, aims to put all costs and benefits 'on screen'.[145] However, one could be forgiven for thinking that conservative estimates could also help balance the assessment, at least as a complement to realistic estimates in cases of uncertainty.

The US approach to cost–benefit analysis is the most developed of the systems investigated here. In the last five years the EU Commission has begun to work more concretely by estimating the costs of all proposed regulation in the EU. However, so far, the EU approach appears to be based mainly on the participation of stakeholders. There is still a wide gap between this and the firm use of economic modelling and the strong and independent role of the OMB in the USA.

7.1.9 The targeting of risks

Like the performance of cost–benefit analyses discussed above, no generally applicable statute dictates that countervailing risks or risk offsets have to be considered by the regulator when issuing measures.[146] In fact, and also as in the case of cost–benefit analyses, some statutes actively appear to prohibit the regulator from considering risk trade-offs, while others require the regulator to perform them.[147]

In its Circular A-4 on Regulatory Analysis, the OMB clearly stated that agencies 'should look beyond the direct benefits and direct costs of [...] rulemaking and consider any important ancillary benefits and countervailing risks'.[148] When doing this, agencies are instructed to monetise the values that expected trade-offs may bring and compare them to the proposed regulation. It is also suggested that agencies should include risk

[145] As has been argued by Sunstein, for example, in Sunstein, *Risk and Reason*, pp. 8 and 29. See also Section 2.3.2, above.

[146] However, attempts were made by the 105th Congress to make the consideration of 'substitution risks' mandatory for regulators through the unsuccessful introduction of the Regulatory Improvement Act 1998, S. 981, 105th Congress, 627 (1998). See, in this regard, Anderson *et al.*, 'Regulatory Improvement Legislation'. For a general discussion on risk trade-offs, targeting and ranking of risks see above, Section 2.3.3.

[147] See, for example, C. Sunstein, 'Cost/benefit Default Principles', *Chicago Working Paper in Law & Economics*, No. 104, 3/6/01, p. 15. Available at: www.law.uchicago.edu/files/files/104.CRS_.Default.REVISED.pdf (accessed 1 March 2010).

[148] Office of Management and Budget, Regulatory Analysis, Circular A-4 To the Heads of Executive Agencies and Establishments, p. 26.

trade-off data in the cost–benefit analysis if they wish. This must be made in a transparent fashion.[149]

The OMB Circular A-4, which aims to clarify the Executive Order relating to the analyses of risks, is not legally binding, as was discussed above. However, the courts have, in some cases, proven willing to allow agencies to base the issuing of regulatory measures upon the analysis of risk trade-offs.[150] For example, in *Corrosion Proof Fittings* the 5th Circuit Court held, in relation to asbestos, that the responsible agency could not solely concentrate on one substance, but would have to consider the risks of other materials that would substitute the risks associated with the asbestos.[151]

Sunstein has summarised the current status of trade-off analysis along the following lines:

> Agencies ought generally to be required to show that they are doing more good than harm. This does not mean that courts should engage in independent review of agency judgments on this score. But it does mean that courts should take a 'hard look' at agency decisions failing to undertake health–health comparisons.'[152]

However, even if such analysis is, in practice, performed by agencies following the OMB guidance, there is no strong, legally enforceable mandate for them to do so. How trade-offs are considered in practice is returned to below in the sections on pesticides and on base stations.

During the 1990s the issue of clearly prioritising risk regulation and ranking the order of dealing with risks became a major issue.[153] Applegate has held that a lack of growth in agency budgets spurred on the development towards a greater emphasis on priority-setting with regard to environmental and health risks.[154] In a situation in which resources were limited, the ranking of priorities became a necessity:

> The agency's first task, upon which the overall efficacy and efficiency of the rest of the process depends, is the allocation of resources among potential regulatory efforts. Agencies must set priorities among the risks within their purview.[155]

[149] *Ibid.* [150] Sunstein, 'Cost/Benefit Default Principles', p. 38.

[151] *Corrosion Proof Fittings*. v. *EPA*, 247 F.2d 12–1 (5th Cir. 1991).

[152] C. Sunstein, 'Health-Health Tradeoffs', *U. Chi. L. Rev.*, 63 (1996), 28.

[153] See, for example Breyer, *Breaking the Vicious Circle*; A. M. Finkel and D. Golding, *Worst Things First? The Debate over Risk-Based National Environmental Priorities* (Resources for the Future, 1994); and Sunstein, *Risk and Reason*.

[154] Applegate, 'Worst Things First?', 277. [155] *Ibid.*, 288.

The performance of risk assessment, cost–benefit analysis and risk trade-off analysis can be used to yield the data needed to establish a rational agenda for the reduction of risks. Such input can be used to measure the impact of precautionary measures and to make sure that the measures issued do, in fact, lower the overall risk level.

The method used for priority-setting in relation to risks has been called comparative risk analysis (CRA).[156] Originally influenced by the EPA Report entitled 'Unfinished Business',[157] CRA aims to 'compare the relative magnitudes of threats posed by environmental factors under EPA jurisdiction with the agency staff, budget, and attention allocated to each factor'.[158] 'Unfinished Business' identified and analysed a large number of risks according to four categories: cancer risks, non-cancer health risks, ecological effects and welfare effects. The risks were then ranked according to how they scored in the different categories. No risks scored high in all categories, but risks that scored at least medium in at least three categories were recommended to be tackled first. The Report laid the groundwork for the systematic approach assumed by the EPA in the years to come.[159] Its conclusions were broadly supported in a Review published by the Science Advisory Board in 1990, which urged the EPA to continue its prioritisation of risks.[160]

During the 1990s the issue of ranking and prioritising risks was a topic of great importance for US agencies (most notably the EPA), and was, together with CBA and RIA, intensely discussed in the literature.[161] Principally, this discussion centred on how the EPA should carry out assessments and rankings. However, in the absence of binding rules or guidelines, it appears that no standardised procedure or strategy for the targeting of risks exists.

156 For an overview of the development of this concept see for example C. W. Kent and F. W. Allen, 'An Overview of Risk-Based Priority Setting at EPA', in Finkel and Golding, *Worst Things First?*.

157 The Environmental Protection Agency, *Unfinished Business: A Comparative Assessment of Environmental Problems*, February 1987.

158 J. K. Hammitt, 'Improving Comparative Risk Analysis', *Duke Envtl. L. & Pol'y F*, 8 (1997), 81.

159 Kent and Allen, 'An Overview of Risk-Based Priority Setting at EPA', 51.

160 The Environmental Protection Agency, *Reducing Risk: Setting Priorities and Strategies for Environmental Protection*, September 1990.

161 See, for example, Finkel and Golding, *Worst Things First*.

7.1.10 Public opinion

That the general public is to be given an opportunity to voice its opinions on proposed administrative decision-making is established in the APA, and called 'notice-and-comment rule-making'.[162] In general, there are two stages in the decision-making process where the public and stakeholders may influence its outcome. First, organised stakeholders with whom a public administration has an established relationship may be consulted during the drafting of a measure. Second, once a full draft measure has been established, it is formally announced and public comments are requested. These comments should be considered by the relevant public body before the measure is formally adopted and included in the Federal Register.[163]

In the above-discussed *Vermont Yankee* case a number of public interest interveners held that the system of notice-and-comment was too weak to allow for efficient participation, and that it thereby 'denied them a meaningful opportunity to participate in the proceedings'.[164] The DC Circuit Court agreed with the interveners without specifying in any detail what more could be required by the regulator.[165] However, on appeal, the US Supreme Court held that the approach advocated by the DC Circuit Court would amount to a procedural requirement not mandated by Congress, and thus considered the limited public consultation provided in the notice-and-comment procedure to be sufficient.[166]

However, as has been pointed out by Applegate, the Court did not preclude that agencies could impose voluntary arrangements pertaining to public involvement. What the Supreme Court stated in *Vermont Yankee* was merely that the review of the courts could not stretch as far as to invent such requirements.[167] The EPA, for example, issued a set of general guidelines for public participation in relation to waste disposal in 1981.[168]

[162] APA 5 U.S.C. 553 (1994).

[163] J. Applegate, 'Beyond the Usual Suspects: The Use of Citizens Advisory Boards in Environmental Decision-making', *Ind. L.J*, 73 (1998), 901.

[164] *Vermont Yankee Nuclear Power Corp.* v. *Natural Resources Defense Council*, p. 643.

[165] *Natural Resources Defense Council, Inc.* v. *United States Nuclear Regulatory Comm'n*, 547 F.2d 633 (D.C. Cir. 1976), p. 654.

[166] *Vermont Yankee Nuclear Power Corp.* v. *Natural Resources Defense Council*, pp. 543–8.

[167] Applegate, 'Beyond the Usual Suspects', 909, note 31.

[168] Environmental Protection Agency, *RCRA Public Participation Policy*, 1981.

In 2003 the EPA issued general guidelines for public participation within the agency's sphere of regulation,[169] which specified that sectoral arrangements, such as those included in the 1981 guidelines on waste regulation, provided minimum requirements for the agency when regulating topics covered by them.[170] The 2003 Guidelines emphasise that, in order for the involvement of both the public and the stakeholders to be meaningful, information must be shared between the EPA and the affected parties. The EPA should establish a budget for each public involvement enterprise, and should, as far as possible, try to accommodate involvement through the setting-up of meetings and panels, and even provide funding and resources for the public and stakeholders.[171] The outcome of public involvement can range from providing information to conducting binding agreements with stakeholders.[172] Even if the guidelines are not formally binding before the courts, they go some way towards showing the intention of the EPA to take the views of interested parties seriously. However, to date concrete examples of deep and far-reaching public involvement can mainly be found at local and regional levels.[173] Thus, no formal role is foreseen for the public to influence decision-making prompted by precautionary considerations. However, it is shown below that the opinion of the public may, possibly, be an important factor when issuing precautionary measures in specific cases.[174]

Finally, it should be pointed out that it remains to be seen whether the passing of the Data Quality Act in 2000 makes it more difficult for the concerns of stakeholders and the general public to be relied upon if such concerns are not based on scientifically certain data.

7.1.11 Conclusion

The precautionary principle, as such, lacks explicit endorsement in US legislation and policy. Despite this fact, activist US courts in the 1970s went to some lengths to carve out a space for precautionary considerations in US regulation. This never amounted to an effective endorsement of a generally applicable precautionary principle, but judgments such as *Ethyl*

[169] Environmental Protection Agency, *Public Involvement Policy of the Environmental Protection Agency*, May 2003. Available at: www.epa.gov/publicinvolvement/pdf/policy2003.pdf (accessed 1 March 2010).

[170] *Ibid.*, p. 1. [171] *Ibid.*, Appendix 1.

[172] *Ibid.*

[173] See, for example, Applegate, 'Beyond the Usual Suspects'.

[174] See, below, Section 7.3.

Corp. and *Reserve Mining* allowed US agencies to act in a precautionary manner without an explicit basis in scientific evidence. Even though this far-reaching development was effectively curtailed by the *Benzene* case and the subsequent emphasis on risk assessment by the executive branch and Congress during the Reagan administration, the possibility of basing decisions on precautionary considerations did not disappear.

The 'hard look' doctrine which was developed by the DC Circuit Court in the 1970s continues to influence the scope of the judicial review, even as the procedure-based approach applied by courts in the 1970s was denounced. For precautionary measures to be enacted, a threshold finding of 'significant risk' must be present. Precisely what is required for this to be the case is unclear, but it appears that mere scientific uncertainty is not sufficient. This means that, when a precautionary measure is challenged, US courts will analyse the scientific evidence brought before them, in order to establish whether or not a risk exists. Here, US courts differ in their approach when compared to the other jurisdictions considered in this book, in which the courts generally allow for a near total deference on the side of the regulator, and the mere existence of scientific uncertainty is generally sufficient to warrant regulation. However, this does not mean that US courts require scientific certainty for regulation to be legal. Instead, it appears that the approach is more akin to that of the interpretation of the SPS Agreement advanced in the WTO Dispute Settlement Body.[175] Like the WTO Panels and the Appellate Body, US courts require a threshold finding of a plausible risk before regulation is permitted. This implicitly mandates the performance of risk assessments to establish whether such a threshold has been met before any regulation can be issued.

In some cases where measures have been taken under varying degrees of uncertainty, claimants have managed to challenge them successfully before the courts. Often, such cases are based on the performance of flawed risk assessments or formal errors on the part of the regulator. However, it appears that it is technically possible to challenge a decision on the merits before US courts. At the very least, it can be said that the scope for such review must be considered greater than in the other legal systems here investigated.

[175] Outside of the scope of Article 5.7 of the SPS Agreement, where precautionary measures are explicitly authorised when the scientific material is so inconclusive that it renders the performance of a risk assessment impossible.

US regulatory measures are preceded by Regulatory Impact Assessments. RIAs include not only the performance of risk assessment and cost–benefit analysis, but also a statement relating to the need for the proposed action by the agency, as well as an examination of alternative approaches. Risk trade-offs should be considered, and there is a strong impetus for ranking risks in order of priority. The executive branch, through its Office of Management and Budget, oversees the rationality and consistency of regulatory proposals, both formally and informally. When the OMB does not agree with the assessment performed by an agency, the measure is sent back for revision. The importance and influence of the OMB is on the rise. The notorious Data Quality Act gives the affected parties an opportunity to file for a 'correction' of the scientific data held by agencies with the OMB. However, to date, the courts have proven reluctant to test the performance of cost–benefit analyses and RIAs in any detail, but have, instead, left its performance up to the executive branch. The same goes for the challenges against regulatory decisions based on the Data Quality Act.

The precautionary principle does not exist as an independent concept in the US legal system. However, this does not mean that precautionary considerations do not play a significant role in US regulation. Compared to Sweden and the UK, risk regulation in the USA is more firmly based on rational concepts such as risk assessment and cost–benefit analysis. Decisions that are not adequately supported by scientific evidence may be overturned by US courts. Thus, the hands of the US regulator are more tied than in Sweden, the UK or the EU, unless Congress has specifically mandated that risks do not need to be established scientifically. At present it would be difficult for the US regulator to base decisions solely on public opposition to a certain activity. The discretion of the regulator simply does not stretch that far. Instead, it has been bound by scientific evidence and rational reasoning. In Sweden the UK and the EU, on the other hand, the regulator is seen as acting on a broader mandate of parliament (or parliament and the Member States in the case of the EU). Consequently, courts are reluctant to overturn decisions taken using this broad discretion.

Contrary to the discussion in Europe, which tends to centre on the increased involvement of stakeholders and the public in precautionary decision-making, many US commentators argue in favour of an increased use of science and procedural requirements. It remains to be seen if the highly controversial Data Quality Act and cases like *Daubert* will move the US system further towards the strict use of scientifically verified data as a basis for regulation. This would inevitably further limit the scope that precautionary considerations play in risk regulation.

7.2 US pesticides regulation

7.2.1 Introduction

Pesticides have been regulated at federal level in the USA since the early 1900s, and within its current legislative structure since the late 1940s. Naturally, the content of the legislation has changed somewhat over the years as the means to assess risks have become more sophisticated, and societal concerns have evolved. However, contrary to public belief, US pesticide regulation has a history that is far more precautionary in nature than most European countries. In the main, this is due to the strict preauthorisation regime set up by statute in combination with the controversial Delaney Clause of 1958. Effectively barring all pesticides with even microscopic carcinogenic properties from authorisation, the Delaney Clause meant that the USA did, in fact, have a zero-risk approach in place for most pesticides for around forty years, until the Delaney Clause was revoked in the mid-1990s. Since the Clause was repealed, the EPA has enjoyed greater discretion to carry out a more nuanced assessment of pesticides, weighing risk assessment results against the potential benefits of authorising a pesticide. To assist the EPA in its tasks, there is a developed structure of scientific bodies which carry out risk assessments and give general, as well as specific, advice.

The legal and institutional structure of pesticide regulation in the USA is outlined and discussed below. As in previous sections, the focus is on whether, and if so, how, precautionary considerations function as a basis for regulation. Important issues in this regard are not only the treatment of scientific evidence and the level of proof for regulation in cases of authorisation, but also the instances in which pesticides are withdrawn from the market. The issue of the costs and benefits of regulation is an important aspect when considering pesticide regulation in the USA, and this is discussed in some detail.

7.2.2 The legal framework

In the USA chemical pesticides have been subject to some degree of regulation ever since the Insecticide Act 1910.[176] Today, the Federal Insecticide,

[176] www.epa.gov/compliance/civil/fifra/index.html (accessed 1 March 2010).

Fungicide and Rodenticide Act (FIFRA) of 1947 is the principal piece of legislation in place in relation to pesticide regulation. Of prime importance here, FIFRA establishes how pesticides are authorised for use on the US market, and how, and under which circumstances, authorisations can be suspended and revoked.[177] Other statutes help complete the picture by dealing specifically with issues of pesticide residues in human food and animal feed.[178]

Pesticides cannot be used or marketed in the USA unless they have been registered following the provisions in FIFRA.[179] The burden of information in FIFRA is put squarely on the registrant, who has to provide evidence to support the assertion that the pesticide does not entail 'unreasonable adverse effects on the environment'.[180] What it is that needs to be registered is not entirely unambiguous. As Miller has pointed out, the term 'pesticide' has been interpreted to mean both an active substance and a product containing one or more active substances.[181] All authorisations, or registrations as they are known in FIFRA, are given for a limited period of time and have to be renewed when they expire.[182] Conditional registrations can be issued by the EPA in some cases where information regarding a substance is temporarily lacking. This may primarily be the case when the substance in question, and the risks associated with it, are known to the EPA.[183] The process may not be initiated in relation to pesticides for which suspected risks have led to the initiation of a special review process.[184]

FIFRA also provides for the revoking of registrations. It makes a separation between instances of 'cancellation' and 'suspension', where the latter constitutes an immediate revocation in a situation in which a product is deemed to pose an 'imminent hazard'.[185] Where a 'cancellation' is considered, a product remains on the market until it is clarified whether or not

[177] For a brief overview of the development of FIFRA see B. L. Johnston, *Environmental Policy and Public Health* (CRC Press, 2007), pp. 262–5.

[178] Food Quality Protection Act 1996 (FQPA), Pub L. 104–70 and Federal Food, Drug and Cosmetic Act 1996 (FFDCA), 21 U.S.C. § 9.

[179] FIFRA, 7 U.S.C. § 136a (a).

[180] FIFRA, 7 U.S.C. § 136a (c) (9) (B) (i) (I), FIFRA; 7 U.S.C. § 136a (c) (5) (C); FIFRA, 7 U.S.C. § 136a (c) (7) (a). The term 'unreasonable adverse effects on the environment' is defined elsewhere in FIFRA and this is returned to in detail under the sections on level and burden of proof.

[181] M. L. Miller, Esq., 'Pesticides', in T. F. P. Sullivan, *Environmental Law Handbook* (Government Institutes, 2003), pp. 652–3.

[182] FIFRA, 7 U.S.C. § 136d (a).

[183] FIFRA, 7 U.S.C. § 136a (c) 7.

[184] Miller, 'Pesticides', p. 655.

[185] FIFRA, 7 U.S.C. § 136d (b) and (c).

it poses 'unreasonable adverse effects on the environment'.[186] Moreover, a product can be suspended during the time that the cancellation procedure is in place. What may trigger these procedures is investigated in detail in the following section.

7.2.3 The institutional framework

The EPA is responsible for implementing and administering FIFRA. This means that the EPA administrators are responsible for receiving and making decisions on the registrations of pesticides, as well as for deciding on whether to initiate and carry out cancellation and suspension procedures.[187] Within the EPA, the task of regulating pesticides belongs to the Assistant Administrator for Prevention, Pesticides, and Toxic Substances.[188] The Assistant Administrator functions as the Head of the Office of Prevention, Pesticides, and Toxic Substances (OPPS).[189]

The EPA is officially an 'independent agency' with its own staff, budget and internal rules. However, it is set up by, and functions under, the executive branch of the US Government, and the US president retains certain important controls over it.[190] Examples of such measures of control include the power of appointments and the issuing of the budget. In both cases, the US Congress needs to concur with the decisions of the president.[191]

The EPA is assisted in its work by an expert Scientific Advisory Panel (SAP) in all cases relating to cancellation/suspension, or the issuing of other regulations.[192] Notably, the SAP does not need to be consulted when issues of registration are considered.[193] It consists of a total of seven scientific experts, who are selected and appointed by the EPA Administrator from a list of twelve nominees, six nominated by the National Institutes

[186] The definition of this term is returned to in detail in the following.

[187] See for example FIFRA, 7 U.S.C. § 136a and d respectively, as well as the EPA FIFRA website at: www.epa.gov/pesticides/ (accessed 1 March 2010).

[188] See EPA Organizational Structure at www.epa.gov/epahome/organization.htm (accessed 1 March 2010).

[189] Information about OPPS is available at http://www.epa.gov/oppts/ (accessed 1 March 2010).

[190] United States Environmental Protection Agency, Reorganisation Plan No. 3 of 1970, Special Message from the President to the Congress about Reorganization Plans to Establish the Environmental Protection Agency and the National Oceanic and Atmospheric Administration, 9 July 1970. Available at: www.epa.gov/history/org/origins/reorg.htm (accessed 1 March 2010).

[191] Reorganisation Plan No. 3 of 1970, Section 1

[192] FIFRA, 7 U.S.C. § 136w (d).

[193] For some of the implications of this see Jasanoff, *The Fifth Branch*, p. 151, note 40.

of Health and six by the National Science Foundation. It is laid down that the members of the SAP should be selected based on 'professional qualifications to assess the effects of the impact of pesticides on health and the environment'. As far as possible, the Panel should comprise members with expertise in the fields of toxicology, pathology, environmental biology and related sciences.[194] Within the framework of the SAP, sub-panels for specific topics and areas of expertise may be established. Sub-panels can consist of members of the SAP, but outside scientists can be appointed by the SAP if this is needed. US Advisory Committees are subject to the Federal Advisory Committee Act (FACA),[195] which establishes congressional oversight over the activities and functions of each advisory committee, the SAP included. The FACA, furthermore, holds that experts serving on committees should be 'fairly balanced in terms of the points of view represented', and that they should not be 'inappropriately influenced' by either the authority under which it serves or outside interests.[196] It thus appears that Congress is intent on safeguarding the independence of US expert committees, as well as the representation of differing interests and viewpoints on the committees themselves. Here, the US system goes further towards establishing such requirements than the other national systems investigated in this book. In both Sweden and the UK the selection of experts mainly falls to the sole discretion of the government or the agency which has had powers delegated to it. Objective criteria such as those included in FACA are rare. In this regard, the US system is more like the EU system of institutionalised scientific advice.[197]

As stated above, the role of the SAP is, according to FIFRA, to advise the decision-maker on scientific aspects of proposed regulation. The courts have rejected the possibility that the EPA Administrator be bound by the conclusions of the expert committee. Instead, they have held that the Administrator is free to disregard the advice given, provided that there is 'justifiable' reason to do so.[198] This issue is returned to in more detail below in relation to the distinction between risk assessment and risk management.

[194] *Ibid.* [195] Federal Advisory Committee Act, Public Law 92–463, 5 U.S.C. App. 1.

[196] *Ibid.*, Section 5 (a) (2) and (3).

[197] For a comparative discussion on the use of scientific bodies in evaluating data relating to pesticides in the US and Europe see R. Brickman, S. Jasanoff and T. Ilgen, *Controlling Chemicals: The Politics of Regulation in Europe and the United States* (Cornell University Press, 1985), pp. 174–80.

[198] Miller, 'Pesticides', p. 677.

Another interesting aspect of FIFRA is that it explicitly demands that all 'major scientific studies' conducted by the EPA should be subjected to peer review by the SAP.[199] This means that the Advisory Committee is not only limited to influencing the outcome of concrete decisions, but it also holds a supervisory role in relation to all scientific work of 'major' importance conducted at the EPA.

7.2.4 Risk assessment and risk management

As stated in the previous Section, the use of risk assessment has increased in the USA in the last twenty-five years. However, the issue of how the process of risk assessment relates to risk management has received surprisingly little attention.[200] In policy documents it is held that even though the roles of the risk assessor and the risk manager are different, it is important that they perform their duties in co-operation.[201] Even though the EPA has held that it considers risk assessment and risk management 'not necessarily separate',[202] it appears in other documents that they are subject to the traditional separation discussed elsewhere in this book.[203]

With regard to pesticide regulation, the risk assessment is mainly performed by the SAP. As discussed in the previous section, referral to the SAP for advice is mandatory in FIFRA during cancellation procedures. However, the advice given by the SAP is not binding on the EPA Administrator, who may disregard it provided that reasons are stated for this. In *Dow* v. *Ruckelshaus*, which involved the cancellation of the pesticide 2,4,5-T, the EPA cancelled the pesticide despite advice from the scientific committee which had maintained that it should be allowed on the market.[204] The 8th Circuit Court held that the Administrator had acted within his discretion when he came to a different conclusion than that of

199 FIFRA, 7 U.S.C. § 136w (e).

200 See above, Section 7.1.5.

201 See Environmental Protection Agency, 'An Evaluation of EPA Risk Assessment Principles and Practices', Staff Paper Prepared for the US Environmental Protection Agency by members of the Risk Assessment Task Force, March 2004. Available at: www.epa.gov/osa/pdfs/ratf-final.pdf (accessed 1 March 2010) and US Environmental Protection Agency, *Risk Characterization Handbook*, December 2000, pp. 20–3. Available at: www.epa.gov/OSA/spc/pdfs/rchandbk.pdf (accessed 1 March 2010).

202 US Environmental Protection Agency, 'An Evaluation of EPA Risk Assessment Principles and Practices', p. 22.

203 See, for example, National Research Council, *Risk Assessment in the Federal Government*, pp. 18–20. See also Section 2.3 above.

204 *The Dow Chemical Company* v. *William D. Ruckelshaus, Administrator Environmental Protection Agency Defendant-Appellant* (8th Circuit 1973) 477 F.2d 1317.

the scientific committee. '[The Administrator] was not required to adopt the Advisory Committee findings', the 8th Circuit Court stated unambiguously.[205] It is also important to note here that the Administrator had issued a number of statements outlining ten reasons for not following the SAP recommendations. Thus, the courts have held that the risk manager, the EPA Administrator in cases regarding pesticides, is not bound by the recommendations of the risk assessor. This emphasises the notion of the difference and separation of tasks between the two levels within the risk regulation framework.

7.2.5 Scientific evidence and the introduction of precautionary measures

In the previous sections it was shown that, since the 1980s, the requirements on the legislator to provide scientific evidence in support of a risk stemming from a product before regulating it have been increasing.[206] The regulation of pesticides, however, provides an important example of an exception to this development. This issue has been particularly publicised with regard to pesticide residues in food. According to the Delaney Clause, which was a 1958 amendment to the Food, Drug, and Cosmetic Act (FDCA), any cancer-causing food additives were barred from appearing in foods regardless of the magnitude of the risk.[207] This zero-risk approach was upheld by courts,[208] but was heavily criticised and difficult for the EPA to uphold in practice.[209] In 1996 Congress decided to revisit the matter and revise the zero-risk approach. The resulting Food Quality Protection Act (FQPA) established that all pesticide residues in food are deemed unlawful, unless the EPA has set a tolerance level for them and the pesticides in question comply with these levels. This meant that the EPA was allowed to decide on where to set the level of regulation, and that it was within its discretion not to regulate pesticides with negligible carcinogenic properties. The 'safety

[205] *Ibid.* [206] See, above, Section 7.1.4.

[207] Miller, 'Pesticides', 693.

[208] See, for example *Lee* v. *Reilly* 968 F.2d 985 (9th Cir. 1992), p. 988, where the court held that the EPA had no discretion to permit carcinogenic foods, regardless of the minimal degree of risk posed by them.

[209] E. Dunkelberger and R. A. Merrill, 'The Delaney Paradox Re-examined: Regulating Pesticides in Processed Foods', *Food & Drug L. J.*, 48 (1993), 411 and A. Montemarano, 'The Delaney Paradox Resurfaces: Regulating Pesticides as Food Additives Under Federal Law', *Rutgers L.J.*, 25 (1993/1994), 433.

level' established by the FQPA was set at 'reasonable certainty' of no harm.[210]

With regard to the allowing of pesticides on the US market, it is useful to separate the issues of registration on the one hand, and suspension and cancellation on the other. As mentioned above, in relation to registration of pesticides, the burden of information is put squarely on the applicant to provide data to support the fact that the pesticide in question does not bring 'unreasonable adverse effects on the environment'.[211] When determining what constitutes 'unreasonable effects on the environment', the EPA should not only consider risks to human health and the environment, but must also take 'into account the economic, social, and environmental costs and benefits' the proposed pesticide brings.[212] This level has been described as a non-zero-risk level, which allows the EPA to decide on registrations on an ad hoc basis, taking risk assessment factors into account, but balancing these with a wider socio-economic analysis.[213] The EPA does not have to wait until scientific uncertainty materialises, but can reject a request for registration as a precautionary measure under conditions of uncertainty, as long as this does not aim to set a zero-risk level.[214] This has been emphasised by courts, which have rejected regulation of *de minimis* toxic risks.[215] In practice, it has been pointed out, time constraints and lack of resources may lead to situations in which the EPA is more likely to approve a registration under uncertainty than arguing that the uncertain risk is 'unreasonable'.[216] However, the EPA emphasises in its own reports the importance of identifying and classifying uncertainty during the risk analysis process,[217] and it has stated that it aims 'to use all information available in an objective, realistic, scientifically balanced way to make decisions'.[218]

When it comes to cancelling and suspending pesticides, the leeway for the EPA is somewhat different. First, as mentioned above, the SAP needs

[210] FQPA, Sec. 405. [211] FIFRA, 7 U.S.C. § 136a (a).

[212] FIFRA 7 U.S.C. § 136 (bb).

[213] J. S. Applegate, 'The Perils of Unreasonable Risk: Information, Regulatory Policy and Toxic Substances Control', *Colum. L. Rev.*, 91:2 (1991), 261.

[214] See, for example, *Ciba-Geigy Corp.* v. *EPA*, 874 F.2d 277, 1989 (5th Cir.), 280, where the court held that FIFRA allows for regulation without the EPA having to show 'actual adverse consequences'.

[215] Applegate, 'The Perils of Unreasonable Risk', 267 and note 68.

[216] M. W. Graf, 'Regulating Pesticide Pollution in California Under the 1986 Safe Drinking Water and Toxic Exposure Act (Proposition 65)', *Ecology L.Q.*, 28 (2001), 663.

[217] US Environmental Protection Agency, 'An Evaluation of EPA Risk Assessment Principles and Practices', 30–51.

[218] *Ibid.*, p. 139.

to be consulted when cancellations are prepared and, even if its verdict is in no way decisive, it highlights the emphasis of scientific advice in this situation. Second, the burden of proof is put more explicitly on the EPA. This is returned to in the following section. If the Administrator finds that a registered pesticide no longer conforms to the requirements in FIFRA on which its registration was based, and that the product poses a 'substantive question of safety',[219] the registration may be cancelled.[220] In *EDF* v. *Ruckelshaus* the DC Circuit Court left it to the general discretion of the Administrator to decide when a 'substantive question of safety' was present. In that case, which regarded the prohibition of the pesticide DDT, the Court recognised that the data relied upon by the Administrator was uncertain in nature, but did not review it in any detail.[221]

The intention to cancel a product has been described as 'a determination [...] that adjudicatory proceedings will commence'.[222] However, during such procedures, which may be lengthy, the product in question remains on the market. In order to provide the EPA with a possibility of removing products from the market at an earlier stage, the possibility of suspension has been included in FIFRA.[223] A suspension of a registration can be ordered when it 'is necessary to prevent an imminent hazard'.[224] Normally, a suspension is issued after a hearing has been held, but it can also be issued in advance of notification to the affected parties in the form of an emergency order.[225] When it comes to the issuing of suspension orders, courts have held that a higher level of evidence must be met by the EPA than is the case for cancellations.[226] The DC Circuit Court held that the level of proof that would have to be met by the Administrator should correspond to a 'likelihood'-requirement, that is, the expected risk from a pesticide would have to be likely to be realised, instead of just pose a 'substantial question of safety'.[227]

219 The DC Circuit found support for this term in the legislative history to FIFRA, see *EDF* v. *Ruckelshaus* (D.C. Circuit, 1971) 439 F. 2d 584, pp. 593–5.

220 FIFRA, 7 U.S.C. § 136d (b).

221 *EDF* v. *Ruckelshaus*, 594.

222 *FTC* v. *Standard Oil Co.*, 449 U.S. 232 (1980).

223 FIFRA, 7 U.S.C. § 136d (c). A suspension is a temporary prohibition of the marketing of a product, pending the results of cancellation procedures.

224 FIFRA, 7 U.S.C. § 136d (c) (1).

225 FIFRA, 7 U.S.C. § 136d (c) (3).

226 *National Coalition Against the Misuse of Peticides, et al.* v. *EPA* (D.C. Circuit, 1989) 867 F.2d 636, p. 644.

227 *Ibid.*, where the Court described the requirement of a risk having to be 'likely to result' in harm. The Court had similarly held in *EDF* v. *EPA* that 'substantial likelihood' of the

As can be deduced from the above, FIFRA does not set any explicit exposure limits which bind the regulator. Instead, every instance of regulation is judged on a case-by-case basis. Precautionary action can be taken under conditions of scientific uncertainty, but the occurrence of scientific uncertainty does not mandate regulatory action on the part of the EPA.[228] Courts have interpreted FIFRA to entail different demands on scientific evidence, depending on which type of action is considered. The level of proof is the highest in cases of emergency suspensions.[229] It should also be pointed out that, in practice, the FQPA requires the EPA to set 'safe' exposure levels for pesticide residues in food. Owing to the fact that cost–benefit balancing is more restricted under FQPA, precautionary regulation and the reliance on uncertain information may be more widespread in such cases.[230] Interestingly, the FQPA introduced the ubiquitous '10x factor', which means that a ten-fold safety factor for infants and children is maintained.[231] In practice, this means that when a 'safe' exposure level has been identified, this has to be multiplied by ten in relation to substances that may be consumed by infants and children. Thus, the '10x factor' is a far-reaching precautionary measure. Interestingly, it is not based on scientific evidence, but, instead, it aims to increase the margin of safety for children even though a risk assessment has established that a product is virtually risk free.

The introduction of the FQPA has also meant a development of the risk assessment techniques employed by the EPA. After its introduction, the EPA routinely includes, for example, aggregate exposure and cumulative effects in its assessments of risks.[232]

7.2.6 *The burden of proof*

As in the case of the EU regulation of pesticides, and as also reflected in the Swedish and UK systems, FIFRA appears to put the onus with regard

realisation of harm would trigger regulation. *EDF* v. *EPA* (D.C. Circuit 1976) 548 F.2d 998, p. 1005.

228 *National Coalition Against the Misuse of Peticides, et al.* v. *EPA*, p. 644–5.

229 *Ibid.*, p. 644.

230 Graf, 'Regulating Pesticide Pollution in California Under the 1986 Safe Drinking Water and Toxic Exposure Act (Proposition 65)', 687, note 100. This issue is returned to in more detail below under the section on cost–benefit analysis.

231 EPA Evaluation Report, 'Changes Needed to Improve Public Confidence in EPA's Implementation of the Food Quality Protection Act', Report No. 2006-P-00003, 19 October 2005, pp. 5–6.

232 *Ibid.*

to pesticides registration on the applicant.[233] Instead of the regulator needing to demonstrate the risk, it is for the regulated party to show that the product is without 'unreasonable risk'.[234] However, it could be argued that the burden borne here by the applicant is a 'false' burden of proof,[235] or at least, that the burden is shared between the applicant and EPA. It is clear that the applicant bears the burden of information. This is the typical situation in relation to pre-authorisation schemes, since the applicant is the party in possession of the scientific evidence on which the registration is to be based. It is also clear that this information forms the basis on which the decision that no 'unreasonable risk' stems from the pesticide must be deduced. However, it is the responsibility of the EPA to identify an 'unreasonable risk' during the registration procedures in order to bar the product from being marketed. If the EPA does not meet this requirement, the pesticide should be registered and allowed to enter the market. If the EPA identifies a risk in an application for registration, the burden falls on the applicant to show that this risk is not unreasonable.[236] Thus, the initial burden of proof rests with the EPA, but the meeting of a relatively low level of proof tips the burden onto the applicant.

In contrast to this, the onus is clearly on the EPA to show that a pesticide poses a 'substantial question of safety' when it intends to initiate cancellation or suspension proceedings. This has also been upheld by courts.[237] In *EDF* v. *Ruckelshaus* the Court held that:

> The legislative history supports the conclusion that Congress intended any substantial question of safety to trigger the issuance of cancellation notices, shifting to the manufacturer the burden of proving the safety of his product.[238]

7.2.7 *Cost–benefit analysis*

One of the main differences with regard to pesticide regulation in the EU, Sweden and the UK, on the one hand, and the USA on the other, is the firm

[233] See, above, Sections 5.2.6 and 6.2.6, respectively.

[234] FIFRA, 7 U.S.C. § 136a (a).

[235] See, above, Section 5.1.7.

[236] The level of proof needed to be met by the EPA here is low, as was discussed in the previous section. The requirements become progressively higher for cancellations and suspensions.

[237] See, for example, *National Coalition Against the Misuse of Pesticides, et al.* v. *EPA*, p. 642.

[238] *EDF* v. *Ruckelshaus*, p. 593; also cited by the Court in *Dow Chemicals* v. *Ruckelshaus*, p. 1324.

demand on, and significance of, the balancing of costs and benefits in the latter.[239] When determining whether a risk is 'unreasonable' or not during the registration procedure, the costs and benefits of the product appear to be as important to the outcome as that of the nature and magnitude of the product's inherent risk.[240] Applegate has held that cost–benefit analysis is a necessary part of pesticide regulation, since it is only the estimate benefits of a product that can motivate anything but a zero-risk approach.[241] If a risk to health or to the environment is not perceived as bringing anything beneficial, why then take the risk at all? This insight may seem trivial, but it helps to explain how Swedish chemicals regulation has become so radical. Since the costs and benefits of products are not considered, even *de minimis* risks are considered sufficient to warrant regulation. Since the Swedish Chemicals Inspectorate (KemI) has confined itself to considering risks alone, and does not juxtapose them with corresponding expected costs and benefits incurred when regulating risks, this also narrows its flexibility. If a risk is found, the product should not be allowed. The flexibility of the EPA, on the other hand, is considerably greater. By balancing the risks, costs and benefits on a case-by-case basis, it has a substantial margin of manoeuvre.[242] Applegate has summarised the issue in the following fashion:

> Likewise, the unreasonable level of risk floats somewhere between harm that is not certain and risk that is not de minimis. Acceptance of non-absolute or relative safety therefore begs the question, how safe? This is where cost enters the risk formula. The unreasonable risk standard expressly incorporates cost-risk-benefit balancing to answer the question raised by relative safety. EPA may not take action without considering the degree of risk involved, the costs associated with regulation, and the benefits that may be achieved by the proposed regulation.[243]

This, however, has not always been the case. Until the Delaney Clause was revoked in the mid-1990s, the EPA was effectively prohibited from taking other considerations than risk into account when deciding on pesticides which might have resulted in residues in food.[244]

The margin of manoeuvre enjoyed by the EPA functions within the confines of what US courts have stipulated. In *National Coalition Against*

[239] See, above, Section 7.1.7.

[240] FIFRA, 7 U.S.C. § 136bb, where 'unreasonable risk' is defined as being contingent upon the 'taking into account the economic, social, and environmental costs and benefits of [the pesticide's] use'.

[241] Applegate, 'The Perils of Unreasonable Risk', 274–5.

[242] *Ibid.*, 276–7.

[243] *Ibid.*, 274–5.

[244] See, above, Sections 7.1.2 and 7.2.1.

Misuse of Pesticides v. *EPA* the DC Circuit Court held that scientifically uncertain risks may, indeed, be offset by the costs incurred by regulation.[245] In *Love* v. *Thomas*, which concerned the EPA emergency suspension of the pesticide dinoseb, the 9th Circuit Court referred to the definition of 'imminent hazard' in FIFRA, and held that, before a suspension can be issued, costs as well as benefits have to be carefully balanced.[246] The Court evaluated the cost–benefit analysis performed by the EPA in relation to dinoseb in detail, and found it to have seriously undervalued the costs for local farmers. It stated:

> With all due respect to the EPA and its overworked staff, such insensitivity to the local economic problems caused by its decision is unbecoming and inappropriate.[247]

However, the Court continued, the EPA could still have been entitled to issue the emergency suspension if a true emergency had been at hand. In the case of emergency suspensions, a balancing of costs and benefits is not required.[248] However, since dinoseb was unlikely to be used by farmers until the next spraying of the potato crop, some six months after the EPA decision, an emergency was not present according to the Court, and therefore the decision appeared 'entirely arbitrary'.[249]

Love v. *Thomas* provides an interesting example of how the courts are not content only to consider whether the EPA has taken a 'hard look' at the costs and benefits, but also take a close look at the details of the balancing, as well as the data reviewed. This illustrates how the reliance on cost–benefit analysis in US pesticide regulation is far more advanced than is the case in the other legal systems compared in this book. It should be pointed out that, even after the demise of the Delaney Clause, courts have suggested that the balancing might be skewed in favour of health protection in cases where cancer is the suspected risk. In *EDF* v. *Ruckelshaus* the Court held that the law 'places a heavy burden on any administrative officer to explain the basis for his decision to permit the continued use of a chemical known to produce cancer in experimental animals'.[250]

245 '[The EPA] might determine after an administrative hearing [...] that scientific uncertainty as to the danger of a specific pesticide (combined perhaps with the economic impact of cancellation on "agricultural commodities, retail food prices and the [...] agricultural economy" 7 U.S.C. § 136d(b) (1982)) indicates that the registration should not be cancelled'. *National Coalition Against Misuse of Pesticides et al.* v. *EPA*, p. 642.

246 *Love* v. *Thomas* (9th Cir. 1988) 858 F.2d 1347, p. 1357.

247 *Ibid.*, p. 1361.

248 Miller, 'Pesticides', 674.

249 *Love* v. *Thomas*, p. 1362.

250 *EDF* v. *Ruckelshaus*, p. 596.

That a 'safer' or similar product already exists on the US market should not lead the EPA to deny an application for registration. However, the fact that this is the case can be factored into the cost–benefit analysis, and, in this way, be taken into account in determining the 'unreasonable' nature of a given risk.[251] Thus, the substitution principle, or product choice principle, is not in effect in the USA, but issues that are similar to it can be factored into the cost–benefit balancing.[252]

7.2.8 Conclusion

As already mentioned in the introduction, the USA has a long history of precautionary regulation in the area of pesticides. Even though the precautionary principle as such is not used, specific provisions, such as the Delaney Clause and the '10x factor' for carcinogenic properties in products intended for consumption by children, are precautionary in nature. Interestingly, these precautionary measures focus on the risk of cancer, while other risks, such as toxicity, are not necessarily treated as harshly. Thus, a choice appears to have been made to target products that may give rise to cancer in particular. With the revocation of the Delaney Clause in the 1990s, the EPA has been given an increased mandate to take issues other than the risk of cancer into account when deciding on the treatment of pesticides.

FIFRA functions as the legislative framework for pesticide regulation in the USA. It is, in many ways, similar to Directive 91/414/EC in place in the EU,[253] in that it places the burden of information squarely on the applicant and makes explicit use of scientific bodies to examine and advise the bureaucratic body empowered to take decisions. Courts have clarified that the EPA is not bound by the advice issued by scientific bodies, but it appears that it has to give its reasons for why it chooses to derogate. At first sight, it appears that the burden of proof in the FIFRA is reversed, and thus placed on the applicant. However, it is argued here that this construction constitutes a 'false' burden of proof, and that the actual burden still rests with the EPA to demonstrate that an 'unreasonable risk' is present during the registration procedure. The level of proof is low and,

[251] See, for example, *EDF* v. *EPA* (D.C. Circuit 1972) 465 F.2d 528, p. 539.

[252] Compare with the situation in Sweden, where the substitution principle is strictly interpreted and closely connected to the precautionary principle. See, for example, Section 5.2.5, above.

[253] Or rather, Directive 91/414/EC is reminiscent of FIFRA, since it preceded the Directive by some forty-five years.

once it is met, the burden is switched onto the applicant to show that no 'unreasonable risk' is present. The level of proof is higher when switched onto the applicant.

Since the revocation of the Delaney Clause, risk is only one among many elements that need to be considered by the EPA when taking decisions under FIFRA. The process of authorising or cancelling authorisations in the USA is often described as a balancing exercise, which indicates that several competing interests must be considered together. Applegate holds that this is a necessary starting point, because, if positive aspects of a product were not considered, there would be nothing to justify anything but a zero-risk approach. Here, it is interesting to compare this with the situation in Sweden, in which the precautionary regulation of pesticides is very far-reaching. This is made possible by the fact that, as Swedish officials have stated, no economic considerations are taken into account when formulating decisions in relation to pesticides.

The US system allows for a more nuanced way of taking decisions. In fact, it requires a more nuanced decision-making process, because it is cost–benefit balancing that will answer the question of whether a product poses an 'unreasonable' risk or not. Furthermore, courts have indicated that they find it within their remit to 'second-guess' the balancing performed by the EPA, at least in cases where they find that cost–benefit analysis is significantly flawed. In *Love* v. *Thomas* the Court investigated the details of the EPA cost–benefit analysis, before it decided to overturn it. Some cases have indicated that the cost–benefit analysis may be skewed in favour of health protection, at least in cases where cancer is the suspected risk. In *EDF* v. *Ruckelshaus* the Court stated that the EPA had a 'heavy burden' to explain why it allowed a pesticide when it had been shown to have carcinogenic properties in animal testing. Finally, in relation to cost–benefit balancing, it should be pointed out that even though the substitution principle is not formally in effect in the USA, the issue of whether a comparable, but less risky, pesticide is already on the market should be allowed to influence the outcome of the balancing. Thus, issues similar to the substitution principle can be considered under FIFRA.

7.3 US regulation of base stations

7.3.1 Introduction

In the chapters on Sweden and the UK it was shown that the public is increasingly concerned with radiation from mobile telephones and base stations. However, in the USA the concern with health risks from radio-frequency radiation is not as widespread or high profile as it is in, for example, the UK. It has been held that the media coverage in the USA with regard to these issues is relatively subdued in comparison with some European countries, and that a well-organised, national network championing health concerns about mobile phones has failed to materialise, as of yet at least.[254] Thus, the US emissions guidelines in place, as well as the statutory instrument which pre-empts state and local authorities from making their own health risk assessments, has received relatively little attention, both in mainstream media and in specialist literature.

The regulation of base stations for mobile telephones is a relatively complicated area, due to the fact that national rules function together with decisions taken at local level during the planning process. The desire for the rapid expansion of mobile phone coverage is often counterbalanced by local planning concerns, including aesthetic, as well as health, considerations. In this section, as in the previous sections, a structure which aims to deal with the constituent parts of the precautionary principle is maintained. Thus, two sections outline the legal and institutional structure in the area, before substantive issues of whether, and if so, how and by whom precautionary considerations can be taken during the process. The central topics here are, of course, not only the issues relating to the level and burden of proof needed for the regulation of base stations, but also the issue of public opinion, which is of central importance to this chapter.

7.3.2 The legal framework

In the USA, as in the other legal systems discussed in this book, several laws and regulations at national, as well as at local, level are concerned with the regulation of base stations. At US federal level, the National Environmental Protection Act (NEPA),[255] the Radiation Control for Health and Safety Act (RCHSA)[256] and the Telecommunications Act[257] make up the legislative framework. At local level, the actual placement of base stations, or cell towers as they are frequently referred to in the USA, is subject to local planning, or 'zoning', rules.

[254] A. Burgess, *Cellular Phones, Public Fears, And a Culture of Precaution* (Cambridge University Press, 2004), pp. 178–80.

[255] National Environmental Protection Act, 42 U.S.C.. 4321.

[256] Radiation Control for Health and Safety Act 1968, 42 U.S.C. 263f.

[257] Telecommunications Act 1996, Pub. L. No. 104–104, 110 Stat. 56.

The most relevant government body in the area of mobile telephony is the Federal Communications Commission (FCC), which had already been established by the Communications Act in 1934.[258] The role of the FCC is to regulate the US radio spectrum, which includes the considering of whether communication equipment significantly affects the quality of the human environment.[259] The institutional structure and tasks of the FCC are dealt with in detail in the following section.

The most significant single piece of legislation in relation to base stations is the Telecommunications Act 1996. It was adopted by Congress, after heavy lobbying,[260] to facilitate the expansion of the US mobile phone network, and to secure a competitive environment for the benefit of consumers.[261] In order to prevent local governments from evaluating environmental and health concerns within the framework of zoning procedures, the Act included a provision which holds that:

> No State or local government or instrumentality thereof may regulate the placement, construction, and modification of personal wireless service facilities on the basis of the environmental effects of radio frequency emissions to the extent that such facilities comply with the [Federal Communications] Commission's regulations concerning such emissions.[262]

The provision explicitly pre-empts local governments from imposing their own assessments, and makes environmental and health considerations a matter for the federal government.[263] In apparent contradiction to this,

[258] The Communications Act 1934 preceded the Telecommunications Act which amended it in 1996.

[259] According to NEPA there is a requirement on all federal agencies to take this into account when issuing regulations. See NEPA Sec. 102, 42 U.S.C. 4332 (c). The fact that the FCC is essentially not designed to make health assessments is returned to in the following section.

[260] For a critical parallel between alleged campaign contributions to members of Congress from telecom actors and the hasty passing of the bill see N. Tinari, 'Cell Phone Towers in Residential Areas: Did Congress Let the Pig in the Parlor with the Telecommunication Act of 1996?', *Temp. L. Rev.*, 73 (2000), 269.

[261] S. Lorde Martin, 'Communication Tower Sitings: the Telecommunications Act of 1996 and the Battle for Community Control', *B.T.L.J.*, 12:2 (1997). Available at: www.law.berkeley.edu/journals/btlj/articles/vol12/index.htm (accessed 1 March 2010).

[262] Telecommunications Act 1996, Sec. 704 (B) (iv).

[263] However, in *Sprint Spectrum L.P.* v. *Mills* the 2nd Circuit Court opened up for more stringent measures to be prescribed by local governments where these act as 'proprietors' rather than 'regulators'. In this case a government-owned school building was subjected to more stringent demands on radiation emissions in a lease agreement. Since the agreement was not considered to be a regulatory act, the arrangement was deemed

the Act also authorises states to 'impose' measures which are necessary to 'protect the public safety and welfare'.[264] However, this statement is subjected to the possibility of pre-emption by the Federal Communication Committee.[265] It has been held that, after the Telecommunications Act came into force, the prospects for local authorities to impose their own assessments with regard to environmental considerations has indeed become bleak.[266] In comparison with Sweden and the UK, the US legislation is strict on this point, where legislation in the former two countries is more ambiguous.

7.3.3 The institutional framework

Since state and local authorities are effectively precluded from establishing health and environmental standards for emissions from base stations, the most important actors in the USA are found at federal level. A number of agencies have a stake in the process, but the 'spider in the web' is the FCC, which has the statutory power to regulate the area.[267] However, since the FCC has admitted that it lacks the expertise to formulate regulatory guidelines for health and environmental exposure to radiation from mobile telephony, it relies heavily on other public and private bodies for scientific input.[268]

The Food and Drug Administration is the government agency that has the general responsibility for regulating radiation emissions from electric products. However, to date it has declined to make real use of its powers in the area of mobile telephony, referring mainly to the lack of scientific certainty with regard to the possible health effects stemming from mobile phone use. It has, however, worked together with industry to seek ways in which the exposure to radio-frequency radiation can be minimised.[269]

by the court to fall outside of the Telecommunications Act, and thus within the discretion of the local government. See *Sprint Spectrum L.P.* v. *Mills* (2nd Cir. 2002) 283 F.3d 404, p. 417–20.

264 Telecommunications Act, Sec. 253 (b). 265 *Ibid.*, Sec. 253 (d).

266 B. M. Gerber, 'Case Comment: City of Rancho Palos Verdes, *California* v. *Abrams*', *Harv. Envtl. L. Rev.*, 30 (2006), 295.

267 Telecommunications Act, Sec. 704 (b).

268 United States General Accounting Office, 'Telecommunications: Research and Regulatory Efforts on Mobile Phone Health Issues', Report to Congressional Requesters, May 2001, AO-01–545, p. 20.

269 L. Grasso, 'Cellular Telephones and the Potential Hazards of RF Radiation: Responses to the Fear and Controversy', *Va. J.L. & Tech.*, 3 (1998), 2.

The Environmental Protection Agency (EPA) also plays a part in the institutional framework for radiation protection and control. Under Reorganisation Plan No. 3 of 1970, it is responsible for advising the president and providing guidance to all federal agencies on matters relating to protective standards for radiation exposure.[270]

In relation to workers, the National Institute for Occupational Safety and Health (NIOSH) and the Occupational Safety and Health Administration (OSHA) have been involved in monitoring and investigating issues relating to exposure to radio-frequency radiation.[271]

The FCC is the only agency that has issued exposure guidelines for radiation from mobile phones. This was done in the form of a non-binding document in the late summer of 1999.[272] The details of these recommendations are returned to in the following sections. The FCC is a federal agency founded in 1934 whose role it is to regulate 'interstate and international communications by radio, television, wire, satellite and cable'.[273] Answering directly to the US Congress, the FCC is directed by five commissioners, who are appointed by the president. One of them is designated as the chairperson, and only three of them can be from the same political party.[274] The agency is divided into six different bureaux and ten offices, where the Office of Engineering and Technology (OET) is the unit which has issued the above-mentioned exposure guidelines. The office has little practical expertise in the area of health and safety regulation,[275] and is generally responsible for advising the Commission on all 'engineering matters'.[276] The fact that health and safety regulation is not the prime focus of the OET can also be deduced from the FCC organisation outline, in which no one seems to be specifically mandated to cover health and safety issues.[277] This is further underscored by the fact that none of the FCC's advisory committees, established under the Federal Advisory Committee Act, are directly concerned with providing scientific input to the

[270] United States General Accounting Office, 'Telecommunications: Research and Regulatory Efforts on Mobile Phone Health Issues', p. 9.

[271] Questions and Answers about Biological Effects and Potential Hazards of Radiofrequency Electromagnetic Fields, OET Bulletin 56, Fourth Edition, August 1999, p. 10.

[272] Questions and Answers about Biological Effects and Potential Hazards of Radiofrequency Electromagnetic Fields.

[273] www.fcc.gov/aboutus.html (accessed 1 March 2010). [274] *Ibid.*

[275] United States General Accounting Office, 'Telecommunications: Research and Regulatory Efforts on Mobile Phone Health Issues'.

[276] www.fcc.gov/oet/ (accessed 1 March 2010).

[277] www.fcc.gov/oet/organization/Welcome.html#sec4 (accessed 1 March 2010).

Committee.[278] As mentioned above, and as is discussed in more detail below, this has led to the FCC being heavily reliant upon other bodies when issuing its advice.

7.3.4 *Risk assessment and risk management*

It was shown above that the reliance on risk assessment as a foundation for US risk regulation has been increasing since the 1980s.[279] In the area of pesticide regulation the institutional structure and separation of the two stages is generally relatively straightforward. In US pesticide regulation an advisory panel is influential in the risk assessment stage and decisions are ultimately taken by the EPA Administrator.[280]

However, the situation in the area of base stations is slightly different, since several institutions have a stake in the regulation, at local level as well as at national level. In the USA the FCC is the institution that has the statutory task of issuing regulations in relation to radiation from base stations. Thus, the FCC is the agency which it is most accurate to associate with risk management tasks in relation to base stations. However, due to the lack of a clear institutional structure, the FCC also has to analyse the scientific evidence, a task more readily associated with the tasks of a risk assessor than a risk manager. This is a task that FCC itself has stated that it feels uncomfortable with, due to the lack of in-house scientific expertise that the agency possesses.[281] Despite this, the FCC has issued recommendations on the basis of its own gathering of data from various other institutions, both public and private.[282] These recommendations are the most authoritative source in place in relation to exposure to radiation stemming from base stations.[283] The FCC guidelines were

[278] Federal Advisory Committees Follow Requirements, but FCC Should Improve Its Process for Appointing Committee Members, Report to the Chairman, Committee on Government Reform, House of Representatives, General Accounting Office, GAO-05–36, 2004. The Intergovernmental Advisory Committee advises the FCC on public health issues, but does not consist of scientific experts. It is returned to in the section on public opinion.

[279] See, above, Section 7.1.4. [280] See, above, Section 7.2.4.

[281] In relation to exposure guidance and testing of handheld devices the FCC has relied on 'one staff specialist in radiofrequency exposure issues' to review test results. This appears a long way from a systematic framework of risk assessment and scientific advice. See United States General Accounting Office, 'Telecommunications: Research and Regulatory Efforts on Mobile Phone Health Issues', p. 4.

[282] For a discussion of the content of this advice, see the following section.

[283] Questions and Answers about Biological Effects and Potential Hazards of Radiofrequency Electromagnetic Fields.

also largely endorsed in the independent study carried out by the General Accounting Office (GAO).[284]

The lack of a clear institutional structure and separation of risk assessment and risk management in the area of regulation of base stations is not unique for the USA. As has been seen above, in both Sweden and the UK the frameworks are stricter in relation to pesticide regulation than in relation to base stations.

7.3.5 Scientific evidence and the introduction of precautionary measures

Two main documents outline the FCC recommendations in relation to the regulation of radiation from base station. The first one, Bulletin 65, deals primarily with exposure from handsets. It establishes that an exposure threshold of 4 SAR (specific absorbtion rate) should be set as the limit for humans, above which 'potentially hazardous exposure may occur'.[285] This, the Bulletin holds, is in line with what 'expert organisations' recommend.[286] The limit, which is similar to that recommended by the International Commission for Non-Ionising Radiation Protection (ICNIRP) and is subsequently in force in Sweden and the United Kingdom as discussed above, offers a slightly stricter level of protection than was the case in earlier recommendations issued by the FCC in 1985. However, this is not to signify that any new scientific evidence has come to light, according to the FCC. Instead, it aims to express a 'precautionary desire to protect sub-groups of the general population who, potentially, may be more at risk'.[287] This appears to be similar to the reasoning provided by the UK authorities when they incorporated the ICNIRP values. The ICNIRP recommendations consist of a SAR value, which is based upon an arbitrary, built-in 'safety margin' around the value at which detrimental effects are identified.[288] Even though the recommendations coincide, Bulletin 65 does not refer to ICNIRP. Instead, the values are based on the conclusions of the National Council on Radiation Protection and Measurement

[284] See United States General Accounting Office, 'Telecommunications: Research and Regulatory Efforts on Mobile Phone Health Issues, Report to Congressional Requesters'.

[285] Evaluating Compliance with FCC Guidelines for Human Exposure to Radiofrequency Electromagnetic Fields, OET Bulletin 65, Edition 97–01, August 1997, p. 8.

[286] *Ibid.* As has been discussed above, the FCC does not consider itself an expert organisation when it comes to environmental and health protection.

[287] *Ibid.* [288] See, above, Sections 5.3.4 and 5.3.5.

(NCRP), the Institute of Electrical and Electronics Engineers, Inc. (IEEE) and the American National Standards Institute (ANSI).[289] Thus, the FCC draws on scientific advice formulated by government bodies (the NCRP[290]), as well as branch organisations (IEEE[291]) and voluntary standard-setting bodies (ANSI[292]) for inspiration.

In August 1999 the FCC responded to the task it had been given by the Telecommunications Act and issued OET Bulletin 56 with guidelines for maximum exposure limits from base stations.[293] These recommendations are built largely on the same conclusions as Bulletin 65, with the difference that the exposure limits are set as maximum field exposure, rather than SAR values.[294] The document was preceded by an extensive consultation procedure with relevant US agencies, such as the Food and Drug Administration (FDA), the OSHA, and the EPA. All the consulted agencies supported the NCRP and ANSI/IEEE guidelines upon which the FCC based their recommendations.[295]

The FCC guidelines separate occupational (or 'controlled') exposure from exposure to which the general public may be subject ('uncontrolled exposure'). The occupational exposure limits are set higher than those established for the general public.[296] As with Bulletin 65 and the ICNIRP guidelines, the FCC recommendations are based on the presumption that scientifically established health risks are a fact at a degree of exposure exceeding an SAR threshold of 4 W/kg.[297] However, ICNIRP recommends different exposure levels in the lower and upper frequency ranges and for localised exposure from certain products such as hand-held wireless telephones. This difference is marginal, and appears to be due to the use of different physics models.[298]

In 2001 US Senators Lieberman and Markey requested the US General Accounting Office (GAO) to conduct an independent survey of mobile phone health issues. This initiative is similar to the UK approach, in which an independent expert group was asked to conduct an investigation into the same matter, outside of the regular institutional structure.[299]

[289] Evaluating Compliance with FCC Guidelines for Human Exposure to Radiofrequency Electromagnetic Fields, p. 7.

[290] www.ncrponline.org (accessed 1 March 2010).

[291] www.ieee.org (accessed 1 March 2010). [292] www.ansi.org (accessed 1 March 2010).

[293] Questions and Answers about Biological Effects and Potential Hazards of Radiofrequency Electromagnetic Fields.

[294] See *ibid.*, Table 1, p. 15. [295] *Ibid.*, p. 12.

[296] *Ibid.*, p. 15. [297] *Ibid.*, p. 11.

[298] www.fcc.gov/oet/rfsafety/rf-faqs.html#Q10 (accessed 1 March 2010).

[299] See above, Section 6.3.3.

Where the UK investigation was mainly motivated by the growing public discomfort with rapid mobile phone technology expansion,[300] the US counterpart appears to have aimed more at a general clarification of the science and legislation affecting the area.[301] The evidence considered in the Report consists of scientific reports issued by international bodies such as the World Health Organisation, as well as by 'expert panels' in the UK, Canada and Australia, which have considered the peer-reviewed scientific assessments in the area.[302] It thus appears that, when drafting the report, the GAO only considered the scientific evidence that had already been considered by other international and national bodies.

The conclusion of the GAO Report is that the evidence does not support the assertion that radiation, within the recommended limits, from handsets or base stations would have any adverse health effects. However, like the Stewart Report in the UK, it emphasises that the evidence existing at this point is not conclusive, and that more research is required to understand the effects of radiation from mobile telephones fully.[303] Unlike the Stewart Report, the GAO Report does not recommend any precautionary measures,[304] although it does go further than the Swedish government assessments in that it highlights the uncertainties inherent in the data in the area.[305] As a response to the uncertainty, the GAO Report recommends increased efforts to carry out research into health effects, both within the framework of the US regulatory structure in the area, and in partnership with industry.[306]

It thus appears that, in the area of mobile phones, mere uncertainty as to health effects is not sufficient for the US government to issue precautionary measures. Some positive scientific evidence of a risk needs to be at hand. In this context, it should be pointed out that some studies, which indicate potential biological effects from extended exposure to low levels

300 *Ibid.*

301 United States General Accounting Office, 'Telecommunications: Research and Regulatory Efforts on Mobile Phone Health Issues', p. 2.

302 *Ibid.*, pp. 9–12. 303 *Ibid.*, pp. 8–9.

304 As was discussed above, the Stewart Report recommended relatively far-reaching precautionary measures as a response to the uncertainty in relation to base stations. This response included a revocation of approval for base stations and a renewed procedure. It also recommended restrictive siting of base stations in the neighbourhood of schools. See above, Section 6.3.5.

305 The Swedish government has instead opted to play down uncertainties in relation to mobile telephony, and has emphasised that radiation stemming from mobile telephone systems is essentially harmless for humans. See above, Section 5.3.

306 United States General Accounting Office, 'Telecommunications: 'Research and Regulatory Efforts on Mobile Phone Health Issues', pp. 12–18.

of radio frequency,[307] had not yet been carried out when the US regulatory bodies issued the bulk of their recommendations. However, no measures have been taken since, either.

In a 1999 challenge before the 2nd Circuit Appeals Court, petitioners argued that the FCC had not adequately considered the uncertain risks of mobile phone radiation when issuing its guidelines.[308] Among other things, the petitioners held that the FCC should have paid more attention to some scientific advice which clearly emphasised potential, but highly uncertain, risks. Furthermore, they held that the FCC was under an obligation developed by the Nuclear Regulatory Commission to keep radiation levels 'as low as reasonably possible'. The Court did not agree with this. Citing earlier case law, it began by stating that the judicial review under the APA had been interpreted to be deferential in nature:

> The agency's action should only be set aside where it relied on factors which Congress has not intended it to consider, entirely failed to consider an important aspect of the problem, offered an explanation for its decision that runs counter to the evidence before the agency, or is so implausible that it could not be ascribed to a difference in view or the products of experts.'[309]

Having thus held that it considered its margin for review narrowly confined, the Court examined the scientific evidence provided by the applicants. It concluded that the evidence in question was not conclusive, and this had been considered by the FCC when drafting its guidelines. That it had chosen not to act in a more precautionary manner and had based its recommendations on the existing uncertainties, the Court did not find to be problematical. It concluded:

> [t]he argument that the FCC should create greater safety margins in its guidelines to account for uncertain data is a policy question, not a legal one.[310]

Thus, it appears that the Court considers it possible for the FCC to rely on the uncertainty if it so wishes, but that such a choice is within its discretion. Here, the situation appears to be similar to the UK situation in which the courts leave precautionary decisions entirely up to the consideration of the relevant government authorities.[311] However, as discussed

[307] See, for example, Blank, 'The Precautionary Principle Must Be Guided by EMF Research', 203–8, and above, Section 5.3.5.

[308] *Cellular Phone Task Force, et al.*, v. *FCC* (2nd Cir. 2000) 205 F.3d 82.

[309] *Ibid.* [310] *Ibid.*

[311] See above, Chapter 6.

above, other US cases in relation to other statutes suggest that courts will review, and in some cases overturn, decisions taken upon the basis of scientifically uncertain evidence.[312]

In some cases local governments have wished to maintain a stricter application of radiation levels than is provided for in the FCC guidelines. Sometimes zoning permits for base stations have been denied on this basis. In order to impose any restrictions on base stations, the action on the part of the state or local government has to be supported by 'substantial evidence'.[313] This does not mean that the evidence needs to be conclusive, but the mere existence of scientific uncertainty is apparently not sufficient. The term 'substantial evidence' has 'a long-standing meaning in federal law'[314] and has been described by the 4th Circuit Court as meaning 'more than a mere scintilla of evidence', but 'less than a preponderance'.[315] The 1st Circuit Court has held that:

> [s]ubstantial evidence is such relevant evidence as a reasonable mind might accept as adequate to support a conclusion.[316]

However, the 4th Circuit Court held in *AT&T Wireless PCS, Inc.* v. *City Council of Virginia Beach* that health concerns are effectively precluded from consideration by local and state authorities by the Telecommunications Act.[317] Thus, the substantial evidence brought by state or local governments must generally regard issues other than health risks. This topic is returned to in the context of public participation below.

7.3.6 *The burden of proof*

As in the previous studies conducted here, the burden of proof is not reversed in the US with regard to the regulation of base stations. As has been shown above, the guidelines issued by the FCC incorporate safety margins in a way that is similar to the ICNIRP values, which are effectively in use in Sweden and in the UK. However, these safety margins do not appear to be precautionary as such; instead, they aim to establish a

312 See above, 7.1.5.

313 Telecommunications Act, Sec. 704 (7) (B) (3) and *ibid.*

314 *Petersburg Cellular Partnership* v. *Boards of Supervisors of Nottoway County*, 205 F.3d 688 (4th Cir. 2000), 694.

315 *AT&T Wireless PCS, Inc.* v. *City Council of Virginia Beach*, p. 430.

316 In *Associated Fisheries of Maine, Inc.* v. *Daley* (1st Cir. 1997) 127 F.3d 104, p. 109.

317 *AT&T Wireless PCS, Inc.* v. *City Council of Virginia Beach* (4th Cir. 1998) 155 F.3d 423, p. 431.

'fence' around the emissions levels at which adverse effects in humans are scientifically proven to appear in order to avoid them being breached accidentally. Since the regulated parties, the providers of mobile phone systems, have not been inclined to challenge the FCC guidelines or their application by state or federal governments, it is difficult to speculate on how these levels could be successfully challenged by the regulated parties. To date, the case law has only concerned instances in which a more precautionary approach than the FCC guidelines has been argued by parties.

7.3.7 Cost–benefit analysis

The Telecommunications Act 1996 is not primarily an instrument for health protection. Instead, its focus is clearly on removing barriers to trade[318] in telecommunications services in order to pave the way for 'competitive markets'.[319] It has been pointed out that '[c]ongress enacted the Telecommunications Act 1996 to meet society's communication demands',[320] rather than as a means to address the issue of radiation from base stations. In the light of this, it would be fair to say that the act is clearly motivated by economic considerations, rather than by a wish to create a radiation-safe environment. Thus, the main focus of the Telecommunications Act is to allow for increased competition between actors on the telecommunications market, and an increasing quality of services and better prices for consumers. This is further emphasised by the fact that the FCC has been given the task of overseeing the Act. As it has itself pointed out, it lacks the competence to formulate radiation protection guidelines on its own.[321] When considering the background and purpose of the Telecommunications Act, it appears that risk considerations play a rather small role in the grand scheme. Thus, it would seem that the scales were already tipped from the outset in favour of the benefits of rapid mobile phone expansion.

When considering the guidelines established by the FCC, and the documents which relate to it, such as Bulletin 65 and the Lieberman/Markey Report, references to cost–benefit analysis are lacking in their

[318] Telecommunications Act, Section 253. [319] *Ibid.*, Par.t II.

[320] T. L. Gustin, 'The Perpetual Growth and Controversy of the Cellular Superhighway: Cellular Tower Siting and the Telecommunications Act of 1969', *Wm. Mitchell L. Rev.*, 23 (1997), 1001.

[321] United States General Accounting Office, 'Telecommunications: Research and Regulatory Efforts on Mobile Phone Health Issues', p. 20.

entirety. Instead, the guidelines are based solely on risk considerations, even though they are based upon scientifically certain effects, rather than on precautionary considerations.[322] Neither risk nor risk trade-offs are dealt with in the guidelines and related documents. This is probably due to the fact that no risk appears to be present, provided that the guidelines are abided by. If cost–benefit or trade-off analyses were carried out when drafting the legislation or the advice mandated by the legislation, these appear informal and almost intuitive in nature.

At both state level and local level, the Telecommunications Act effectively pre-empts state and local governments from making other assessments than those suggested in the FCC guidelines. This means that they are prevented from making their own assessments of risk levels. Thus, independent assessments of the costs and benefits, as well as of trade-offs inherent in potential regulation, are also beyond their reach. Only if 'substantial evidence' is presented can they act more restrictively than has been prescribed by the federal government.[323]

7.3.8 Public opinion

As has been stated repeatedly above, the FCC guidelines in place in the USA are based solely on scientifically proven risks. When establishing these limits, no record shows that attention was paid to any potential opposition to mobile phone technology or radiation on the part of the public. In relation to this, it has been pointed out that public opposition to mobile phone technology in the USA is at a significantly lower level than, for example, in the UK, where public interest groups are well organised and vocal.[324] Since the Telecommunications Act pre-empts local and state governments from taking more precautionary decisions than is provided for in the FCC guidelines, the role of public opinion at local and state level is also severely limited. However, the possibilities that do exist for citizens to oppose the establishment of base stations in their vicinity will be dealt with in some detail here.

While the opposition to base station development in some other countries is more centrally organised and often based on health concerns,[325] public opposition in the US seems to be ad hoc, and based upon what has

[322] Questions and Answers about Biological Effects and Potential Hazards of Radiofrequency Electromagnetic Fields, pp. 9–16.

[323] Telecommunications Act, Sec. 704 (B) (II) (iii).

[324] Burgess, *Cellular Phones, Public Fears, And a Culture of Precaution*, pp. 178–80.

[325] *Ibid.*

been described as a 'nimby' (not in my back yard) mentality.[326] Often the main reason for opposition to the development of base stations is not the uncertain health risks associated with them, but, instead, the effect that the establishment of them may have on property values. Here, aesthetic concerns are likely to be more prominent than the potential risks,[327] and they are not regulated in any detail in the Telecommunications Act, although it appears that, in order for such concerns to constitute a basis for a denial of a zoning application, 'a local zoning board's aesthetic judgment must be grounded in the specifics of the case'.[328] Thus, general bans on the establishment of base stations seem to be impossible to maintain even when they are based on an aesthetic rationale.

In *Cellular Telephone Co.* v. *The Town Board of Oyster Bay* the issue of what weight can be given to public concerns with health risks from base stations when taking zoning decisions was dealt with in some detail.[329] When deciding on the placement of mobile phone towers, the Town Board of Oyster Bay in New York held public hearings in order for the citizens of the town to have input in the process. At the hearings, AT&T presented evidence as to the need for the development in order to safeguard uninterrupted mobile phone coverage in the area. It also presented scientific evidence with regard to the radio frequency emissions that the development would give rise to in order to mitigate the fears of adverse health effects.[330] Representatives of the citizens of Oyster Bay spoke out against the proposed developments and cited, primarily, the risks of adverse health effects, including cancer, as their reason for opposing the plans. In February 1997 the Town Board denied the applicant's request for zoning permission, and primarily cited 'safety reasons' as the basis for its decision. In addition to this, the Board also furnished a number of other reasons, including the failure to provide parking, the need for a full environmental impact assessment and aesthetic concerns.[331]

With regard to the health concerns, the Court held that previous case law had established that 'generalised health concerns of citizens are insufficient to rise to the level of substantial evidence',[332] and that the

[326] 'Not In My Backyard'. See Gustin, 'The Perpetual Growth and Controversy of the Cellular Superhighway', 1006–1007.

[327] See, for example, K. Hafner, First Come Cellphone Towers, then the Babel, *New York Times*, Section 1, Page 1, 1 May 2005.

[328] *Southwestern Bell Mobile Sys.* v. *Todd* (1st Cir. 2001) 244 F.3d 51, p. 61.

[329] *Cellular Telephone Co.* v. *The Town Board of Oyster Bay* (2nd Cir 1999) 166 F.3d 490.

[330] *Ibid.* [331] *Ibid.*

[332] *Ibid.* Here, the Court cited the judgment in *Illinois RSA No. 3, Inc.* v. *County of Peoria*, 963 F. Supp. 732 (C.D.Ill. 1997), p. 745

Telecommunications Act precludes 'health concerns from radio emissions' being considered.[333] However, the Court held that aesthetic considerations could, potentially, be grounds for denying a planning permission, even though, in this case, it found the referral in question to be unsupported by evidence. Furthermore, and interestingly, it also touched upon the issue of whether declining property values can function as grounds for denying a permit. In the end, it declined to give a clear ruling on the interesting issue of whether declining property values as a result of the fear of health risks from base stations would be permissible under the Telecommunications Act. AT&T held that allowing for such reasoning would only function as a proxy to usher in reasoning based upon health considerations which are not allowed. In the end, the Court found the reasoning of the citizens in the public hearings to be too general in nature to fulfil the substantive evidence requirement and did not rule on the issue as such.[334]

In summary, the Court in *AT&T Wireless Services* v. *The Town Board of Oyster Bay* did not exclude that public concerns could potentially be taken into account. But it did clearly state that such considerations would have to be supported by a significant body of evidence in order for it to 'rise to the level of "substantial"'.[335] It also appears that concerns other than health risks are more likely to be effective when challenging a zoning decision since the Telecommunications Act explicitly authorises the FCC to issue advice on this.[336] Thus, the situation in the USA is somewhat different to that in the UK, where public fears appear to be increasingly considered. Instead, it is closer to the Swedish situation, in which the centrally established guidelines are upheld consistently by courts even in the face of public resistance.

In a more recent case from the 2nd Circuit Court, it is highlighted that the applicant bears the burden of showing that the establishment of a base station is necessary, at least in the face of public opposition. In this case the applicant had failed to show why the proposed establishment of a base station would be necessary in order to maintain full mobile phone coverage in the area. The Court placed weight on the fact that other service

333 *Ibid.* Here, the Court cited the judgment in *AT&T Wireless PCS, Inc.* v. *City Council of Virginia Beach*, p. 431 note 6.

334 *Ibid.* 335 *Ibid.*

336 See, for example, *AT&T Wireless PCS, Inc.* v. *City Council of Virginia Beach*, p. 431 note 6, where the 4th Circuit Court had to work around the health concerns issue in order to find a way to allow for public opposition to influence the City Council.

providers managed to afford this without building on the contested site.[337] It is interesting to note that courts are seemingly expanding the possibilities of opposing the establishment of base stations, but only provided that the reason for the opposition is not one of health concerns.

7.3.9 Conclusion

The regulation of base stations in the USA is an example of the conflict that arises between the desire at federal level to push through the rapid expansion of mobile phone technology while maintaining the responsibility for planning decisions at state and local levels. This is a situation that is familiar from the previous case studies presented above. The US example, however, distinguishes itself in that a statutory instrument explicitly pre-empts local governments from taking a more precautionary stance than that presented by the FCC in the national guidelines. In Sweden the de facto situation is similar, due to the fact that the Radiation Protection Agency has issued, in itself non-binding, advice, which has consistently been upheld in the Swedish courts. In the UK it appears that local governments may have somewhat more of a possibility of maintaining more stringent regulation based on public opinion.

The Telecommunications Act 1996 was the instrument which gave the FCC the task of developing national emissions guidelines for mobile phones. This the FCC has done in two Bulletins, also issued in the late 1990s. The content of the guidelines are almost identical to the guidelines issued by ICNIRP, which are in place in Sweden and in the UK. Both the FCC and the ICNIRP guidelines are based solely on scientifically proven adverse effects in humans. The break-off point in both sets of guidelines is 4 SAR. With regard to what levels of frequency can be permitted to be emitted from base stations in order for this level to be avoided, the ICNIRP guidelines are marginally stricter in some frequency ranges due to the use of different physics models.

The FCC has repeatedly held that it feels uncomfortable with the task of issuing health and safety guidelines, since it is not primarily a health and safety agency, and thus lacks the necessary expertise. When studying the organisation of the Office of Engineering and Technology, which is responsible for drafting the guidelines within the FCC, the lack of expertise is striking. Owing to this fact, the FCC has relied on other government

[337] *Omnipoint Communications Inc.* v. *The City of White Plains* (2nd Cir. 2005) 430 F.3d 529, p. 536.

agencies, such as the FDA and branch organisations, when formulating its advice. Interestingly, it has not referred to the ICNIRP, the internationally recognised body which is active in the area. As in the UK, an independent report commissioned by a branch of the government (the US Congress and a ministry in the UK respectively) has evaluated the current state of the scientific research. Both reports highlighted the uncertainties regarding the radiation from mobile phones. However, the US report stopped short of recommending any precautionary measures, while the UK equivalent went to great lengths to advocate a precautionary approach based on the same set of data. In Sweden no such independent report has been commissioned. The advice present in Sweden is drafted solely by the Radiation Protection Agency.

With regard to the content of the guidelines in place at national level in the USA it can be concluded that they are similar to those recommended by the ICNIRP and in place in most EU countries, including Sweden and the UK, as discussed earlier. Like the ICNIRP guidelines, they cannot be considered to be based on the precautionary principle, since they do not aim to address the uncertainties in the scientific material on which they are based. Instead, they focus solely on protecting the public against the scientifically verified risks from radio-frequency emissions. It is true that they, like the ICNIRP values, recommend a 'safety margin' around the scientifically certain risks. However, the purpose of this does not appear to be to take the uncertain nature of the data in question into account, but, instead, to function as a 'fence' so that the risks do not accidentally materialise.

As already mentioned, federal statute pre-empts state and local governments from opposing the establishment of base stations based on health concerns. This approach has been upheld in the case law. However, some cases indicate that courts may permit a relatively far-reaching consideration of aesthetic concerns and the effects of the establishment of base stations on property values by local governments. This could lead to a situation in which an active public opposed to the establishment of base stations could successfully challenge mobile phone providers on such grounds. So far, the courts have avoided ruling on the sensitive issue of how dwindling property values based on the fear, rational or not, that radiation from base stations may be a valid basis on which to oppose developments. If this is, indeed, deemed to be a valid ground, it opens a significant 'back door' for local governments to take precautionary considerations into account when making zoning decisions. Courts have already stated that the burden of proof is on the mobile phone service providers to show the necessity of the establishment of base stations for continuous mobile phone coverage, at least when the establishment of the base stations is faced with public opposition. On the whole, the case law indicates that the courts are inclined to afford local governments at least some leeway to oppose the establishment of base stations. Here, the US situation is closer to how the UK courts have dealt with the issue than to the Swedish situation, in which it is excessively difficult for local governments to oppose the expansion of mobile phone services.

In conclusion, the regulation of base stations in the USA is not explicitly based on the precautionary principle. The exposure limits in the USA are comparable to the ICNIRP guidelines, which have been endorsed in the EU, the UK and Sweden. These limit values are not precautionary in nature, but are based only on scientifically certain effects. However, local governments may have the possibility of issuing more precautionary planning decisions based upon public opposition to base stations. This would be similar to the situation in the UK, but is in contrast to the Swedish situation, in which the national guidelines must be abided by and local governments appear more limited in their power to refuse the establishment of base stations.

8

Conclusions

Since its inclusion in the Treaty of Maastricht in 1992 and, in particular, since the food scares of the last years of the 1990s, the precautionary principle has been held up as a fundamental principle of environmental and health regulation in the EU. Frequent reference is made to the precautionary principle, both in relation to the internal market and in international trade relations. In a European Union which is increasingly concerned with the negative effects of human development and economic progress, a legal instrument which allows for early intervention is generally perceived as necessary and reasonably uncontroversial.

However, this book has shown that the use of the precautionary principle, by the EU institutions as well as by the Member States, is far from uncontroversial. Similar situations are not treated in a comparable fashion, which makes it increasingly difficult to foresee how and when precautionary measures will be applied, and it is often difficult for the affected parties to enjoy effective redress before the relevant courts. Furthermore, EU Member States increasingly justify derogation not only from harmonised rules but also from the basic free movement provisions by invoking the precautionary principle. This practice runs the risk of undermining the careful compromises and agreements which underlie the EU, and, as a consequence, a fragmentation of the internal market.

This book has analysed whether it is possible to discern a definition of the precautionary principle, as well as whether there are systematic and coherent elements behind the issuing of precautionary measures at international, European and national levels. In so doing, a number of factors have been identified as being instrumental for the outcome of precautionary decision-making. These factors have been studied separately in each chapter, and they include the definition of the precautionary principle in each particular legal system, and in the particular policy area studied, the scope of judicial review, the separation of risk assessment and risk management, the role of scientific evidence, the placement of the burden of proof, and the role of cost–benefit analysis, risk trade-off analysis, and the ranking of risks. Where it has appeared to have been influential for the application of the precautionary principle, the role of the public in decision-making has also been considered.

8.1 International law

8.1.1 International environmental law

In international law the precautionary principle is still a reasonably new and unclear concept. Even though reference to a 'precautionary principle' or a 'precautionary approach' has been included in a number of, primarily, sectoral international agreements in the area of the environment, the International Court of Justice (ICJ) and other international tribunals have proven reluctant to acknowledge that it has any legal effect except in specific agreements. Principally, it has been pointed out that, to date, it has been difficult to identify the consistent practice and *opinio juris* which are necessary for a concept to assume the status of international custom. De Sadeleer[1] has argued that the precautionary principle could be considered to be an 'emerging principle' of international law. However, the role and status of such a concept remains uncertain.

Even if the status of the precautionary principle in international law must be considered weak from a legal point of view, from a political viewpoint the inclusion of precautionary considerations in international agreements has been of great importance. Most prominently, the triple negative definition of the precautionary principle included in Principle 15 of the Rio Declaration is the definition of the principle most frequently referred to by most commentators and decision-makers.

8.1.2 WTO law

As in 'traditional' international law, both the Panels and the Appellate Body of the World Trade Organisation (WTO) have proven reluctant to afford the precautionary principle any weight as an independent legal concept. Instead, they have consistently held that the precautionary principle finds reflection in Article 5.7 of the SPS Agreement, which allows Members to issue protective measures without basing them on risk assessments in instances where the scientific data are insufficient to perform

[1] N. De Sadeleer, Environmental Principles: From Political Slogans to Legal Rules, (Oxford University Press, 2002).

a risk assessment. In all other instances, the requirement laid down in Article 5.1 of the SPS Agreement to base measures on risk assessments is absolute.

Despite this, when studying the Panel and Appellate Body reports in detail, it appears that the above does not exhaust the possibilities of taking precautionary measures under the SPS Agreement. Article 5.7 merely functions as a last resort when the scientific evidence is so incomplete that a meaningful risk assessment would be impossible to perform. The Appellate Body stated in *EC-Hormones I* that minority scientific opinions may be relied upon in order to justify precautionary measures, provided that they come from reputable sources. Thus, in instances of conflicting scientific evidence, a member does not need to follow the majority scientific opinion, but is free to rely on scientific evidence which suggests a more precautionary approach. The WTO Panels will scrutinise the scientific evidence put before them. In fact, the review performed by WTO Panels is the deepest carried out by any judicial body in the legal systems investigated in this book.

As demonstrated in, for example, *Japan-Apples*, the Panels will look for 'sufficient scientific evidence' in order to support a trade-restrictive measure issued by a member. In order to establish whether this level has been met, Panels will make a detailed study of the scientific evidence presented to them, and, in some cases, even procure their own scientific evaluation of it *ex officio*. A scientifically credible chain between a product or an activity and the argued risk in question needs to be established, even if it cannot be scientifically proven. This chain will be examined in detail by the WTO Panels. On appeal, the Appellate Body is restricted to considering formal issues and will not 'second-guess' the scientific assessment performed by the Panels.

When considering the above, the scope for WTO Members to act in a precautionary manner is greater than is generally assumed. It is indeed possible for Members to take measures based on inconclusive evidence and still abide by WTO rules. But it is not sufficient to base a measure upon mere suspicion or fear. A scientific link needs to be presented in order for the measures to stand. This link does not have to be scientifically proven, and it does not need to represent the majority scientific view. In *EC-Hormones I*, it was made clear that the scientific evidence needed to be specific to the risk from which the Member aimed to protect itself. So far, no Member has successfully managed to maintain precautionary measures, but it should also be pointed out that, so far, only a very limited amount of cases relating to the precautionary principle have been considered by the WTO Dispute Settlement Body.

8.2 EU law

The application of the precautionary principle in the EU is a highly complex issue. To begin with, the principle is codified in Article 191 TFEU. As a consequence, it is to function as a principle which underpins all environmental policy in the EU. In its case law the European Court of Justice (ECJ) has extended its scope to include issues of human health protection. There are abundant examples of how the precautionary principle influences EC legislation. Some highly publicised examples are the regulation of genetically modified organisms (GMOs) and the comprehensive package for the regulation of chemicals (REACH). In these instances, the precautionary principle has led to the adoption of legislative frameworks which rely on systems of preauthorisation and *ex ante* controls.

In a Communication of 2000 the Commission intended to clarify the application of the precautionary principle in the EU. It emphasised that scientific evidence and expertise should be the basis for regulation. Furthermore, it held that a weighing of costs and benefits should precede regulation, and that the burden of proof could be allowed to be reversed on a case-by-case basis. However, the non-binding document does not propose a comprehensive framework to be applied when precautionary decisions are considered, and it did not touch upon the issue of how to approach situations in which Member States wish to rely on the precautionary principle in order to avoid harmonising measures.

In its case law the EU courts have adopted a deferential approach to issues surrounded by scientific uncertainty. In most cases it affords seemingly unlimited discretion to the EU institutions to which the tasks of regulation or implementation have been given. In the voluminous and sometimes contradictory *Pfizer* judgment, the General Court (formerly Court of First Instance (CFI)) aimed to follow the vague framework established by the Commission Communication on the Precautionary Principle. It highlighted the importance of risk assessment and cost–benefit analysis, but stopped short of establishing any binding tests against which such requirements could be tested. Instead, it held that, in order to avoid regulation, a challenging party would have to 'prove conclusively' that the risk suggested by an EU body entrusted with a regulatory task was unlikely to materialise. By requiring the challenging party to show the non-existence of a risk, the CFI raised the level of proof to the impossible. It follows logically from the concept of scientific uncertainty that it is not possible to demonstrate conclusively the existence or non-existence of a risk. If it were possible to prove conclusively the existence of the risk, it would not be uncertain. In fact, it is widely recognised that it is impossible to show any fact conclusively since every scientific observation is surrounded by varying degrees of uncertainty. Therefore, the conclusion of the *Pfizer* judgment leads to a situation in which an EU institution can adopt any decision it sees fit in cases of scientific uncertainty, as long as it acts within the margins of its discretion. The EU courts, if they uphold the conclusions established in the judgment, refuse to make any qualitative requirements on decisions taken under scientific uncertainty. It should be mentioned that, in the more recent judgment in *Sweden* v. *Commission*, the CFI disregarded its limited review in order to overturn a Commission decision to authorise the pesticide paraquat based on a different interpretation of scientific evidence. This is a surprising and paradoxical approach considering the previous consistent case law of the ECJ and the General Court in relation to judicial review. So far, this is the only case in which the limited review doctrine has been disregarded by the European Courts. It remains to be seen if the case signals a deliberate and consistent shift in the policy of the ECJ and the General Court.

The limited review doctrine creates a situation in which the original burden of proof is indeed on the EU institutions to support precautionary measures. However, what needs to be shown is that the issue in question is surrounded by scientific uncertainty. Since, to date, the EU courts have not shown an inclination to examine scientific evidence in any detail, little is required in terms of evidence to demonstrate the presence of scientific uncertainty. Thus, the level of proof to be met is very low. When scientific uncertainty has been established, the courts do not investigate whether cost–benefit and risk trade-off analyses have been taken into account. Instead, the use of such instruments remains completely at the discretion of the relevant institutions. There is no specific requirement for the decisions taken to be efficient or effective.

The worrying conclusion of the EU case law, so far, is that effective redress is virtually impossible for individuals or undertakings affected by precautionary decisions emanating from the European institutions. In order to remedy this, the ECJ needs to step out of the shadow of its limited review and establish a system against which precautionary measures can be tested. If it fails to do so, the application of the precautionary principle risks will remain arbitrary and impossible to challenge. In order to establish a more foreseeable, effective and efficient system of precautionary regulation, some lessons can be learned from the US system. This issue is returned to below.

A further problem with the application of the precautionary principle in the EU is the issue of how the Member States are allowed to make use of the principle as a 'shield' against harmonisation measures. In harmonised areas Member States are increasingly making use of the 'safeguard clauses' built into harmonising measures, and in non-harmonised areas Member States often successfully base their justifications for derogation from the basic free movement provision in Article 34 TFEU (formerly Article 28 EC) on a precautionary interpretation of facts under Article 36 TFEU (formerly Article 30 EC) or the mandatory requirements established in the case law of the ECJ. In these instances the ECJ has also proven deferential when it comes to challenges of Member State practice by the Commission. In principle, the burden of proof is on the Member States when they wish to derogate from treaty provisions or harmonising measures. However, since the ECJ rests when scientific uncertainty has been established, the level of proof required by the Member States is very low. This leads to a situation in which it is relatively easy for Member States to maintain or introduce national rules which hinder intra-EU trade. As stated above, a continuation of this practice will inevitably lead to a fragmentation of the internal market.

The precautionary principle clearly enjoys the status of a legal principle in the EU. It has been interpreted by the EU courts to allow for regulation under conditions of scientific uncertainty. An in-depth analysis of the relevant case law indicates that EU institutions or Member States are not obliged to use the precautionary principle under scientific uncertainty. Rather it is of an enabling character, that is, once scientific uncertainty is present, the precautionary principle may be invoked.[2] No qualitative requirements are made on precautionary decisions. Instead, it appears to establish that, in the absence of scientific certainty, any decision may be justified.

8.3 The national legal systems

The three case studies on the use of the precautionary principle by Sweden, the UK and the USA in the regulation of pesticides and base stations for

[2] However, in the recent *Sweden* v. *Commission*, the CFI overturned a decision where it considered the Commission not to have acted precautionarily enough when granting an authorisation of the pesticide paraquat. It remains to be seen if this judgment is relevant outside of Directive 91/414/EC on plant protection products, as well as if the approach will be upheld in the future case law of the ECJ.

mobile telephones illustrate that most states are unlikely to use the principle as a rational instrument for reducing the overall risk level of their society. Instead, precautionary considerations are sometimes used to justify predetermined policy choices which do not always aim toward overall risk reduction, but appear to be motivated by macro-economic interests and public and media pressure.

The states which have been investigated were chosen to give as broad as possible a scope of how the precautionary principle is being applied. Sweden and the UK are interesting since they are both Member States of the EU which have a different approach to the precautionary principle, and thus serve to illustrate the conflicts between different approaches within the EU. The USA was chosen since it is not a member of the EU, and is often in open opposition to the EU approach to the precautionary principle, as well as the use of it as a legal principle. However, when one goes beyond the rhetoric, it is clear that the USA also allows for precautionary decision-making, and in some instances goes further than the EU and its Member States.

The policy areas investigated were chosen because both pesticides and base stations are examples of involuntary, uncertain risks which are subject to regulation in one way or another. Of particular interest is the fact that precautionary considerations, or the precautionary principle, have been used in fundamentally different ways in relation to similar risks in different policy areas within a single state, as well as in relation to the same policy areas in different states.

With regard to the authorisation of pesticides in the EU, the area is harmonised within the framework of Directive 91/414/EC. The Directive does not make any reference to the precautionary principle or precautionary considerations in general. Instead, a pesticide should be allowed on the market if it is considered that its 'normal use' does not produce 'harmful' effects on human health or 'unacceptable influence' on the environment. The approach to pesticides by individual Member States is of pivotal importance in EU pesticide regulation since the risk assessment as well as the authorisation of pesticides is Member State driven through the use of comitology procedures under Decision 1999/468/EC. Thus, although the area is harmonised at EU level, Member States are highly influential when deciding on whether to allow a specific pesticide or not. If Member States apply the precautionary principle differently in their internal policies towards pesticides, this is likely to create conflicts when evaluating pesticides at EU level.

There is no comprehensive EU policy in place in relation to base stations for mobile telephony, but Council Recommendation 1999/519/EC suggested that Member States apply the internationally agreed-upon ICNIRP standard as an appropriate level of protection from low-frequency radiation. However, in practice the area is not harmonised and Member States are free to apply whichever policies they see fit in relation to radiation from mobile phones and base stations.

8.3.1 Sweden

In Sweden the precautionary principle is codified in the Swedish Environmental Code and is supposed to underpin all government regulation in the areas of environment and health. It is not more closely defined, but can best be described as a broad principle which establishes a wide duty of care, rather than only aiming to allow for regulation in the face of uncertainty. Connected to the precautionary principle is the substitution principle which entails an individual responsibility to opt always for the use of a less hazardous product when one is available. Due to the strict and mandatory nature of the precautionary principle included in the Swedish Environment Code, on paper this is one of the most far-reaching versions of the principle in place in any legal system.

When considering the reasoning in the relevant preparatory works, judicial review by courts in administrative cases in Sweden appears to be deeper than in the EU and in many other European countries. Courts are not formally hindered from reviewing the facts of a case under review and no 'reasonableness' or 'manifest error' threshold needs to be met in order for the courts to overturn an administrative decision. However, in practice courts have proven reluctant to review the substance of decisions taken by agencies and the government using their discretion. However, it is possible that the system of environmental courts recently introduced in Sweden may increase the depth of review. The courts incorporate scientific and technical expertise in the tribunals, and there is some indication that they intend, at least in some cases, to make full use of the possibilities of judicial review.

Sweden does not conduct a clear separation of risk assessment and risk management. In cases in which agencies have had decision-making powers delegated to them, the scientific analyses and the management decisions are often taken by the same people. 'Common sense' plays a large role in the decision-making. The applicant must, in most cases, carry out the scientific assessment in order to have a product authorised.

When issuing a precautionary measure, the initial burden of proof rests with the regulator. The level of proof required is low, and some support exists for the notion that a decision not to allow a product must be based on scientific evidence and may not be theoretical. Once this low level of proof has been met by the regulator, it shifts on to the party facing regulation and is subsequently high. Commentators have held that what is required is to disprove the chain of evidence provided by the regulator, rather than proving that the activity or the product in question is safe. This seems to lead to a situation in which the level of proof required is slightly lower than that required by the ECJ.

According to the Swedish Environmental Code, all measures issued must be 'reasonable'. The article seems to encompass what the courts previously referred to as the requirement that measures must be 'environmentally justifiable', as well as meet some form of proportionality requirement. The term 'environmentally justifiable' functioned as a form of cost–benefit requirement which aimed to hinder measures to improve the environment for which the benefits were so small that they could not justify the high costs associated with them. It is important to note that this was not developed as a traditional cost-efficiency requirement, but was, instead, used to prevent cases of extreme differences between costs and benefits. The reasonableness requirement too is not based on cost-efficiency reasoning. Instead, it appears to aim to focus on the regulated party and its economic ability to sustain regulation. Thus, economically successful undertakings have been deemed to be able to accept stricter regulation than weaker ones.

Swedish policy on environment and health regulation is laid down in sixteen broad environmental quality objectives, which have been adopted by the Swedish Parliament. These goals are to be reached within one generation. The environmental quality objectives are, in turn, divided into partial goals, which have been given deadlines for completion. This system provides for a broad, but relatively detailed, system of prioritising risks. The prioritisation is not based on rational concepts such as risk assessment and cost–benefit analysis, but appears to be political in nature.

In the area of pesticide regulation, the version of the precautionary principle outlined in the Environmental Code is used to its full extent. Swedish chemicals policy is highly restrictive, and an environmental policy goal established by the Swedish parliament holds that all man-made substances which may have detrimental effects should be phased out of the natural environment by 2020. This is a remarkably absolute version of the principle.

The Swedish Chemicals Agency (KemI) is the agency entrusted with issuing authorisations and detailed rules regarding pesticides. No formal separation of risk assessment and risk management is in effect; instead, decisions are often taken in informal deliberations between the civil servants in the agency.

In its assessments of pesticides KemI considers the risk of the intrinsic properties of each substance, and not the risk posed by the substance in normal use. This leads to a situation in which a substance with which a risk is associated is not deemed fit for use even if it is abundantly clear that the substance does not pose a risk under normal conditions of use. The strict application of the precautionary principle has caused Sweden significant problems in complying with EU risk assessments, which tend to be based on the normal use of substances. As an example, in one case the plaintiff argued that it would be unreasonable to have to show that the substance in question (mancozeb) was safe for use at high doses, since the normal use of the product would only concern significantly lower doses. KemI and the Swedish courts were unmoved by these pleas. They reasoned that it was not sufficient to show that a pesticide was without risk when used under normal circumstances. Instead, each of the substances which are contained in the pesticide must be shown to be risk-free. Such reasoning ultimately leads to a situation where only zero-risk products are tolerated.

Furthermore, the substitution principle, in practice, constitutes a particular manifestation of the precautionary principle, since it means that if 'safer' pesticides of equal effect as the one in question exist on the market, the pesticide in question should not be allowed. Thus, it establishes comparative assessments, instead of absolute requirements with regard to what level of risk should be tolerated. Such moving targets are difficult to comply with, and may lead to situations of considerable legal uncertainty.

With regard to cost–benefit balancing in Swedish law; it appears that, when evaluating substances, KemI only considers the risk and not benefits associated with the products in question. This enables it to apply a very restrictive policy towards pesticides. If the benefits of a product are not considered, it is virtually impossible to justify anything but a zero-risk strategy. Thus, the proportionality requirement in Swedish environmental regulation is reserved for situations in which it is not 'environmentally motivated' to prohibit a certain product or activity. The requirements for this to be the case are very high.

The area concerning base stations in Sweden is regulated in a number of different statutes, comprising the Environmental Code, the Radiation

Protection Act and the Planning and Building Act. As was stated above, all areas relating to environmental and health protection in Sweden are covered by the precautionary principle included in the Environmental Code. According to the Radiation Protection Act, the Radiation Safety Authority (formerly the Radiation Protection Agency (SSI)) is responsible for drawing up detailed rules for protection from radiation, and these rules should incorporate precautionary considerations. Finally, the siting of base stations is decided by local authorities according to the Planning and Building Act.

The Radiation Protection Agency has issued a General Advice which outlines the Swedish position on protection from radiation from base stations. The General Advice is based on the EU Council Recommendation of 1999, which, in turn, incorporates the Guidelines of the International Commission for Non-Ionising Radiation Protection (ICNIRP). The ICNIRP guidelines are not precautionary in nature, but are based upon the scientifically verifiable thermal effects of radiation. They do not consider scientifically uncertain issues such as biological and non-thermal effects of radiation. Thus, the Swedish approach to radiation from base stations cannot be said to be precautionary in nature.

In a communication to local governments, a Swedish agency has unequivocally stated that the precautionary principle should not be applied with regard to base stations at local level. Thus, local government is prevented from assuming a more precautionary stance than that decided at national level. This approach has been upheld by the courts in the scarce case law existing in the area. To date, the courts have not been inclined to investigate the scientific evidence underlying challenges of the system currently in place. Instead, they have afforded full discretion to the Swedish Radiation Safety Authority in deciding on the appropriate level of protection. Cost–benefit considerations do not play a formal role in the regulation of base stations.

Thus, the non-precautionary approach assumed in relation to base stations appears to be the inverse of the precautionary approach assumed in relation to pesticides. The reasons for this may include a long history in Sweden of strict regulation of chemicals, and a possibly exaggerated awareness of the risks of chemicals in the Swedish population. However, it is impossible to overlook the economic implications of the regulation of the two respective policy areas. Sweden has a very small chemical industry and a small agricultural sector which is not highly dependent on pesticides. On the other hand, the Swedish telecoms sector is at the core of Swedish economic growth. Full mobile phone coverage is a highly publicised goal of the Swedish government and Ericsson is a world leader in mobile phone technology and a symbol of Swedish industry in the twenty-first century. Consequently, it is understandable that the regulation of pesticides is likely to be significantly easier and more acceptable to the public and industry than the regulation of base stations. However, the result of the above is that national economic interests lead to a different approach in relation to comparable risks. This reliance on 'the precautionary principle' to justify the regulation of one type of risk but to ignore a comparable one raises several concerns. Among the most serious is the result that the precautionary principle can be used as a general clause to justify any type of regulation that the government sees fit to enact. Connected to this are concerns relating to the effective redress of affected parties, as well as the lack of efficiency in the government's risk agenda.

8.3.2 The UK

In the UK the precautionary principle is not codified in any general statute as it is in Sweden. It is only recently that the UK government has begun to give importance to the precautionary principle in policy documents. However, these documents do little to clarify the content and scope of the principle.

The strictly limited review afforded to UK courts in administrative cases means that little guidance can be found in the case law. Some early cases which dealt with precautionary considerations considered that, as far as the principle was applicable in the UK, it would be restricted to interpreting measures taken to comply with EU law. This stance was due to the fact that Article 191 TFEU (formerly Article 174 EC) was the only legal instrument in which UK courts could find the principle mentioned. However, in more recent cases the UK government has invoked the precautionary principle, and, when challenged, UK courts have invariably found this practice to be within the government's discretion.

In policy documents it has been held that, even if the application of the precautionary principle is within the discretion of the government, decisions based on it should be taken following the performance of cost–benefit and risk trade-off analyses. Some support for this can also be found in the case law. Advanced models are suggested for making such comparisons. These include 'willingness to pay' and 'willingness to accept' calculations, as well as 'quality adjusted life years'. Whether, and if so, how, the UK government has made use of these instruments in relation to the precautionary principle is unclear. The documents in which they are suggested are not binding and no other rules indicate that the performance of a cost–benefit analysis would be a formal requirement. The courts have not tested the application of the principle against any formal requirements. The same is true with regard to public involvement in decision-making: its role is emphasised in policy documents, but is not made legally binding.

Following this approach, no general statute obliges the UK government, or the agencies entrusted with regulatory tasks, to act in a precautionary manner in cases of uncertainty in the area of pesticide regulation. However, there is nothing to prevent them from taking precautionary measures when they see fit. In practice, it appears that different decisions may be taken regarding different pesticides with comparable risks. For example, the pesticide lindane, whose carcinogenic properties are widely accepted as being highly uncertain, was subject to precautionary action, whereas paraquat, with its well-documented toxic properties, was declared safe for use.

The agency responsible for pesticide regulation in the UK is the Pesticide Safety Directorate (PSD), which, in practice, has strong ties with the Department for Environment, Food and Rural Affairs (DEFRA). Scientific advice is usually given by the Advisory Committee on Pesticides, but is not binding on the PSD or the DEFRA. Interestingly, the case law has shown that the PSD can principally control the scope of scientific material to be assessed by the advisory committee. As a result, it cannot be established that a strict institutional separation of risk management and risk assessment is upheld in practice.

It was held above that the UK government has emphasised the usefulness and importance of cost–benefit and risk trade-off analyses, as well as the involvement of the general public in decision-making. However, if these issues are indeed considered, this is done in an informal manner. In addition, in the area of pesticide regulation the UK government has avoided laying down formal requirements against which its decisions can be checked.

With regard to base stations, the general framework is broadly similar to that in place in Sweden. At national level, the Health Protection Agency (HPA) functions as an independent agency, which does not answer directly to the Ministry in the way that the PSD does in the field of pesticide regulation. The HPA issues recommendations with regard to protection from non-ionising radiation. It is assisted in this task by the Advisory Group on Non-Ionising Radiation. However, the most influential document in the area to date is the 'Stewart Report', which was drafted by the Expert Group on Mobile Phones. In the Stewart Report independent experts commissioned by the government recommended that far-reaching precautionary measures be issued in relation to base stations. Even though the UK government did not fully endorse these recommendations, government policy has proven more open to discussions on the possible negative health effects of radiation from base stations than the Swedish categorical refusal to categorise the issue as surrounded by uncertainty at all. The recommendations of the ICNIRP are currently the standard applied in relation to protection from non-ionising radiation in the UK, as they are in Sweden.

Furthermore, and also as in Sweden, concrete decisions regarding the actual siting of base stations are taken at local level by local planning boards. However, and unlike the situation in Sweden, it appears that local governments have the possibility of making use of some degree of independence in relation to the level of protection chosen at local level. Where Swedish planning boards are explicitly bound to respect the national radiation guidelines, the meagre and largely contradictory case law existing in this area in the UK suggests that local planning boards may issue more precautionary measures than have been established nationally. It is still unclear exactly what UK local planning boards need to show in order to issue such precautionary measures, but it appears that public opinion plays an important role in establishing whether base stations should be restricted to local level. In some cases it has been emphasised that even public opinion which 'has no scientific or logical reason' can function as material consideration for a local planning board when taking a decision on whether to allow a base station. However, no requirements are made on the formal weighing of the costs and benefits or the risk trade-off analyses, and it is unclear what role such considerations play in practice.

8.3.3 The USA

The USA is generally viewed as being a staunch opponent of the precautionary principle. The principle lacks a place in US statutes, and, in recent years, increased demands on scientific evidence and the importance of cost–benefit analysis have given the impression that precaution is not part of the US regulatory agenda. Some commentators have argued that precaution is not a part of the US mindset, and the *New York Times* asked itself the question whether precaution, after all, is not just something for Europeans.

However, in reality precautionary decision-making has an interesting and developed history in US regulation. Some commentators have even stated that the US Congress has a clear 'precautionary preference' when issuing regulation in relation to health and environmental protection. In the 1970s groundbreaking pieces of environmental legislation such as the Clean Air Act were broadly interpreted by activist US courts to entail strongly precautionary elements.

One highly influential aspect of the US system in the field of risk regulation is the comparably extensive review of regulatory measures performed by US courts. In the 1970s activist US courts went to some lengths to carve out a space for precautionary considerations in US regulation. This never amounted to an effective endorsement of a generally applicable precautionary principle, but judgments such as *Ethyl Corp.* and *Reserve Mining* allowed US agencies to act in a precautionary manner without explicit basis in scientific evidence. Even though the 'hard look' doctrine developed by the DC Circuit Court in the 1970s was curtailed by the change in the regulatory agenda brought about by the Reagan administration in the early 1980s, it still continues to influence the scope of judicial review.

For precautionary measures to be enacted, a threshold finding of 'significant risk' must be present. Exactly what is required for this to be the case is unclear, but it appears that mere scientific uncertainty is not sufficient to justify regulation. This means that US courts must analyse the scientific evidence brought before them, in order to establish whether a risk actually exists. Here, the US courts differ in their approach when compared to the other jurisdictions described above, in which courts generally allow for a near total deference on the side of the regulator, and the mere existence of scientific uncertainty is generally sufficient to warrant regulation. However, this does not mean that US courts require scientific certainty for regulation to be legal. Instead, it appears that the approach is more akin to that of the interpretation of the SPS Agreement advanced in WTO dispute settlement. Courts will actively investigate the risk assessment performed by the relevant agencies to discern whether it has been carried out in an objective and rational fashion, or whether other formal errors have been committed. This leads to a situation in which it is at least plausible that an affected party could substantively challenge a regulatory measure taken under conditions of uncertainty.

US regulatory measures should be preceded by Regulatory Impact Assessments (RIAs). RIAs include the performance of risk assessment and cost–benefit analysis, but also a statement relating to the need for the proposed action by the agency, as well as an examination of alternative approaches. Risk trade-offs should be considered, and there is a strong impetus for ranking risks in order of priority.

The executive branch of the US government retains important powers of control through the Office of Management and Budget (OMB), which can ask agencies to review assessments with which it does not agree. The powers of the OMB have been increased further with the introduction of the highly criticised Data Quality Act, which allows for the possibilities for citizens to demand the 'correction' of data held by government which is not scientifically 'sound'.

Pesticides regulation is one area of US legislation in which the influence of, if not a coherent precautionary principle, then at least precautionary considerations, in the broad sense, is particularly clear. For decades the Delaney Clause meant that a zero-risk approach was taken towards the residues of carcinogenic properties in food. In practice, this meant that a remarkably precautionary approach was introduced in relation to pesticides. Interestingly enough, this approach focused solely on the risks of cancer; other risks, such as toxicity, were treated in a more lenient fashion.

Since the Delaney Clause was revoked in the 1990s, the US approach to pesticides regulation has become more nuanced. Cancer risks are no longer explicitly treated in a more restrictive manner than other risks, such as toxicity, for example.

The Federal Insecticide, Fungicide and Rodenticide Act (FIFRA) functions as the legislative framework for pesticide regulation in the USA. It is, in many ways, similar to Directive 91/414 in place in the EU, in that it places the burden of information squarely on the applicant, and makes explicit use of scientific bodies to examine and advise the bureaucratic body empowered to take decisions. Courts have clarified that the EPA is not bound by the advice issued by scientific bodies.

Under the FIFRA pesticides with 'unreasonable risks' should not be permitted. This means that several competing interests must be considered together when taking a decision on authorising or cancelling authorisations under the FIFRA. This includes risks, but also the economic impact of regulation. It has been pointed out that this is a necessary starting point, because, if positive aspects of a product were not considered, there would be nothing to justify anything but a zero-risk approach. It is interesting, here, to compare this with the situation in Sweden, where the precautionary regulation of pesticides is very far-reaching. This is made possible by the fact that, as Swedish officials have stated, no economic considerations are taken into account when formulating decisions in relation to pesticides.

The FIFRA requires a more nuanced decision-making process, because it is a cost–benefit balance that will answer the question of whether a product poses an 'unreasonable' risk or not. Furthermore, courts have indicated that they find it within their remit to 'second-guess' the balancing performed by the EPA, at least in cases in which they find that cost–benefit analysis is significantly flawed. In *Love* v. *Thomas* the court investigated the details of the cost–benefit analysis carried out by the Environmental Protection Agency (EPA) before it decided to overturn it. Some cases have indicated that the cost–benefit analysis may be skewed in favour of health protection, at least in cases in which cancer is the suspected risk. In *EDF* v. *Ruckelshaus* the court stated that the EPA has a 'heavy burden' to explain why it allows a pesticide if it has been shown to have carcinogenic properties in animal testing. Finally, in relation to cost–benefit balancing, it should be pointed out that, even if the substitution principle is not formally in effect in the USA, the issue of whether a comparable but less risky pesticide is already on the market should be allowed to influence the outcome of the balancing. Thus, issues comparable to the substitution principle can be considered under the FIFRA. The possibilities for public participation in the regulation of pesticides are limited in the USA, just as in the cases discussed above.

In the USA, as in the other states investigated here, particularly Sweden, a conflict exists between the wish to expand mobile phone coverage as quickly as possible, while, at the same time, respecting local planning rules and the concerns of individuals. The US example, however, distinguishes itself, in that a statutory instrument explicitly pre-empts local government from taking a more precautionary stance than that presented by the relevant government body, the Federal Communications Commission (FCC) in the national guidelines. In Sweden the situation is similar, due to the fact that the Radiation Protection Agency has issued advice, albeit in itself non-binding advice, which has been upheld consistently in the Swedish courts. In the UK it appears that local governments may have a somewhat greater possibility of maintaining more stringent regulation based on public opinion.

The Telecommunications Act 1996 is the instrument which gave the FCC the task of developing national emissions guidelines for mobile phones. This the FCC has done in two Bulletins, also issued in the late 1990s. The content of the guidelines are almost identical to the guidelines issued by ICNIRP, which are in place in Sweden and the UK. Both the

FCC and the ICNIRP guidelines are based solely on scientifically proven adverse effects in humans.

Interestingly, the FCC has repeatedly stated that it feels uncomfortable with the task of issuing health and safety guidance, since it lacks the appropriate expertise to do so. In practice, it relies heavily on other government agencies when formulating advice in this regard. As in the UK, an independent report commissioned by a branch of the government (the US Congress and a ministry in the UK respectively) has evaluated the current state of the scientific research in the area of non-ionising radiation. Both reports highlighted the uncertainties with regard to radiation from mobile phones. However, the US report stopped short of recommending any precautionary measures, whereas the UK equivalent went to great lengths to advocate a precautionary approach based on the same set of data. In Sweden no such independent report has been commissioned.

Even if local governments are prevented by statute from assuming a more precautionary approach than that suggested by the FCC, some cases indicate that courts may permit a relatively far-reaching consideration of aesthetic concerns and the effects of the establishment of base stations on property values by local governments. This could lead to a situation in which an active public opposed to the establishment of base stations could successfully challenge mobile phone providers on such grounds. So far, courts have avoided ruling on whether the sensitive issue of how dwindling property values based on the fear, be it rational or not, of radiation from base stations may be a valid basis upon which to oppose developments. If this is, indeed, deemed to be a valid ground, it opens a significant 'back door' for local governments to take precautionary considerations into account when making zoning decisions. Courts have already stated that the burden of proof is on mobile phone service providers to show the necessity of the establishment of base stations for continuous mobile phone coverage, at least when the establishment of the base station is faced with public opposition.

8.4 The different types of the precautionary principle

In summary, there are at least three different ways in which the precautionary principle is currently applied. First, the principle may be viewed as a fundamental principle which obliges governments to act in a precautionary manner. This is the version which can be found in the Swedish Environmental Code. It is prima facie strict in application and places a low level of proof on the party wishing to invoke the principle. Generally, this version of the principle presupposes a deferential approach taken by the relevant courts, and the level of proof needed to be met to challenge precautionary measures is very high.

Second, the precautionary principle can be viewed as enabling governments to act in a precautionary manner. In the UK system, for example, it is for the government to use its virtually unlimited discretion to issue precautionary measures where it sees fit. The courts in such systems are extremely deferential and a successful challenge of a precautionary measure is very difficult. The application of the precautionary principle in the EU to date has, in the main, resembled this version.

Third, the USA and the WTO systems are examples of systems in which the precautionary principle can only be applied after procedural steps have been taken. These include the establishment of a risk, which does not need to be certain, but must at least be feasible, and the balancing of the costs and benefits of regulation. Courts acting within this version will take a 'hard look' at the decisions taken to establish that the decision-maker has, indeed, taken the relevant information into account when issuing a regulation. If not, the decision can be overturned.

As has been shown in this book, none of the systems described are upheld consistently in practice. In the end, the precautionary principle appears to be applied differently in different situations. In systems in which a procedural framework is lacking, it appears that the precautionary principle may, in some cases, be used to further socio-economic goals, rather than to aim at rational overall risk reduction. Precautionary measures in the EU, Sweden and the UK are not issued on the basis of rational criteria of overall risk prevention and efficiency. There is no obligation to compare similar risks with each other, or to enact effective and efficient regulation. Such systems are open to a number of criticisms. First, by not subjecting the issuing of precautionary measures to rational criteria such as cost–benefit and risk trade-off analyses, there is a risk that public money may be wasted and citizens may not be protected adequately against real risks. Second, decisions taken under conditions of uncertainty are often extremely difficult to challenge effectively in practice before the courts. Thus, the affected parties do not have access to effective redress. The unfortunate result of the above is that the precautionary principle as applied in the EU and the Member States investigated here can be used as a vessel that can be used to usher through virtually any regulation deemed necessary on subjective grounds.

8.5 Which way forward?

In order to remedy the unsatisfactory situation outlined above, the EU and its Member States should glance across the Atlantic. Even though the US system of risk regulation is far from perfect, it offers a more rational alternative to the lack of a system currently in place in the EU and some of its Member States. By subjecting the issuing of precautionary measures to strict and objective Regulatory Impact Assessments, a balancing of the pros and cons of regulation can be carried out, and more effective and efficient risk-reducing choices can be made.

Choosing the appropriate level of protection is an eminently political choice, which is, and should be, influenced by societal values and ethics. However, since public funds are limited even when it comes to reducing risks to the public and the environment, citizens should expect and require that government policies be aimed at purchasing the highest level of protection for the money available.

This can be done by the EU and the Member State governments issuing formal, legally binding guidelines on how the precautionary principle should be used. Such guidelines should focus on the importance of independent scientific advice showing at least a threshold finding of a plausible link between an activity and a risk. Such a link should not have to be scientifically certain, but needs to be assessed in each individual case. Thus, the initial burden of proof in order to issue precautionary measures necessarily rests with the party wishing to invoke restrictive measures. As has been demonstrated above, this is not different from the allocation of the burden of proof currently in place in all systems investigated here. What is different is that a regulating party should have to show that a risk is plausible by taking the relevant scientific advice into account before the activity is regulated. Consequently, a party wishing to challenge the precautionary measures should not have to 'show conclusively' that the risk will not materialise. Such a high level of proof is impossible for the regulated party to meet. Instead, it should be sufficient to show that the risk is not plausible in order for courts to overturn the decision.

Furthermore, before issuing a precautionary measure, the regulating party should be formally bound to consider the costs and benefits associated with regulation. If such a balance is not carried out, it is impossible for the regulator to know if the intended regulation will indeed be efficient. Connected to this is the importance of analysing the trade-offs involved in a regulatory measure. By regulating one product, the market may, indeed, be opened up for other products which may pose even greater risks than the regulated product. If this is the case, the proposed regulation is meaningless at best, and may even increase the overall level of risk in a society. Following the outcome of cost–benefit analyses should not be binding for the regulator, but it would be reasonable to require the regulator to state its reasons for choosing to disregard rational criteria when issuing regulation.

Once a choice has been made on which risks to regulate, they should be ranked in order of importance. Such a ranking should include the respective costs and benefits brought about by regulation as well by lack of regulation. The ranking of risks allows the regulator to compare regulatory choices, but it is also a powerful communication tool. By summarising the information available and the regulatory choices decided on, measures can be presented to citizens in a coherent and transparent fashion. This is likely to provide a good incentive for public debate and increased understanding of regulatory choices in the general public. In addition, it would be reasonable for the regulator to provide an explanation if it chooses not to follow the established ranking when targeting risks.

Finally, by creating legally binding, formal requirements such as the ones briefly outlined above, it becomes possible for parties that are affected by precautionary regulation to foresee the decisions of governments. This creates a necessary minimum of legal certainty in the application of the law in this field. Furthermore, it becomes effectively possible for affected parties to challenge precautionary measures which are not taken in conformity with the requirements.

Good decision-making is decision-making that is based on as complete an understanding of both the facts and the consequences of the proposed decisions as is possible. Under conditions of scientific uncertainty, it is clear that a complete understanding of the scientific facts is not possible. However, the scientific facts are only one aspect of the decision. The issues of the efficiency and the effectiveness of the proposed regulation based on different scenarios must still be performed, and are of comparable importance to the science underlying the decision.

In summary, the key to the precautionary principle is to know '[w]hen do we have enough information to act as if something is a cause'.[3] It is here submitted that only when all the available information is collected and put on the table can we begin to understand if a product or an activity is a cause of a risk or not. In relation to the precautionary principle,

[3] D. Kriebel, 'How Much Evidence is Enough? Conventions of Causal Inference', *Law & Contemp. Probs*, 72.(2009), 121.

fundamental information includes scientific evidence, knowledge of costs and benefits of the regulation in question, and knowledge of the risk trade-offs involved in the regulation in question. Such information should always be gathered and should also be made available to the public in order for the best decisions to be formulated, and to enable citizens to be informed and be given the opportunity to make their opinions heard with regard to the regulatory choices made.

Furthermore, to act 'precautionarily' cannot mean that one specific risk is targeted while other more hazardous options are left unregulated. Instead, in order to act in a truly precautionary manner, risks need to be considered together, as a whole. A 'truly precautionary' decision is one that decreases the overall level of risk in society, not one that targets an individual risk without considering other risks. Thus, in order to act in a 'truly precautionary' manner, it is not sufficient to focus on one single risk, but the full spectrum of risks needs to be considered at every instance when regulation is considered. Risks should be targeted in such a manner that the worst risks are tackled first in the most economically efficient manner.

Finally, and perhaps most importantly, in a world of globalised markets, individual states are continually going to be required to justify trade restrictive measures. If individual states are allowed to apply the precautionary principle to justify economic policy choices, this will undermine the system of European and international trade which has become fundamental both to our continuing economic growth and to the world as we know it. Therefore, not only national governments, but also the EU institutions, and the ECJ in particular, as well as the WTO, need to be able to check national regulations against the same formal rules, as outlined above. Failure by Member States to show why regulation is necessary and proportional, as well as effective and efficient, should lead to the measures being revoked.

The precautionary principle does, indeed, have its place in the EU as well as at international level. But it must be a principle which is applied in a coherent, predictable, efficient and effective fashion.

It has been shown above that precautionary measures are issued by responsible governments in certain cases regardless of whether they subscribe to a 'precautionary principle' or not. Further codification of the precautionary principle, in the EU as well as internationally and nationally, should focus on the procedural application of the principle as outlined above, rather than on its definition.

BIBLIOGRAPHY

Ackerman, F. and L. Heinzerling, *Priceless: On Knowing the Price of Everything and the Value of Nothing*, The New Press, 2004.

Advisory Committee on Pesticides, *A Guide to Pesticide Regulation in the UK and the Role of the Advisory Committee on Pesticides*, ACP 19 (311/2005). Available at: www.pesticides.gov.uk/acp.asp (accessed 1 March 2010).

Minutes of the 290th Meeting of the Advisory Committee on Pesticides (ACP) on 4 March 2002. Available at: www.pesticides.gov.uk/acp.asp?id=269 (accessed 1 March 2010).

Applegate, J. S., 'The Perils of Unreasonable Risk: Information, Regulatory Policy and Toxic Substances Control', *Col L. Rev.*, 91:2 (1991).

'Worst Things First: Risk, Information, and Regulatory Structure in Toxic Substances Control', *Yale J. on Reg.*, 9 (1992).

'A Beginning and not an End in Itself: The Role of Risk Assessment in Environmental Decision-Making', *U. Cin. L. Rev.*, 63 (1994/1995).

'Beyond the Usual Suspects: The Use of Citizens Advisory Boards in Environmental Decision-making', *Ind. L.J.*, 73 (1998).

'The Precautionary Preference: An American Perspective on the Precautionary Principle', *Human and Ecological Risk Assessment*, 6:3 (2000).

Anderson, E., *Value in Ethics and Economics*, Harvard University Press, 1993.

Anderson, F., M. A. Chirba-Martin, E. D. Elliott, C. Farina, E. Gellhorn, J. D. Graham, C. Boyden Gray, J. Holmstead, R. M. Levin, L. Noah, K. Rhyne and J. B. Wiener, 'Regulatory Improvement Legislation: Risk Assessment, Cost-Benefit Analysis, and Judicial Review', *Duke Env. L. & Pol'y F.*, 11 (2000).

Antonopolou, L. and P. van Meurs, 'The Precautionary Principle within European Union Public Health Policy: The Implementation of the Principle under Conditions of Supranationality and Citizenship', *Health Policy*, 66 (2003).

Arrow, K. I., 'Know a Hawk from a Handsaw', in Szenberg, M. (ed.), *Eminent Economists: Their Life and Philosophies*, Cambridge University Press, 1992.

Asselt, M. B. A. van, *Perspectives on Uncertainty and Risk: The PRIMA Approach to Decision Support*, Kluwer, 2000.

Asselt, M. B. A. van and E. Vos, 'The Precautionary Principle and the Uncertainty Paradox', *Journal of Risk Research*, 9 (2006), 4.

Atik, J. and D. A. Wirth, 'Science and International Trade – Third Generation Scholarship', *B.C Int'l & Comp. L. Rev.*, 26 (2003).

Babich, A., 'Too Much Science in Environmental Law', *Colum. J. Envl. L.*, 28 (2003).

Beck, U., *Risk Society: Towards a New Modernity*, Sage Publishers, 1992.

World Risk Society. Blackwell Publishers Ltd, 1999.

Bengtsson, B., *Speciell fastighetsrätt*, Iustus, 1999.

Miljöbalkens återverkningar, Norstedts, 2001.

Bergkamp, L., 'Understanding the Precautionary Principle (Parts I and II)', *Env. Liability*, 10 (2002).

Bergkvist, P., 'PM inför beslut om fortsatt godkännande för EBDC-medel' (not published).

Bergqvist, U., *et al.*, 'Exponering för radiofrekventa fält och mobiltelefoni', SSE rapport: 09 (April 2001) (2001).

Bernasconi-Osterwalder, N. *et al.*, *Environment and Trade: A Guide to WTO Jurisprudence*, Earthscan, 2006.

Bernstein, P. L., *Against the Gods: The Remarkable Story of Risk*, John Wiley & Sons Inc., 1996.

Bisgaard, C., 'Assessing the Standard of Review for Trade-Restrictive Measures in the Sanitary and Phytosanitary Agreement', in E. Brown Weiss and J. H. Jackson, *Reconciling Environment and Trade*, Transnational Publishers, 2001.

Blank, M., 'The Precautionary Principle Must Be Guided by EMF Research', *Electromagnetic Biology and Medicine*, 25 (2006).

Bodansky, D., 'Scientific Uncertainty and the Precautionary Principle', *Environment* 33 (1991) 4–5.

'Commentary: The Precautionary Principle', *Environment*, 34 (1992), 4–5.

Bohanes, J., 'Risk Regulation in WTO Law: A Procedure-based Approach to the Precautionary Principle', *Colum. J. Transnat. L.*, 40 (2001–2002).

Bossche, P. van den, *The Law and Policy of the World Trade Organization*, Cambridge University Press, 1st edn 2005, 2nd edn 2008.

Bossche, P. van den and D. Prevost, 'The Agreement on the Application of Sanitary and Phyto-sanitary Measures', in P. Mcrory, A. Appleton and M. Plummer (eds.), *The World Trade Organisation: Legal, Economic and Political Analysis*, Springer, 2005.

Breyer, S., *Breaking the Vicious Circle: Towards Effective Risk Regulation*, Harvard University Press, 1993.

Brickman, R., S. Jasanoff and T. Ilgen, *Controlling Chemicals: The Politics of Regulation in Europe and the United States*, Cornell University Press, 1985.

Burgess, A., *Cellular Phones, Public Fears, And a Culture of Precaution*, Cambridge University Press, 2004.

Button, C., *The Power to Protect: Trade, Health and Uncertainty in the WTO*, Hart, 2004.

Calster, G. van, *International and EU Trade Law: The Environmental Challenge*. Cameron/May, 2000.

Cameron, D. and J. Abouchar, 'The Status of the Precautionary Principle in International Law', in D. Freestone and E. Hey (eds.), *The Precautionary Principle in International Law: The Challenge of Implementation*, Kluwer Law International, 1996.

Cameron, J., 'The Precautionary Principle in International Law', in T. O'Riordan, J. Cameron and A. Jordan (eds.), *Reinterpreting the Precautionary Principle*, Cameron May, 2001.

Cane, P., *Administrative Law*, Oxford University Press, 2004.

Carson, R., *Silent Spring*, Houghton Mifflin, 1962.

Cecot, C., R. Hahn, A. Renda and L. Schrefler, 'An Evaluation of the Quality of Impact Assessment in the European Union with Lessons for the US and the EU', AEI-Brookings Joint Center for Regulatory Studies, Working Paper, December 2007.

Chittenden, F., T. Ambler and D. Xiao, 'Impact Assessment in the EU', in S. Weatherill, *Better Regulation*, Hart Publishing, 2007.

Christoforou, T., 'Science, Law and Precaution in Dispute Resolutions on Health and Environmental Protection: What Role for Scientific Experts?', in J. Bourrinet and S. Maljean-Dubois, *Le commerce international des organismes génétiquement modifies*, Aix-Marseilles/Paris, 2002.

Coggon, D., 'The Precautionary Principle and Pesticide Regulation', Chairman's Paper for ACP Open Meeting 2001. Available at: www.pesticides.gov.uk/acp.asp?id=490 (accessed 1 March 2010).

Covello, V. T., 'Risk Communication', in P. Calow, *Handbook of Environmental Risk Assessment and Management*, Wiley Blackwell, 1998.

Cranor, C. F., 'Asymmetric Information, the Precautionary Principle, and Burdens of Proof', in C. Raffensperger and J. Tickner (eds.), *Protecting Public Health and the Environment: Implementing the Precautionary Principle*, Island Press, 1999.

Croley, P. and J. H. Jackson, 'WTO Dispute Procedures, Standard of Review and Deference to National Governments', *AM. J. Int'l L.*, 90 (1996).

Croley, S., 'White House Review of Agency Rulemaking: An Empirical Investigation', *U. Chi. L. Rev.*, 70 (2003).

Damon, W. A., 'The Alkali Act: Its Administration and Relation to Chemical Industry', *Journal of the Society of Chemical Industry*, 54:50 (2007), 1070–5. Published online 29 May 2007.

Danish Environmental Assessment Institute, *Getting Proportions Right: How Far Should the EU Impact Assessments Go?*, Environmental Assessment Institute, 2006.

'Procedures for Derogations from the Principle of Approximation of Laws under Article 95 EC', *CMLRev*, 40 (2003).

'The Precautionary Principle in European Community Health and Environmental Law', *ELJ*, 12:2 (2006).

Implementing the Precautionary Principle: Approaches from the Nordic Countries, the EU and the USA, Earthscan, 2007.

'The Precautionary Principle in European Community Health and Environment Law: Sword or Shield for the Nordic Countries?', in N. de Sadeleer, *Implementing the Precautionary Principle: Approaches from the Nordic Countries, EU and USA*, Earthscan, 2007.

Department for Environment: Food and Rural Affairs UK. *A Guide to Risk Assessment and Risk Management for Environmental Protection*, HMSO, 1995.

News Release, 19 April 2002. Available at: www.hse.gov.uk/press/2002/e02076.htm (accessed 1 March 2010).

Figure 2.1. Available at:www.defra.gov.uk/environment/quality/risk/eram-guide/(accessed 1 March 2010).

Guidelines for Environmental Risk Assessment and Management. Available at:www.defra.gov.uk/environment/quality/risk/eramguide/ (accessed 1 March 2010).

Dinham, B., 'Why Paraquat Should Be Banned', *Outlook on Pest Management*, December 2004.

Douglas, M. and Wildawsky, A., *Risk and Culture*, University of California Press, 1992.

Douma, W., 'The Precautionary Principle; Its Application in International, European and Dutch Law', Proefshchrift Rijksuniversiteit Groningen, 6 January 2003.

Dunkelberger, E. and R. A. Merrill, 'The Delaney Paradox Re-examined: Regulating Pesticides in Processed Foods', *Food & Drug L. J.*, 48 (1993).

Duriodé, B., 'Plastic Panics: European Risk Regulation in the Aftermath of BSE', in J. Morris (ed.), *Rethinking Risk and the Precautionary Principle*, Butterworth-Heinemann, 2000.

'The True Cost of Precautionary Chemicals Regulation', *Risk Analysis*, 23:2 (2003).

Dworkin, R., *A Matter of Principle*, Harvard University Press, 1985.

Ehlerman, C.-D., 'The Internal Market Following the Single European Act', *CMLRev*, 24 (1987).

Ehrlich, P. R. and A. Ehrlich, *The End of Affluence: A Blueprint for your Future*, Ballantine Books, 1974.

Ekelöf, P. O., *Rättegång* IV, Norstedts, 2009.

Ellis, E. (ed.), *The Principle of Proportionality in the Laws of Europe*, Oxford University Press, 1999.

Eriksson, M. Band, 'Placera masten rätt: Miljö och naturskyddskrav på 3G-utbyggnaden', *Svenska Naturskyddsföreningen Rapport*, Svenska Natuskyddsföreningen 2003.

Fidler, D. P., *International Law and Public Health*, Transnational Publishers Inc., 2000.

Finkel, A. M. and D. Golding, *Worst Things First? The Debate over Risk-Based National Environmental Priorities*, Resources for the Future, 1994.

Fischoff, B., 'Risk Perception and Communication Unplugged: Twenty Years of Process', *Risk Analysis*, 15:2 (1995), 133–45.

Fisher, E., 'Is the Precautionary Principle Justiciable?', *J. Env. L.*, 13:3 (2001).

'The Rise of the Risk Commonwealth and the Challenge for Administrative Law', *Public Law* (2003), 455–78.

Risk Regulation and Administrative Constitutionalism, Hart Publishing, 2007.

Fisher, E., *et al.*, *Implementing the Precautionary Principle*, Edward Elgar Publishing, 2006.

Fitzmaurice, M., *Contemporary Issues in International Environmental Law*, Edgar Elgar Publishing, 2009.

Flick, C., 'The Movement for Smoke Abatement in 19th-Century Britain', *Technology and Culture*, 21:1 (1980), 29–50.

Ford, R., 'Planning For a New Telecommunications Network in the UK', *J.P.L.*, September (2001).

Freestone, D. and E. Hey (eds.), *The Precautionary Principle in International Law: The Challenge of Implementation*, Kluwer Law International, 1996.

Funtowicz, S. O. and J. R. Ravetz, *Uncertainty and Quality in Science for Policy*, Kluwer, 1990.

'Science for the Post-Normal Age', *Futures*, 25:7 (1993), 739–55.

Furedi, F. 'Precautionary Culture and the Rise of Possibilistic Risk Assessment', *Erasmus Law Review*, 2:2 (2009).

Geiser, K., 'Establish a General Duty of Precaution in Environmental Protection Policies in the United States: A Proposal', in C. Raffensperger and J. Tickner (eds.), *Protecting Public Health and the Environment: Implementing the Precautionary Principle*, Island Press, 1999.

Gerber, B. M., 'Case Comment: City of Rancho Palos Verdes, California v. Abrams', *Harv. Envtl. L.* Rev., 30 (2006).

Gocha, N., Rohm and Haas Comments on DeZäta Granulat Registration withdrawal by KemI, 18 January 1995 (not published).

Goldstein, B. D., 'The Interface Between Science and Law', *Colum. J. Envtl. L.*, 14 (1989).

Goldstein, B. D. and R. S. Carruth, 'Science in the Regulatory Process: Implications of the Precautionary Principle for Environmental Regulation in the United States: Examples from the Control of Hazardous Air Pollutants in the 1990 Clean Air Act Amendments', *Law & Contemp. Prob.*, 66 (2003).

Graf, M. W., 'Regulating Pesticide Pollution in California Under the 1986 Safe Drinking Water and Toxic Exposure Act (Proposition 65)', *Ecology L.Q.*, 28 (2001).

Graham, J. D., 'Legislative Approaches to Achieving More Protection Against Risk at Less Cost', *U Chi L F* 13, (1997).

'The Perils of the Precautionary Principle: Lessons from the American and European Experience', Speech before the Heritage Foundation, Washington DC, 20 October 2003. Available at: www.heritage.org/Research/Regulation/hl818.cfm (accessed 1 March 2010).

'The Evolving Regulatory Role of the U.S. Office of Management and Budget', *Review of Environmental Economics and Policy*, 1:2 (2007), 171–91.

Graham, J. D. and J. K. Hartwell, 'The Risk Management Approach', in: J. D. Graham and J. K. Hartwell (eds.), *The Greening of Industry*, Harvard University Press, 1997.

Graham, J. D. and S. Hsia. 'Europe's Precautionary Principle: Promise and Pitfalls', *Journal of Risk Research*, 5:4 (2002).

Graham, J. D. and J. B. Wiener, *Risk versus Risk: Tradeoffs in Protecting Health and the Environment*, Harvard University Press, 1995.

Grando, M. T., 'Allocating the Burden of Proof in WTO Disputes: A Critical Analysis, *Journal of International Economic Law*, 9:3 (2006).

Grasso, L., 'Cellular Telephones and the Potential Hazards of RF Radiation: Responses to the Fear and Controversy', *Va. J.L. & Tech.*, 3 (1998).

Grimeaud, D. Hoofdstuk II. 'The Precautionary Principle in International Environmental and Trade Law', in M. Faure and E. Vos (eds.), *Juridische afbakening van het voorzorgbeginsel: mogelijkheden en grenzen*, *Achtergrondstudie voor de Nederlandse Gezondheidsraad*, No. A03/03, 2003.

Gustin, T. L., 'The Perpetual Growth and Controversy of the Cellular Superhighway: Cellular Tower Siting and the Telecommunications Act of 1969', *Wm. Mitchell L. Rev.*, 23 (1997).

Hafner, K., 'First Come Cellphone Towers, then the Babel', *New York Times*, Section 1, Page 1, 1 May 2005.

Hahn R. W. and J. A. Hird, 'The Costs and Benefits of Regulation: Review and Synthesis', *Yale J. on Reg.*, 8 (1991).

Hahn, R. W. and C. R. Sunstein, 'A New Executive Order for Improving Federal Regulation? Deeper and Wider Cost-Benefit Analysis', *Chicago Working Paper in Law & Economics*, No. 150, 04/2002, p. 1. Available at: www.law.uchicago.edu/Lawecon/wp1–50.html (accessed 1 March 2010).

Haigh, N., 'The Introduction of the Precautionary Principle into the UK', in T. O'Riordan and J. Cameron (eds.), *Interpreting the Precautionary Principle*, Earthscan, 1994.

Hajer, M., *Environmental Discourse*, Clarendon Press, 1995.

Hammitt, J. K., 'Improving Comparative Risk Analysis', *Duke Envtl. L. & Pol'y F*, 8 (1997).

Hanekamp, J. C., 'Neither Acceptable nor Certain – Cold War Antics for 21st Century Precautionary Culture', *Erasmus Law Review*, 2:2 (2009).

Hart, H. L. A., *The Concept of Law*, Clarendon Press, 1961.

Health Protection Agency. AGNIR reports. Available at: www.hpa.org.uk/web/HPAweb&HPAwebStandard/HPAweb_C/1195733801791 (accessed 1 March 2010).

Heinzerling, L., 'Regulatory Costs of Mythic Proportions', *Yale Law Journal*, 107 (1981).

Hilson, C., 'Planning Law and Public Perceptions of Risk: Evidence of Concern or Concern Based on Evidence', *J.P.L* (2004).

HM Treasury, *The Green Book, Appraisal and Evaluation in Central Government*, 2003. Available at: www.hm-treasury.gov.uk/d/green_book_complete.pdf (accessed 1 March 2010).

Managing Risks to the Public: Appraisal Guidance, 2005. Available at: www.hm-treasury.gov.uk/consult_greenbook_index.htm (accessed 1 March 2010).

Horn, H. and P. C. Mavroidis, 'Burden of Proof in Environmental Disputes in the WTO: Legal Aspects', *EELR*, 18:2 (2009).

House of Commons, Science and Technology Committee Publications, Third Report, printed on 2 September 1999, Annex 2. Available at: www.parliament.the-stationery-office.co.uk/pa/cm199899/cmselect/cmsctech/489/48902.htm (accessed 1 March 2010).

Scientific Advice, Risk and Evidence Based Policy Making, 7th Report of Session 2005–2006, Vol. I, Authority of the House of Commons, 2006. Available at: www.publications.parliament.uk/pa/cm200506/cmselect/cmsctech/900/900-i.pdf (accessed 1 March 2010).

'Government Policy on the Management of Risk', 5th Report of Session 2005–2006, Vol. I and II: Minutes of Evidence, Authority of the House of Lords, 2006. Available at: www.publications.parliament.uk/pa/ld200506/ldselect/ldeconaf/183/183ii.pdf (accessed 1 March 2010).

Howse, R., 'Democracy, Science, and Free Trade: Risk Regulation on Trial at the World Trade Organization', *Mich. L. Rev.*, 98 (2000), 2333–8.

Independent Expert Group on Mobile Phones. Homepage. Available at: www.iegmp.org.uk (accessed 1 March 2010).

'Mobile Phones and Health', 11 May 2000, 1.30–1.48. Available at: www.iegmp.org.uk/report/text.htm (accessed 1 March 2010).

International Commission on Non-Ionizing Radiation Protection, 'Guidelines for Limiting Exposure to Time-Varying Electric, Magnetic, and Electromagnetic Fields (up to 300 GHZ)', *Health Physics*, 74:4 (1998).

International Program on Chemical Safety (INCHEM) Poison Information Monograph (PIM) Nr. 399. Available at: www.inchem.org/documents/pims/chemical/pim399.htm (accessed 1 March 2010).

Jaeger, C. C., O. Renn, E. A. Rosa and T. Webler, *Risk, Uncertainty and Rational Action*, Earthscan, 2001.

Jans, J. 'Proportionality Revisited', *Legal Issues of Economic Integration*, 27:3 (2000), 239–65.

Jasanoff, S. *The Fifth Branch: Science Advisors as Policymakers*, Harvard University Press, 1990.

Science at the Bar, Harvard University Press, 1995.

Johnston, B. L., *Environmental Policy and Public Health*, CRC Press, 2007.

Johnston, J. C., 'A Game Theoretic Analysis of Alternative Institutions for Regulatory Cost-Benefit Analysis', *U. Pa. L Rev.*, 150:5 (2002).

'Paradoxes of the Safe Society: A Rational Actor Approach to the Reconceptualization of Risk and the Reformation of Risk Regulation', *U. Pa. L Rev.*, 151:5 (2003).

Jordan, A., 'The Precautionary Principle in the European Union', in T. O'Riordan, J. Cameron and A. Jordan (eds.), *Reinterpreting the Precautionary Principle*, Cameron May, 2001.

Jowell, J. 'Of Vires and Vacuums: The Constitutional Context of Judicial Review', in C. Forsyth (ed.), *Judicial Review and the Constitution*, Hart Publishing. 2000.

Kagan, E. 'Presidential Administration', *Harv. L. Rev.*, 114 (2001).

Kahneman D. and A. Tversky, 'Prospect Theory: An Analysis of Decision under Risk', *Econometrica*, 47:2 (1979).

Kannan, P. M., 'The Precautionary Principle: More than a Cameo Appearance in United States Environmental Law', *Wm. & Mary Envtl. L & Pol'y Rev*, 31 (2007).

Karlsson, M., 'The Precautionary Principle, Swedish Chemicals Policy and Sustainable Development', *Journal of Risk Research*, 9:4 (2006), 344–8.

Kelly, C. R., 'Power, Linkage and Accommodation: The WTO as an International Actor and Its Influence on Other Actors and Regimes', *Berkely J. Int'l L.*, 24 (2006).

KEMI Swedish Chemicals Agency, Principles for Identifying Unacceptable Pesticides, KemI Report, No. 4/ 92, National Swedish Chemical Inspectorate, 1992.

Statement of 16 September 2003. Available at: www.kemi.se/templates/News____3090.aspx (accessed 1 March 2010).

Kent, C. W. and F. W. Allen, 'An Overview of Risk-Based Priority Setting at EPA', in A. M. Finkel and D. Golding, *Worst Things First? The Debate Over Risk-Based National Environmental Priorities*, Resources for the Future, 1994.

Keynes, J. M. 'The General Theory of Employment', *The Quarterly Journal of Economics* 51 (1937).

King, J. 'Application of Risk Assessment in Policy and Legislation in the European Union and in the United Kingdom', in P. Calow, *Handbook of Environmental Risk Assessment and Management*, Blackwell Science, 1998.

Kiss, A. C. and D. Shelton., *International Environmental Law*, Transnational Publishers, 1994.

Knight, F., *Risk, Uncertainty and Profit*, Houghton Mifflin Co., 1933.

Krämer, L., *EC Environmental Law*, Sweet and Maxwell, 2000.

Kriebel, D., 'How Much Evidence is Enough? Conventions of Causal Inference', *Law & Contemp. Probs.*, 72 (2009).

Kuhn, T., *The Structure of Scientific Revolutions*, University of Chicago Press, 1962.

Ladeur, K.-H. 'The Introduction of the Precautionary Principle into EU Law: A Pyrrhic Victory for Environmental and Public Health Law? Decision-making Under Conditions of Complexity in Multi-level Political Systems', *CMLR*, 40:6 (2003).

Lavranos, N., 'The Epilogue in the MOX Plant Dispute: An End Without Findings', *EELR*, 18:3 (2009).

Lehrberg, B., *Praktisk juridisk metod*, Iustus Förlag, 2001.

Lemons, J., K. Schrader-Frechette and C. Cranor , 'The Precautionary Principle: Scientific Uncertainty and Type I and Type II Errors', *Foundations of Science*, 2:2 (1997).

Lindblom, P. H., *Miljöprocess Del II*, Iustus, 2002.

Loewenberg, S., 'Precaution is for Europeans', *New York Times*, Section 4, p. 14, 18 May 2003.

Löfstedt, R., 'The Changing Character of Regulation: A Comparison of Europe and the United States', *Risk Analysis*, 21 (2002).

'Risk Communication and Management in the Twenty-First Century', 7:3 *International Public Management Journal* 335, 2004.

'The Swing of the Regulatory Pendulum in Europe: From Precautionary Principle to (Regulatory) Impact Analysis', *Journal of Risk and Uncertainty*, 28:2 (2004).

Risk Management in Post-Trust Societies, Palgrave Macmillan, 2005.

'The Plateau-ing of the European Better Regulation Agenda: An analysis of Activities Carried out by the Barroso Commission', *Journal of Risk Research*, 10:4 (2007).

Löfstedt, R. and R. Fairman, 'Scientific Peer Review to Inform Regulatory Decision Making: A European Perspective', *Risk Analysis*, 26:1 (2006), 25–31.

Löfstedt R. and D. Vogel 'The Changing Character of Consumer and Environmental Regulation: A Comparison of Europe and the United States', *Risk Analysis*, 21:3 (2001).

Lorde Martin, S., 'Communication Tower Sitings: the Telecommunications Act of 1996 and the Battle for Community Control', *B.T.L.J.*, 12:2 (1997). Available at: www.law.berkeley.edu/journals/btlj/articles/vol12/index.htm (accessed 1 March 2010).

MacCormack, G., 'Scandinavian Realism', *The Juridical Review*, 1970.

Madeley, J., 'Paraquat – Syngenta's controversial herbicide', SSNC Report 2002, pp. 10–12. Available at www.chemsec.org/images/stories/news_publications/rap-inter-syngenta-paraquat.pdf (accessed 1 March 2010).

Majone, G., 'The Precautionary Principle and Regulatory Impact Analysis'. Paper presented at the International Seminar on Regulatory Impact Analysis. Rome 15 June 2001.

Marchant, G. E. and K. L. Mossman, *Arbitrary and Capricious: The Precautionary Principle in the European Union Courts*, American Enterprise Institute Press, 2004.

Marr, S., *The Precautionary Principle in the Law of the Sea: Modern Decision Making in International Law*, Martinus Nijhoff Publishers, 2003.

McElveen, J. C. and C. Amatea, 'Risk Symposium: Legislating Risk Assessment', *U. Cin. L. Rev.*, 63 (1995).

McNelis, N., 'EU Communication on the Precautionary Principle', *JIEL*, 3:3 (2000).

Meadows, D. *et al.*, *The Limits to Growth: A report for the Club of Rome's project on the predicament of mankind*, Universe Books, 1974.

Meddelandeblad från Socialstyrelsen till kommunala miljö- och hälsoskyddsnämnder, länsstyrelser och andra berörda myndigheter, June 2008. Available at: www.fas.se/upload/dokument/särskilda_områden/eloverkanslighet/2008-1-11_2008112.pdf (accessed 1 March 2010).

Meuwese, A. C. M., 'Inter-institutionalising EU Impact Assessment', in S. Weatherill, *Better Regulation*, Hart Publishing, 2007.

Michanek, G., 'Att väga säkert och vikten av att säkra', in E. M. Basse *et al.* (eds.), *Fågelperspektiv på rättsordningen: Vänbok till Staffan Westerlund*, Iustus Förlag, 2002.

'Chapter 7: Sweden', in N. de Sadeleer (ed.), *Implementing the Precautionary Principle: Approaches form the Nordic Countries, EU and USA*, Earthscan, 2007.

Miles, T. J. and C. R. Sunstein, 'The Real World of Arbitrariness Review', John M. Olin Law & Economics Working Paper No. 368, November 2007.

Miljömål: Environmental Objectives Portal, Summary of environmental policy goals. Available at: www.miljomal.nu (accessed 1 March 2010).

Miljøstyrelsen, Vedrørende Dithane DG, 9 June 1993 (available from the author).

Miller, M. L. Esq., 'Pesticides', in T. F. P. Sullivan, *Environmental Law Handbook*, Government Institutes, 2003.

Montemarano, A., 'The Delaney Paradox Resurfaces: Regulating Pesticides as Food Additives Under Federal Law', *Rutgers L.J.*, 25 (1993/1994).

Mooney, C., *The Republican War on Science*, Basic Books, 2005.

Morris, J., *Rethinking Risk and the Precautionary Principle*, Butterworth-Heinemann, 2000.

Moss, D. A., *When All Else Fails: Government as the Ultimate Risk Manager*, Harvard University Press, 2002.

Nash, J. R., 'Standing and the Precautionary Principle', *Colum. L. Rev.*, 108 (2008).

National Radiological Protection Board, Report by the Board of NRPB, 'Mobile Phones and Health', Doc. of the NRPB, 15:5 (2004).

National Research Council, *Risk Assessment in the Federal Government: Managing the Process*, National Academy Press, 1983.

The New Oxford Dictionary, Oxford University Press, 1998.

Newsweek, 'The Battle Over Caution', *Newsweek*, 13 December 2003. Available at: www.newsweek.com/id/60863 (accessed 1 March 2010).

Nilsson, A., *Att byta ut kemikalier*, Nerenius och Santerus, 1997.

'The Precautionary Principle in Swedish Chemicals Law and Policy', in N. de Sadeleer, *Implementing the Precautionary Principle: Approaches from the Nordic Countries, EU and USA*, Earthscan, 2007.

Nilsson, R., 'Science and Politics in the Regulation of Chemicals in Sweden', in M. Gough (ed.), *Politicizing Science: The Alchemy of Policymaking*, Hoover Institutions Press, 2003.

O'Brien, M., *Making Better Environmental Decisions: An Alternative to Risk Assessment*, MIT Press, 2000.

O'Riordan, C. and A. Jordan, *The History and Contemporary Significance of the Precautionary Principle, in Interpreting the Precautionary Principle*, Earthscan, 1994.

Organisation for Economic Co-operation and Development (OECD) , Council Recommendation C(90)164 on Integrated Pollution Prevention and Control, 31 January 1991.

Pak, M., 'Note & Comment: An IQ Test for Federal Agencies? Judicial Review of the Information and Quality Act Under the APA', *Wash. L. Rev.*, 80 (2005).

Palmeter, D., 'The WTO Standard of Review in Health and Safety', in G. A. Bermann and P. C. Mavroidis, *Trade and Human Health and Safety*, Cambridge University Press, 2006.

Patterson, A., *Risk Regulation and Scientific Expertise in the United Kingdom: The Precautionary Principle in Public Policy*, Mellen Press, 2008.

Pauwelyn, J., 'The WTO Agreement on Sanitary and Phytosanitary (SPS) Measures As Applied in the First Three SPS Disputes', *JIEL*, 2:4 (1999).

'The Use of Experts in WTO Dispute Settlement', *ICLQ*, 51 (2002), 328–9.

'Expert Advice in WTO Dispute Settlement', in G. A. Bermann and P. C. Mavroidis, *Trade and Human Health and Safety*, Cambridge University Press, 2006.

Peng, J. *et al.*, 'The Herbicide Paraquat Induces Dopaminergic Nigral Paoptosis through Sustained Activation of the JNK Pathway', *J. Bio. Chem.*, 279:31 (2004).

Pescatore, P., 'Some critical remarks on the "Single European Act"', *CMLRev.*, 24 (1987).

Prevost, D., 'Opening Pandora's Box: The Panel's Findings in the EC-Approval and Marketing of Biotech Products Dispute', *Legal Issues of Economic Integration*, 34:1 (2007), 73–8.

Raul, A. C. and J. Z. Dwyer, 'Science in the Regulatory Process: "Regulatory Daubert": A Proposal to Enhance Judicial Review of Agency Science by Incorporating Daubert Principles into Administrative Law', *Law & Contemp. Prob.*, 667 (2003).

Regeringskansliet. Official English Translations of the Environmental Code. Available at: www.regeringen.se/content/1/c4/13/48/385ef12a.pdf (accessed 1 March 2010).

The Swedish Judicial System – A Brief Presentation, published by the Swedish Ministry of Justice, available at: www.regeringen.se/content/1/c6/01/62/79/023155e4.pdf (accessed 1 March 2010).

Renda, A., *Impact Assessment in the EU: State of the Art and the Art of the State*, Center for European Policy Studies, 2006.

Rohrmann, B. and O. Renn. 'Risk Perception Research – An Introduction', in O. Renn and B. Rohrman, *Cross-cultural Risk Perception: A Survey of Empirical Studies*, Kluwer, 2000.

Rotmans, J. and van Asselt, M. B. A., 'Uncertainty Management in Integrated Assessment Modeling: Towards a Pluralistic Approach', *Environmental Monitoring and Assessment*, 69:2 (2001).

The Royal Society, *Risk: Analysis, Perception, Management*, The Royal Society 1992.

Rozin, P., 'Technological Stigma: Some Perspectives from the Study of Contagion', in J. Flynn, P. Slovic and H.. Kunverther (eds.), *Risk, Media, and Stigma: Understanding Public Challenges to Modern Science and Technology*, Earthscan, 2001.

Ruessmann, L. A., 'Reflections on the WTO Doha Ministerial Conference: Putting the Precautionary Principle in its Place: Parameters for the Proper Application of a Precautionary Approach and the Implications for Developing Countries in Light of the Doha WTO Ministerial', *Am. U. Int'l L. Rev.*, 17 (2002), 925–6.

Sadeleer, N. de, *Environmental Principles: From Political Slogans to Legal Rules*, Oxford University Press, 2002.

Sandin, P., 'Dimensions of the Precautionary Principle', *Human and Ecological Risk Assessment*, 5 (1999).

'A Paradox Out of Context: Harris and Holm on the Precautionary Principle', *Cambridge Quarterly of Healthcare Ethics*, 15 (2006).

Sandin, P., M. Peterson, S.-O. Hansson, C. Rudén, and A. Juthe, 'Five Charges Against the Precautionary Principle', *Journal of Risk Research*, 5:4 (2002).

Schierow, L-J., 'Environmental Risk and Cost-Benefit Analysis: A Review of Proposed Legislative Mandates, 1993–1998', CRS Report for Congress, 22 January 1999.

Science and Environmental Health Network (SEHN), *Precautionary Principle FAQs*. Available at: www.sehn.org/ppfaqs.html (accessed 1 March 2010).

Scientific Committee for Animal Nutrition (SCAN) , Opinion on the immediate and longer-term risk to the value of Streptogramins in Human Medicine posed by the use of Virginiamycin as an animal growth promoter of 10 July 1998.

Scott, J., 'The European Regulation of GMOs and the WTO', *Colum. J. Eur. L.*, 9 (2003).

The WTO Agreement on Sanitary and Phytosanitary Measures, Oxford University Press, 2007.

'Of Kith and Kine (and Crustaceans): Trade and Environment in the EU and WTO', available at: www.jeanmonnetprogram.org/papers/99/990307.html (accessed 1 March 2010).

Scott, J. and E. Vos, 'The Juridification of Uncertainty: Observations on the Ambivalence of the Precautionary Principle within the EU and the WTO', in Ch. Joerges and R. Dehousse , R. (eds.), *Good Governance in Europe's Integrated Market*, Oxford University Press, 2002.

Sedman, R. M. and P. W. Hadley, 'Risk Assessment and Risk Management: Mending the Schism', *Risk – Issues in Health & Safety*, 3 (1992).

Shapiro, S. and R. L. Glicksman, 'Congress, The Supreme Court and the Quiet Revolution in Administrative Law', *Duke L. J.*, 819 (1988), 819–20.

Shrader-Frechette, K., 'Scientific Method, Anti-Foundationalism and Public Decision', in E. Löfstedt and L. Frewer (eds.), *The Earthscan Reader in Risk and Modern Society*, Earthscan, 1998.

Slovic, P., 'Perceptions of Risk', *Science*, 236:4799 (1987), 280–5.

'Informing and Educating the Public about Risk', in P. Slovic, *The Perception of Risk*, Earthscan, 2000.

'Perceived Risk, Trust and Democracy', in P. Slovic, *The Perception of Risk*, Earthscan, 2000.

The Perception of Risk, Earthscan, 2000.

Slovic, P., B. Fischoff and S. Lichtenstein, 'Facts and Fears: Understanding Perceived Risk', in P. Slovic, *The Perception of Risk*, Earthscan, 2000.

'Rating the Risks', in P. Slovic, *The Perception of Risk*, Earthscan, 2000.

Starr, C., 'Social Benefit Versus Technological Risk', *Science*, 165 (1969).

Stirling, A. 'On Science and Precaution in the Management of Technological Risk'. An Esto Project, Prepared for the European Commission – JRC Institute Prospective Technological Studies Seville, May 1999 (available from author).

Strategy Unit Report, 'Risk: Improving Government's Capability to Handle Risk and Uncertainty', November 2002, pp. 74–92. Available at: www.cabinetoffice.gov.uk/media/cabinetoffice/strategy/assets/su%20risk%20summary.pdf (accessed 1 March 2010).

Sunstein, C., *After the Rights Revolution: Reconceiving the Regulatory State*, Harvard University Press, 1990.

'The Cost-Benefit State', *Chicago Working Paper in Law & Economics*, No. 039, 05/1996, Available at: www.law.uchicago.edu/files/files/39.sunstein.pdf (accessed 1 March 2010).

'Cost/benefit Default Principles', Chicago Working Paper in Law and Economics, No. 104, 3/6/01, p. 15. Available at: www.law.uchicago.edu/files/files/104.CRS_.Default.REVISED.pdf (accessed 1 March 2010).

'Health-Health Tradeoffs', *U. Chi. L. Rev.*, 63 (1996).

Risk and Reason: Safety, Law and the Environment, Cambridge University Press, 2004.

The Laws of Fear: Beyond the Precautionary Principle, Cambridge University Press, 2005.

'On the Divergent American Reactions to Terrorism and Climate Change', *Colum. L. Rev.*, 107 (2007).

Svärd, S. O. and T. Rahm, *Prövning enligt miljöskyddslagen*, Gebers Förlag, 1975.

Swedish Ministry for Foreign Affairs (EU Legal Secretariat), Letter To the Registrar at the Court of Justice of the European Communities, 26 February 2004 (available from the author).

Thiruchelvam, M. *et al.*, 'The Nigrostriatal Dopaminergic System as a Preferential Target of Repeated Exposures to Combined Paraquat and Maneb: Implications for Parkinson's Disease', *The Journal of Neuroscience*, 20:24 (2000).

'Overexpression of Superoxide Dismute or Glutathione Peroxidase Protects against the Paraquat + Maneb-induced Parkinson Disease Phenotype', *J. Biol. Chem.*, 280:23 (2005).

Thornton, J. and S. Beckwith, *Environmental Law*, Sweet & Maxwell, 2004.

Tickner, J and C. Raffensberger, 'The Precautionary Principle in Action: A Handbook', Science and Environmental Health Network, 13. Available at: www.mindfully.org/Precaution/Precaution-In-Action-Handbook.pdf (accessed 1 March 2010).

Tinari, N., 'Cell Phone Towers in Residential Areas: Did Congress Let the Pig in the Parlor with the Telecommunication Act of 1996?', *Temp. L. Rev.*, 73 (2000).

Tengs, T. O., M. E. Adams, J. S. Pliskin, D. G. Safran, J. E. Siegel, M. C. Weinstein and J. D. Graham, 'Five-hundred life-saving interventions and their cost-effectiveness', *Risk Analysis* 13 (1995).

Trouwborst, A. *Evolution and Status of the Precautionary Principle in International Law*, Aspen Publishers, 2002.

'The Precautionary Principle in General International Law: Combating the Babylonian Confusion', *RECIEL*, 16:2 (2007).

'Prevention, Precaution, Logic and Law: The Relationship Between the Precautionary Principle and the Preventative Principle in International Law and Associated Questions', *Erasmus Law Review*, 2:2 (2009), 106–27.

United States Environmental Protection Agency, 'An Evaluation of EPA Risk Assessment Principles and Practices', Staff Paper Prepared for the U.S. Environmental Protection Agency by members of the Risk Assessment Task Force, EPA/100/B-04/001.

Guidelines for Ensuring and Maximizing the Quality, Objectivity, Utility, and Integrity of Information Disseminated by the Environmental Protection Agency, EPA/260R-02–008. Available at www.epa.gov/quality/information-guidelines/documents/EPA_InfoQualityGuidelines.pdf (accessed 1 March 2010)

Reorganization Plan No. 3 of 1970, Special Message from the President to the Congress about Reorganization Plans to Establish the Environmental Protection Agency and the National Oceanic and Atmospheric Administration, 9 July 1970. Available at: www.epa.gov/history/org/origins/reorg.htm (accessed 1 March 2010).

RCRA Public Participation Policy, 1981.

Unfinished Business: A Comparative Assessment of Environmental Problems, February 1987. Available at http://yosemite.epa.gov/water/owrccatalog.nsf/9da204a4b4406ef885256ae0007a79c7/d8555a5ca86d824a85256b06007256d1!OpenDocument (accessed 1 March 2010).

Reducing Risk: Setting Priorities and Strategies for Environmental Protection, September 1990. Executive summary available at www.epa.gov/history/topics/risk/01.htm (accessed 1 March 2010).

Risk Characterization Handbook, December 2000, available at www.epa.gov/OSA/spc/pdfs/rchandbk.pdf (accessed 1 March 2010).

'Late Lessons From Early Warnings: The Precautionary Principle 1896–2000', Environmental Issue Report No. 22, 2002. Available at: http://reports.eea.europa.eu/environmental_issue_report_2001_22/en (accessed 1 March 2010).

Public Involvement Policy of the Environmental Protection Agency, May 2003. p. 1 and Appendix 1. Available at: www.epa.gov/publicinvolvement/pdf/policy2003.pdf (accessed 1 March 2010).

'An Evaluation of EPA Risk Assessment Principles and Practices', Staff Paper Prepared for the US Environmental Protection Agency by members of the Risk Assessment Task Force, March 2004, available at: www.epa.gov/osa/pdfs/ratf-final.pdf (accessed 1 March 2010).

'Changes Needed to Improve Public Confidence in EPA's Implementation of the Food Quality Protection Act', Report No. 2006-P-00003, 19 October 2005.

United States General Accounting Office, Telecommunications: Research and Regulatory Efforts on Mobile Phone Health Issues, Report to Congressional Requesters, AO-01-545, 2001 Available at www.gao.gov/new.items/d01545.pdf (accessed 1 March 2010).

OMB's Role in Reviews of Agency's Draft Rules and the Transparency of Those Reviews, GAO-03-929, 2003. Available at: www.gao.gov/new.items/d03929.pdf (accessed 1 March 2010).

Viscusi, K., *Rational Risk Policy*, Oxford University Press, 1998.

'Monetizing the Benefits of Risk and Environmental Regulation', *Fordham Urb. L. J.*, 33 (2006), 1003–4.

Vogel, D., 'Ships Passing in the Night: GMOs and the Politics of Risk Regulation in Europe and the United States', Working Paper No 2001/16, Robert Schuman Centre for Advanced Studies, European University Institute, 2001.

von Moltke, K., 'The Vorsorgeprinzip in West German Environmental Policy', in *Royal Commission On Environmental Pollution*, Twelfth Report, HMSO 1988.

Vos, E., 'EU Food Safety Regulation in the Aftermath of the BSE Crisis', *Journal of Consumer Policy*, 23 (2000).

'Antibiotics, the Precautionary Principle and the Court of First Instance', *Maastricht Journal*, 11:2 (2004).

Wahlström, B. 'The Precautionary Approach to Chemicals Management: A Swedish Perspective', in C. Raffensperger and J. Tickner (eds.), *Protecting Public Health and the Environment: Implementing the Precautionary Principle*, Island Press, 1999.

Walker, W. E., P. Harremoes, J. Rotman, J. P. van der Sluijs, M. B. A. van Asselt, P. Janssen, and M. P. Krayer von Krauss, 'Defining Uncertainty: A Conceptual Basis for Uncertainty Management in Model-Based Decision Support', *Integrated Assessment*, 4:1 (2003).

Wall Street Journal, 'Jolly Green Justices', Editorial, 3 April 2007, p. A14.

Weale, A., 'Ecological Modernisation and the Integration of European Environmental Policy', in J. D. Lifferink, *European Integration and Environmental Policy*, Belhaven Press, 1993.

Weber, M., *The Theory of Social and Economic Organization*, Henderson and Parson, 1947.

Weiss, R., 'Law is Nemesis of Regulation', *Washington Post*, 16 August 2004, p. A01

Westerlund, S., *Lag om kemiska produkter: Kommentar*, C. E. Fritzes Förlag, 1985.

Miljöskyddslagen: En analytisk lagkommentar, Åmyra Förlag, 1990.

Wiener, J. 'Precaution in a Multi-Risk World', Duke Law School Public Law and Legal Theory Working Paper No. 23, December 2001.

Wiener, J. and Rogers, M. D., 'Comparing precaution in the United States and Europe', *Journal of Risk Research*, 5:4 (2002).

Wildawsky, A. 'Trial and Error versus Trial Without Error', in J. Morris, *Rethinking Risk and the Precautionary Principle*, Butterworth-Heinemann, 2000.

Winickoff, D. *et al.*, 'Adjudicating the GM Food Wars: Science, Risk, and Democracy in World Trade Law', *Yale J. Int'l Law*, 30 (2005).

Wood, S. G., S. Q. Wood and R. A. Wood, 'The Precautionary Principle? An American Assessment', in M. Paques, *Le Principe de precaution en droit administratif/Precautionary Principle and Administrative Law, National and International Reports, International Academy of Comparative Law*, Bruylant, 2007.

Zander, J., 'The Green Guarantee in Article 95 of the EC Treaty: Two Recent Cases', *Journal of Environmental Law*, 16:1 (2004).

Zwanenberg, P. and E. Millstone, 'Mad Cow Disease' 1980s–2000: How Reassurances Undermined Precaution', in *Late Lessons From Early Warnings: The Precautionary Principle 1896–2000*, Environmental Issue Report No. 22, 2002. Available at: http://reports.eea.europa.eu/environmental_issue_report_2001_22/en/ (accessed 1 March 2010).

INDEX

For EU product safety concerns, contact us at Calle de José Abascal, 56–1°,
28003 Madrid, Spain or eugpsr@cambridge.org.

www.ingramcontent.com/pod-product-compliance
Lightning Source LLC
LaVergne TN
LVHW011530200926
848484LV00001B/217